D1250749

Shape Interrogation
for Computer
Aided Design and
Manufacturing

Springer
Berlin
Heidelberg
New York
Barcelona
Hong Kong
London
Milan
Paris
Tokyo

Nicholas M. Patrikalakis
Takashi Maekawa

Shape Interrogation for Computer Aided Design and Manufacturing

With 165 Figures, 8 in color

 Springer

Nicholas M. Patrikalakis

MIT Rm 5-428
77 Massachusetts Avenue
Cambridge, MA 02139-4307
USA

e-mail: nmp@mit.edu

Takashi Maekawa

MIT Rm 5-426A
77 Massachusetts Avenue
Cambridge, MA 02139-4307
USA

e-mail: tmaekawa@mit.edu

Catalog-in-Publication Data applied for

Die Deutsche Bibliothek - CIP-Einheitsaufnahme
Patrikalakis, Nicholas M.:
Shape interrogation for computer aided design and manufacturing/
Nicholas M. Patrikalakis; Takashi Maekawa.-Berlin; Heidelberg; New York;
Barcelona; Hong Kong; London; Milan; Paris; Tokyo: Springer, 2002
ISBN 3-540-42454-7

Mathematics Subject Classification (2000): 53A04, 53A05, 14Q05, 14Q10, 65G30

ISBN 3-540-42454-7 Springer-Verlag Berlin Heidelberg New York

This work is subject to copyright. All rights are reserved, whether the whole or part of the material is concerned, specifically the rights of translation, reprinting, reuse of illustrations, recitation, broadcasting, reproduction on microfilm or in any other way, and storage in data banks. Duplication of this publication or parts thereof is permitted only under the provisions of the German Copyright Law of September 9, 1965, in its current version, and permission for use must always be obtained from Springer-Verlag. Violations are liable for prosecution under the German Copyright Law.

Springer-Verlag Berlin Heidelberg New York
a member of BertelsmannSpringer Science+Business Media GmbH

http://www.springer.de

© Springer-Verlag Berlin Heidelberg 2002
Printed in Germany

The use of general descriptive names, registered names, trademarks, etc. in this publication does not imply, even in the absence of a specific statement, that such names are exempt from the relevant protective laws and regulations and therefore free for general use.

Typeset by the authors using a Springer TeX macro package
Cover design: *design & production* GmbH, Heidelberg

SPIN 10794601   46/3142db - 5 4 3 2 1 0 – Printed on acid-free paper

FLORIDA GULF C.
UNIVERSITY LIBRARY

Preface

Objectives and Features

Shape interrogation is the process of extraction of information from a geometric model. Shape interrogation is a fundamental component of Computer Aided Design and Manufacturing (CAD/CAM) systems and was first used in such context by M. Sabin, one of the pioneers of CAD/CAM, in the late sixties. The term *surface interrogation* has been used by I. Braid and A. Geisow in the same context. An alternate term nearly equivalent to shape interrogation is *geometry processing* first used by R. E. Barnhill, another pioneer of this field. In this book we focus on shape interrogation of geometric models bounded by free-form surfaces. Free-form surfaces, also called sculptured surfaces, are widely used in scientific and engineering applications. For example, the hydrodynamic shape of propeller blades has an important role in marine applications, and the aerodynamic shape of turbine blades determines the performance of aircraft engines. Free-form surfaces arise also in the bodies of ships, automobiles and aircraft, which have both functionality and attractive shape requirements. Many electronic devices as well as consumer products are designed with aesthetic shapes, which involve free-form surfaces.

When engineers or stylists design geometric models bounded by free-form surfaces, they need tools for shape interrogation to check whether the designed object satisfies the functionality and aesthetic shape requirements. This book provides the mathematical fundamentals as well as algorithms for various shape interrogation methods including nonlinear polynomial solvers, intersection problems, differential geometry of intersection curves, distance functions, curve and surface interrogation, umbilics and lines of curvature, geodesics, and offset curves and surfaces.

The book can serve as a textbook for teaching advanced topics of geometric modeling for graduate students as well as professionals in industry. It has been used as one of the textbooks for the graduate course "Computational Geometry" at the Massachusetts Institute of Technology (MIT). Currently there are several excellent books in the area of geometric modeling and in the area of solid modeling. This book provides a bridge between these two areas. Apart from the differential geometry topics covered, the entire book is based on the unifying concept of recasting all shape interrogation problems to the solution of a nonlinear system.

Structure and Outline

Chapter 1 presents a brief overview of analytical methods for the representation of curves and surfaces in a computer environment. We focus on the parametric representation of curves and surfaces, commonly used in CAD systems for shape specification. We next review the theory of Bernstein polynomials and associated algorithms and their application in the definition and manipulation of Bézier curves and surface patches. Finally in this chapter, we review the theory of B-spline basis functions and associated algorithms and their application in the definition and manipulation of B-spline and Non-Uniform Rational B-spline curves and surface patches. In our development of Bernstein polynomials and B-spline basis functions and the associated curve and surface representations, we do not provide detailed proofs as they are already contained in other existing books on geometric modeling, on which we rely for instructional purposes.

Chapters 2 and 3 provide an overview and introduction into the classical elementary differential geometry of explicit, parametric and implicit curves and surfaces, necessary for the development of the more advanced differential geometry topics that are presented in Chaps. 6, 8, 9 and 10. Much of the material of Chaps. 2 and 3 (except the treatment of curvatures of implicit surfaces) can be generally found in various forms in existing books on differential geometry and is included for convenience of the reader and completeness of our development.

Chapter 4 focuses on the development of geometrically motivated solvers for nonlinear equation systems and the related numerical robustness (reliability) issues. Much of the shape interrogation problems defined and solved in this book can be reduced to solving systems of n nonlinear polynomial equations in l unknowns, each of which varies within a known interval. Much of the development is based on the Interval Projected Polyhedron (IPP) algorithm, developed in our Design Laboratory at MIT in the early nineties. Some shape interrogation problems involve more general nonlinear functions including radicals of polynomials. These are also converted to nonlinear polynomial systems of higher dimensionality via an auxiliary variable method. The fundamental feature of the IPP algorithm is that it allows recasting of continuous shape interrogation problems encountered in geometric modeling and processing of free-form shapes into the discrete problem of computing convex hulls of a set of points in a plane and their intersections with other convex hulls along a particular axis. In this way, a bridge between the largely disparate fields of geometric modeling of free-form shapes (largely based on numerical analysis and approximation theory) and discrete computational geometry (largely based on the theory of algorithms and combinatorics) is established. Another fundamental feature of the IPP algorithm, is the use of rounded interval arithmetic motivated by questions of numerical robustness or reliability, which have high importance in CAD/CAM systems. Interval methods are a special branch of numerical analysis, with great potential for applications in geometric modeling and processing problems. Interval methods have not yet been used extensively in practice, because, if they are applied naively, they lead to interval growth that reduces the possible achievable precision in a numerical computation. However, when combined with geometric modeling algorithms based on convex combinations (as the de Casteljau algorithm), they lead to very minor interval growth and permit effective and high accuracy solutions in practice. The IPP algorithm robustly eliminates subregions of the domain which do not contain roots, thereby allowing effective bracketing of the roots of the nonlinear system within a given box with size typically much smaller than the actual accuracy of the results of current CAD/CAM systems.

Chapter 5 presents the first major shape interrogation problem analyzed in this book. Intersection is a fundamental operation in the creation of geometric models encoded in the Boundary Representation form of solid modeling. Intersection is also very useful in geometric processing for visualization, analysis and manufacturing of solid models. We present a unified methodology for solving intersection problems, which reduces all such problems to solving a system of nonlinear polynomial equations which in turn can be solved using the method of Chap. 4. We also present a novel classification of intersection problems by virtue of their dimensionality, the type of geometric representations involved, and the number system used in problem statement and solution. The point to point, point to curve, point to surface, curve to curve, curve to surface and surface to surface intersection problems are treated in some detail. Various special cases of interest, where the geometric entities involved (points, curves and surfaces) are represented implicitly or parametrically in terms of polynomials, are treated in some depth.

Chapter 6 is motivated by efficient tracing of intersection curves of two surfaces which intersect either transversely or tangentially, and presents the first, second and higher order derivatives of these entities for use in developing efficient and robust tracing algorithms. The surfaces involved may be parametric or implicit in any combination.

Chapter 7 presents methods for the computation of the stationary points of distance functions between points, parametric curves and parametric surface patches (in any combination). The curves and surfaces may be defined by general piecewise rational polynomials. The resulting problems are again reduced to solving systems of nonlinear equations which can be solved using the IPP algorithm developed in Chap. 4. Distance functions are closely related to intersection problems and are also useful in many other applications including feature recognition via medial axis transforms, animation, collision detection, and manufactured object localization and inspection.

Chapter 8 addresses a variety of curve and surface interrogation methods involving their position vectors and several higher order derivatives. Particular emphasis is placed on robust extraction of stationary points of curvature maps and the consequent application in robust contouring of such maps.

Again the problem reduces to solving systems of nonlinear equations which can be solved using the IPP algorithm developed in Chap. 4. The interrogation methods analyzed in this chapter have many applications in aesthetic and functional surface design and analysis, in fairing of oscillatory shapes, in meshing of surface patches and in machining automation.

Chapter 9 discusses the problems of umbilics and lines of curvature as methods of shape interrogation and identification. Umbilics are computed via solution of a nonlinear polynomial system following the IPP algorithm of Chap. 4. Curvature lines are computed via integration of a system of differential equations via an adaptive numerical process with specialized treatment near umbilics. The stability problem of umbilics under perturbation of the underlying surface is also analyzed for use in surface identification and feature recognition problems.

Chapter 10 addresses yet another shape interrogation problem involving the geodesics of parametric and implicit surfaces. The classical geodesic equations are reviewed and numerical methods for the effective computation of geodesics between two points on a surface or a point and a curve on a surface are presented. The numerical methods involve iterative solution of a nonlinear boundary value problem via either shooting or relaxation methods. Geodesics have applications in feature recognition via medial axis transforms, in path planning in robotics (for distance minimization), in representation of geodesic offsets for design and in manufacturing.

Chapter 11, the final chapter of this book, focuses on the problem of offset (or parallel) curves and surfaces. Offsets have important applications in NC machining, feature recognition via medial axis transforms and in tolerance region specification. The definition and computation of singularities (and especially self-intersections) of planar offset curves and offset surfaces is treated in depth. The methods developed are in part analytical, and in part numerical relying on the IPP algorithm of Chap. 4 and on integration of systems of nonlinear differential equations. The related concepts of Pythagorean hodographs, general offsets and pipe surfaces, which build on the theory of offset curves and surfaces, are also reviewed and analyzed in some detail.

Problems that instructors can use in developing their own courses are provided immediately after Chap. 11. Many of these problems have been used in our graduate course at MIT.

Errors

A book of this size is likely to contain omissions and errors. If you have any constructive suggestions or find errors, please communicate them to N. M. Patrikalakis, MIT Room 5-428, 77 Massachusetts Avenue, Cambridge, MA 02139-4307, USA (e-mail: nmp@mit.edu), and T. Maekawa, MIT Room 5-426A, 77 Massachusetts Avenue, Cambridge, MA 02139-4307, USA (e-mail: tmaekawa@mit.edu).

Acknowledgements
We wish to recognize the following former and current students who have
helped in the development of this book: Panos G. Alourdas, Christian Bliek,
Julie S. Chalfant, Wonjoon Cho, Donald G. Danmeier, H. Nebi Gursoy, An-
dreas Hofman, Chun-Yi Hu, Todd R. Jackson, Kwang Hee Ko, George A.
Kriezis, Hongye Liu, John G. Nace, P. V. Prakash, Guoling Shen, Evan C.
Sherbrooke, Stephen Smyth, Krishnan Sriram, Seamus T. Tuohy, Marsette
A. Vona, Guoxin Yu and Jingfang Zhou. We also wish to acknowledge Ste-
phen L. Abrams for his assistance with software development and Fred Baker
for editorial assistance.

We also thank Chryssostomos Chryssostomidis, David C. Gossard, Mal-
colm Sabin, Takis Sakkalis, Nickolas S. Sapidis, Franz-Erich Wolter and Xiuzi
Ye and several anonymous referees selected by Springer for useful discussions
and their comments.

We also wish to acknowledge MIT's funding of this book development
from the Bernard M. Gordon Engineering Curriculum Development Fund
via the Dean of the School of Engineering and via additional support from
the Department of Ocean Engineering.

We, finally, dedicate this book to our families, our wives Sandra Jean
and Yuko and our children Alexander, Andrew, Nikki, and Takuya, whose
love, patience, understanding and encouragement made this lengthy project
possible.

Cambridge, MA, June, 2001   *Nicholas M. Patrikalakis*
                                             Takashi Maekawa

Contents

1. **Representation of Curves and Surfaces** 1
   1.1 Analytic representation of curves 1
      1.1.1 Plane curves 1
      1.1.2 Space curves 3
   1.2 Analytic representation of surfaces 4
   1.3 Bézier curves and surfaces 6
      1.3.1 Bernstein polynomials 6
      1.3.2 Arithmetic operations of polynomials in Bernstein form 7
      1.3.3 Numerical condition of polynomials in Bernstein form . 9
      1.3.4 Definition of Bézier curve and its properties 12
      1.3.5 Algorithms for Bézier curves 13
      1.3.6 Bézier surfaces 18
   1.4 B-spline curves and surfaces 20
      1.4.1 B-splines 20
      1.4.2 B-spline curve 21
      1.4.3 Algorithms for B-spline curves 24
      1.4.4 B-spline surface 29
   1.5 Generalization of B-spline to NURBS 30

2. **Differential Geometry of Curves** 35
   2.1 Arc length and tangent vector 35
   2.2 Principal normal and curvature 39
   2.3 Binormal vector and torsion 43
   2.4 Frenet-Serret formulae 47

3. **Differential Geometry of Surfaces** 49
   3.1 Tangent plane and surface normal 49
   3.2 First fundamental form I (metric) 52
   3.3 Second fundamental form II (curvature) 55
   3.4 Principal curvatures 59
   3.5 Gaussian and mean curvatures 64
      3.5.1 Explicit surfaces 64
      3.5.2 Implicit surfaces 65
   3.6 Euler's theorem and Dupin's indicatrix 68

4. Nonlinear Polynomial Solvers and Robustness Issues 73
    4.1 Introduction 73
    4.2 Local solution methods................................. 74
    4.3 Classification of global solution methods.................. 76
        4.3.1 Algebraic and Hybrid Techniques.................. 76
        4.3.2 Homotopy (Continuation) Methods 78
        4.3.3 Subdivision Methods............................. 78
    4.4 Projected Polyhedron algorithm 78
    4.5 Auxiliary variable method for nonlinear systems with square
         roots of polynomials 88
    4.6 Robustness issues...................................... 90
    4.7 Interval arithmetic.................................... 92
    4.8 Rounded interval arithmetic and its implementation 95
        4.8.1 Double precision floating point arithmetic 95
        4.8.2 Extracting the exponent from the binary representation 98
        4.8.3 Comparison of two different $unit-in-the-last-place$
               implementations 101
        4.8.4 Hardware rounding for rounded interval arithmetic ... 102
        4.8.5 Implementation of rounded interval arithmetic 103
    4.9 Interval Projected Polyhedron algorithm 105
        4.9.1 Formulation of the governing polynomial equations ... 105
        4.9.2 Comparison of software and hardware rounding 106

5. Intersection Problems 109
    5.1 Overview of intersection problems 109
    5.2 Intersection problem classification 111
        5.2.1 Classification by dimension 112
        5.2.2 Classification by type of geometry 112
        5.2.3 Classification by number system.................. 114
    5.3 Point/point intersection 114
    5.4 Point/curve intersection 114
        5.4.1 Point/implicit algebraic curve intersection 114
        5.4.2 Point/rational polynomial parametric curve intersection 117
        5.4.3 Point/procedural parametric curve intersection 121
    5.5 Point/surface intersection 121
        5.5.1 Point/implicit algebraic surface intersection.......... 121
        5.5.2 Point/rational polynomial parametric surface intersec-
               tion 122
        5.5.3 Point/procedural parametric surface intersection 125
    5.6 Curve/curve intersection 126
        5.6.1 Rational polynomial parametric/implicit algebraic curve
               intersection (Case D3) 126
        5.6.2 Rational polynomial parametric/rational polynomial
               parametric curve intersection (Case D1) 130

5.6.3   Rational polynomial parametric/procedural parametric and procedural parametric/procedural parametric curve intersections (Cases D2 and D5) 131

5.6.4   Procedural parametric/implicit algebraic curve intersection (Case D6) 133

5.6.5   Implicit algebraic/implicit algebraic curve intersection (Case D8) 133

5.7   Curve/surface intersection 134

5.7.1   Rational polynomial parametric curve/implicit algebraic surface intersection (Case E3) 135

5.7.2   Rational polynomial parametric curve/rational polynomial parametric surface intersection (Case E1) 135

5.7.3   Rational polynomial parametric/procedural parametric and procedural parametric/procedural parametric curve/surface intersections (Cases E2/E6) 136

5.7.4   Procedural parametric curve/implicit algebraic surface intersection (Case E7) 136

5.7.5   Implicit algebraic curve/implicit algebraic surface intersection (Case E11) 137

5.7.6   Implicit algebraic curve/rational polynomial parametric surface intersection (Case E9) 137

5.8   Surface/surface intersections 137

5.8.1   Rational polynomial parametric/implicit algebraic surface intersection (Case F3) 138

5.8.2   Rational polynomial parametric/rational polynomial parametric surface intersection (Case F1) 147

5.8.3   Implicit algebraic/implicit algebraic surface intersection (Case F8) 151

5.9   Overlapping of curves and surfaces...................... 155

5.10 Self-intersection of curves and surfaces 157

5.11 Summary... 159

6. **Differential Geometry of Intersection Curves** 161

6.1   Introduction 161

6.2   More differential geometry of curves 162

6.3   Transversal intersection curve 164

6.3.1   Tangential direction 164

6.3.2   Curvature and curvature vector 165

6.3.3   Torsion and third order derivative vector 167

6.3.4   Higher order derivative vector..................... 168

6.4   Intersection curve at tangential intersection points 170

6.4.1   Tangential direction 171

6.4.2   Curvature and curvature vector 173

6.4.3   Third and higher order derivative vector 176

6.5   Examples... 177

6.5.1   Transversal intersection of parametric-implicit surfaces 177
6.5.2   Tangential intersection of implicit-implicit surfaces ... 179

7. Distance Functions 181
7.1   Introduction 181
7.2   Problem formulation 182
    7.2.1   Definition of the distances between two point sets.... 182
    7.2.2   Geometric interpretation of stationarity of distance
            function 184
7.3   More about stationary points 185
    7.3.1   Classification of stationary points.................. 185
    7.3.2   Nonisolated stationary points 190
7.4   Examples... 192

8. Curve and Surface Interrogation 195
8.1   Classification of interrogation methods 195
    8.1.1   Zeroth-order interrogation methods 196
    8.1.2   First-order interrogation methods 197
    8.1.3   Second-order interrogation methods................. 200
    8.1.4   Third-order interrogation methods................... 205
    8.1.5   Fourth-order interrogation methods 208
8.2   Stationary points of curvature of free-form parametric surfaces 210
    8.2.1   Gaussian curvature 210
    8.2.2   Mean curvature 213
    8.2.3   Principal curvatures 214
8.3   Stationary points of curvature of explicit surfaces 215
8.4   Stationary points of curvature of implicit surfaces 221
8.5   Contouring constant curvature 223
    8.5.1   Contouring levels................................. 223
    8.5.2   Finding starting points........................... 223
    8.5.3   Mathematical formulation of contouring 225
    8.5.4   Examples 227

9. Umbilics and Lines of Curvature........................... 231
9.1   Introduction 231
9.2   Lines of curvature near umbilics 232
9.3   Conversion to Monge form............................... 237
9.4   Integration of lines of curvature 242
9.5   Local extrema of principal curvatures at umbilics 244
9.6   Perturbation of generic umbilics 250
9.7   Inflection lines of developable surfaces.................... 256
    9.7.1   Differential geometry of developable surfaces 256
    9.7.2   Lines of curvature near inflection lines 262

10. Geodesics 265
  10.1 Introduction 265
  10.2 Geodesic equation 266
      10.2.1 Parametric surfaces 266
      10.2.2 Implicit surfaces................................... 270
  10.3 Two point boundary value problem 272
      10.3.1 Introduction 272
      10.3.2 Shooting method 273
      10.3.3 Relaxation method 274
  10.4 Initial approximation 275
      10.4.1 Linear approximation 275
      10.4.2 Circular arc approximation 277
  10.5 Shortest path between a point and a curve 278
  10.6 Numerical applications 281
      10.6.1 Geodesic path between two points 281
      10.6.2 Geodesic path between a point and a curve 282
  10.7 Geodesic offsets 284
  10.8 Geodesics on developable surfaces 287

11. Offset Curves and Surfaces 293
  11.1 Introduction 293
      11.1.1 Background and motivation......................... 293
      11.1.2 NC machining 293
      11.1.3 Medial axis 299
      11.1.4 Tolerance region 306
  11.2 Planar offset curves.................................... 307
      11.2.1 Differential geometry 307
      11.2.2 Classification of singularities 308
      11.2.3 Computation of singularities 311
      11.2.4 Approximations 312
  11.3 Offset surfaces 316
      11.3.1 Differential geometry 316
      11.3.2 Singularities of offset surfaces 318
      11.3.3 Self-intersection of offsets of implicit quadratic surfaces 319
      11.3.4 Self-intersection of offsets of explicit quadratic surfaces 328
      11.3.5 Self-intersection of offsets of polynomial parametric
               surface patches..................................... 335
      11.3.6 Tracing of self-intersection curves................... 343
      11.3.7 Approximations 345
  11.4 Pythagorean hodograph 349
      11.4.1 Curves 349
      11.4.2 Surfaces... 351
  11.5 General offsets 352
  11.6 Pipe surfaces 353
      11.6.1 Introduction 353

11.6.2 Local self-intersection of pipe surfaces 355
11.6.3 Global self-intersection of pipe surfaces 356

Problems 367

A. Color Plates 377

References 381

Index 405

1. Representation of Curves and Surfaces

We first introduce three forms to represent geometric objects mathematically. They are the *parametric, implicit* and *explicit* forms. Implicit and explicit forms are often referred to as *nonparametric* forms. Then we briefly review the representation of curves and surfaces in Bézier and B-spline form and treat the special properties associated with each.

1.1 Analytic representation of curves

1.1.1 Plane curves

A plane curve can be expressed in the *parametric* form as

$$x = x(t), \quad y = y(t) , \tag{1.1}$$

where the coordinates of the point (x, y) of the curve are expressed as functions of a parameter t within a closed interval $t_1 \leq t \leq t_2$. The functions $x(t)$ and $y(t)$ are assumed to be continuous with a sufficient number of continuous derivatives. The parametric curve is said to be of *class r*, if the functions have continuous derivatives up to the order r, inclusively [206]. In vector notation the parametric curve can be specified by a vector-valued function

$$\mathbf{r} = \mathbf{r}(t) . \tag{1.2}$$

Another method of representing a curve analytically is to impose one condition on a variable point (x, y) by an equation of the form

$$f(x, y) = 0 . \tag{1.3}$$

This is an *implicit* equation for a plane curve. When $f(x, y)$ is linear in variables x and y, (1.3) represents a straight line. If $f(x, y)$ is of the second degree in x and y (i.e. $ax^2 + 2bxy + cy^2 + 2dx + 2ey + h = 0$), (1.3) represents a variety of plane curves called *conic* sections [79]. The implicit equation for a plane curve can also be expressed as an intersection curve between a parametric surface and a plane. We will discuss this formulation in Chap. 5.

The *explicit* form can be considered as a special case of parametric and implicit forms. If t can be expressed as a function of x or y, we can easily eliminate t from (1.1) to generate the explicit form

$$y = F(x) \quad or \quad x = G(y) . \tag{1.4}$$

This is always possible at least locally when $\frac{dx}{dt} \neq 0$ or $\frac{dy}{dt} \neq 0$ [412]. Conversely if we set x or y in (1.4) to be equal to the parameter t we obtain the parametric form (1.1). Also if the implicit equation (1.3) can be solved for one variable in terms of the other, we also obtain (1.4). This is always possible at least locally when $\frac{\partial f}{\partial y} \neq 0$ or $\frac{\partial f}{\partial x} \neq 0$ [166].

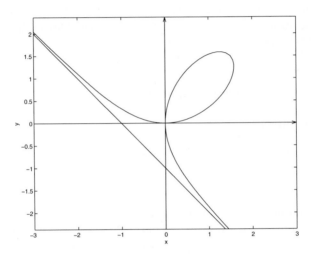

Fig. 1.1. Folium of Descartes

Example 1.1.1. Figure 1.1 shows the Folium of Descartes, introduced by R. Descartes in 1638, with its asymptotic line [227]. It can be expressed in parametric form

$$\mathbf{r}(t) = \left(\frac{3t}{1+t^3}, \frac{3t^2}{1+t^3} \right)^T , \quad -\infty < t < \infty \ (t \neq -1) , \tag{1.5}$$

where superscript T denotes transpose of a vector. For $t < -1$ the curve is located in the fourth quadrant and approaches the origin as t goes to $-\infty$. For $-1 < t < 0$ the curve is located in the second quadrant, and $t = 0$ corresponds to the origin. In the first quadrant it forms a loop moving counter-clockwise as t increases from 0 to $+\infty$. Eliminating t from (1.5), the Folium of Decartes can be also expressed in an implicit form

$$f(x, y) = x^3 + y^3 - 3xy = 0 . \tag{1.6}$$

We can easily trace the curve using the parametric equation (1.5) by evaluating $x(t)$ and $y(t)$ for a discrete sampling of t, while such tracing is more
difficult when using the implicit equation (1.6). However, determining if a
point (x_0, y_0) lies on the curve is easier when using the implicit rather than
the parametric equation of the curve. For example, we can verify that the
point $(\frac{3}{2}, \frac{3}{2})$ lies on the curve by substituting $x = \frac{3}{2}$ and $y = \frac{3}{2}$ into implicit
form and deducing that $f(\frac{3}{2}, \frac{3}{2}) = 0$. However, it is more complex to deduce
this using the parametric form. We first set $x(t) = \frac{3}{2}$ which yields a cubic
equation $t^3 - 2t + 1 = 0$. The roots of the cubic equation are 1, $\frac{-1 \pm \sqrt{5}}{2}$. Then
we substitute each root into $y(t)$ to see if it becomes equal to $\frac{3}{2}$. An alternate
way to do this involves the theory of resultants from algebraic geometry that
we will see in Sect. 5.4.2.

1.1.2 Space curves

The parametric representation of space curves is:

$$x = x(t), \quad y = y(t), \quad z = z(t), \quad t_1 \le t \le t_2 . \tag{1.7}$$

The implicit representation for a space curve can be expressed as an intersection curve between two implicit surfaces

$$f(x, y, z) = 0 \cap g(x, y, z) = 0 , \tag{1.8}$$

or parametric and implicit surfaces

$$\mathbf{r} = \mathbf{r}(u, v) \cap f(x, y, z) = 0 , \tag{1.9}$$

or two parametric surfaces

$$\mathbf{r} = \mathbf{p}(\sigma, t) \cap \mathbf{r} = \mathbf{q}(u, v) . \tag{1.10}$$

The differential geometry properties of the intersection curves between implicit surfaces are discussed in Sects. 2.2 and 2.3 as well as in Chap. 6 together
with the intersection curves between parametric and implicit, and two parametric surfaces. In Sect. 5.8 algorithms for computing the intersections (1.8),
(1.9) and (1.10) are discussed.

   If t can be expressed as a function of x, y, or z, we can eliminate t from
the parametric form (1.7) to generate the explicit form. Let us assume t is a
function of x, then we have

$$y = Y(x), \quad z = Z(x) . \tag{1.11}$$

This is always possible at least locally when $\frac{dx}{dt} \ne 0$ [412]. Also if the two
implicit equations $f(x, y, z) = 0$ and $g(x, y, z) = 0$ can be solved for two of
the variables in terms of the third, for example y and z in terms of x, we
obtain the explicit form (1.11). This is always possible at least locally when
$\frac{\partial f}{\partial y}\frac{\partial g}{\partial z} - \frac{\partial f}{\partial z}\frac{\partial g}{\partial y} \ne 0$ [412]. Therefore the explicit equation for the space curve
can be expressed as an intersection curve of two cylinders projecting the curve
onto xy and xz planes.

1.2 Analytic representation of surfaces

Similar to the curve case there are mainly three ways to represent surfaces, namely parametric, implicit and explicit methods. In *parametric* representation the coordinates of a point (x, y, z) of the surface patch are expressed as functions of the parameters u and v in a closed rectangle:

$$x = x(u, v), \quad y = y(u, v), \quad z = z(u, v), \quad u_1 \leq u \leq u_2, \quad v_1 \leq v \leq v_2 . \quad (1.12)$$

The functions $x(u, v)$, $y(u, v)$ and $z(u, v)$ are continuous and possess a sufficient number of continuous partial derivatives. The parametric surface is said to be of *class* r, if the functions have continuous (partial) derivatives up to the order r, inclusively. In case the class is not explicitly given, it is assumed that the functions have infinitely many derivatives. In vector notation the parametric surface can be specified by a vector-valued function

$$\mathbf{r} = \mathbf{r}(u, v) . \quad (1.13)$$

An *implicit surface* is defined as the locus of points whose coordinates (x, y, z) satisfy an equation of the form

$$f(x, y, z) = 0 . \quad (1.14)$$

When (1.14) is linear in variables x, y and z, it represents a plane. If (1.14) is of second degree in the variables x, y, z, it represents *quadrics* [79]

$$ax^2 + by^2 + cz^2 + dxy + eyz + hxz + kx + ly + mz + n = 0 . \quad (1.15)$$

Some of the quadric surfaces such as elliptic paraboloid, hyperbolic paraboloid and parabolic cylinder have explicit forms (see Fig. 8.9). Paraboloid of revolution is a special case of elliptic paraboloid where the major and minor axes are the same. The rest of the quadrics have implicit forms including ellipsoid, elliptic cone, elliptic cylinder, hyperbolic cylinder, hyperboloid of one sheet and two sheets, where the hyperboloid of revolution is a special form. The natural quadrics, sphere, circular cone and circular cylinder, which are special cases of ellipsoid, elliptic cone and elliptic cylinder, are widely used in mechanical design and CAD/CAM systems. Also they result from standard manufacturing operations such as rolling, turning, filleting, drilling and milling [149]. According to a survey conducted by the Production Automation Project group at the University of Rochester in the mid 1970's, 80-85% of mechanical parts were adequately represented by planes and cylinders, while 90-95% were modeled with the addition of cones [434, 363, 149].

If the implicit equation (1.14) can be solved for one of the variables as a function of the other two, say z is solved in terms of x and y, we obtain an explicit surface

$$z = F(x, y) . \quad (1.16)$$

This is always possible at least locally when $\frac{\partial f}{\partial z} \neq 0$ [166]. And if the two variables u, v of the parametric form can be solved in terms of x and y, we can substitute $u = u(x, y)$ and $v = v(x, y)$ into $z = z(u, v)$ which yields an explicit form. This is possible when $\frac{\partial x}{\partial u}\frac{\partial y}{\partial v} - \frac{\partial x}{\partial v}\frac{\partial y}{\partial u} \neq 0$ [76]. Conversely when the explicit form $z = F(x, y)$ is given, the parametric form is derived by setting $x = u$, $y = v$, $z = F(u, v)$. Thus, the explicit form can be considered as a special case of implicit and parametric forms.

Example 1.2.1. Let us consider a hyperbolic paraboloid surface patch in the parametric form:

$$x = u + v, \quad y = u - v, \quad z = u^2 - v^2, \quad 0 \leq u, v \leq 1.\qquad(1.17)$$

Since we can easily solve for u and v in terms of x and y as $u = \frac{x+y}{2}$ and $v = \frac{x-y}{2}$, the explicit form is obtained as

$$z = xy, \quad 0 \leq x + y \leq 2, \; 0 \leq x - y \leq 2.\qquad(1.18)$$

Table 1.1. Representations of curves and surfaces

Geometry	Parametric	Implicit	Explicit
Plane curves	$x = x(t),\ y = y(t)$ $t_1 \leq t \leq t_2$	$f(x, y) = 0$ or $\mathbf{r} = \mathbf{r}(u, v) \cap$ plane	$y = F(x)$
Space curves	$x = x(t),\ y = y(t),$ $z = z(t),\ t_1 \leq t \leq t_2$	$f(x, y, z) = 0 \cap g(x, y, z) = 0$ or $\mathbf{r} = \mathbf{r}(u, v) \cap f(x, y, z) = 0$ or $\mathbf{r} = \mathbf{p}(\sigma, t) \cap \mathbf{r} = \mathbf{q}(u, v)$	$y = Y(x) \cap$ $z = Z(x)$
Surfaces	$x = x(u, v),$ $y = y(u, v),$ $z = z(u, v),$ $u_1 \leq u \leq u_2,$ $v_1 \leq v \leq v_2$	$f(x, y, z) = 0$	$z = F(x, y)$

Table 1.1 summarizes the three representation forms for plane curves, space curves and surfaces. Table 1.2 compares the three representations [119, 116]. It is clear from the tables that the parametric form is the most versatile method among the three and the explicit is the least. Furthermore, the explicit form can always be easily converted to parametric form. Therefore we will mainly focus on the parametric and implicit forms throughout this book. Methods to fit and manipulate free-form shapes in implicit form are more complex than those for the parametric form both with respect to computation and geometric intuition. However, a considerable body of research aimed at alleviating precisely this obstacle has been published over the last fifteen years, see for example [373, 299, 16]. In this book we do not cover implicit surface fitting and design methods.

Table 1.2. Comparison of different methods of curve and surface representation

Disadvantages		
Explicit	Implicit	Parametric
• Infinite slopes are impossible if $f(x)$ is a polynomial. • Axis dependent (difficult to transform). • Closed and multivalued curves are difficult to represent.	• Difficult to fit and manipulate free form shapes. • Axis dependent. • Complex to trace.	• High flexibility complicates intersections and point classification.
Advantages		
Explicit	Implicit	Parametric
• Easy to trace.	• Closed and multivalued curves and infinite slopes can be represented. • Point classification (solid modeling, interference check) is easy. • Intersections/offsets can be represented.	• Closed and multivalued curves and infinite slopes can be represented. • Axis independent (easy to transform). • Easy to generate composite curves. • Easy to trace. • Easy in fitting and manipulating free-form shapes.

1.3 Bézier curves and surfaces

Good introductory books on Bézier/B-spline curves and surfaces are provided by Faux and Pratt [116], Mortenson [276], Ding and Davies [75], Rogers and Adams [348], Beach [21], Nowacki et al. [289] and Lee [231], while for a more comprehensive mathematical introduction to B-splines, Bezier and B-spline curves and surfaces, the reader should refer to textbooks by Yamaguchi [455], Hosaka [173], Risler [346], Farin [92], Hoschek and Lasser [175], Piegl and Tiller [314] and Gallier [121].

1.3.1 Bernstein polynomials

The *Bernstein polynomials* are defined as

$$B_{i,n}(t) = \frac{n!}{i!(n-i)!}(1-t)^{n-i}t^i, \qquad i = 0,\ldots,n \,. \tag{1.19}$$

They form a basis for polynomials (see Sect. 4.4) and have several properties of interest:

- *Non-negativity:* $B_{i,n}(t) \geq 0, \quad 0 \leq t \leq 1, \quad i = 0, \ldots, n$.
- *Partition of unity:* $\sum_{i=0}^{n} B_{i,n}(t) = (1 - t + t)^n = 1$ (by the *binomial* theorem).
- *Symmetry:*

$$B_{i,n}(t) = B_{n-i,n}(1 - t) . \tag{1.20}$$

- *Recursion:* $B_{i,n}(t) = (1 - t)B_{i,n-1}(t) + tB_{i-1,n-1}(t)$ with $B_{i,n}(t) = 0$ for $i < 0$, $i > n$ and $B_{0,0}(t) = 1$.
- *Linear precision:*

$$t = \sum_{i=0}^{n} \frac{i}{n} B_{i,n}(t) , \tag{1.21}$$

which implies that the monomial t can be expressed as the weighted sum of Bernstein polynomials of degree n with coefficients evenly spaced in the interval [0,1]. This property is used extensively in Chaps. 4 and 5.
- *Degree elevation:* The basis functions of degree n can be expressed in terms of those of degree $n + 1$ [106] as:

$$B_{i,n}(t) = \left(1 - \frac{i}{n+1}\right) B_{i,n+1}(t) + \frac{i+1}{n+1} B_{i+1,n+1}(t) , \tag{1.22}$$

where $i = 0, 1, \cdots, n$. Or more generally in terms of basis functions of degree $n + r$ [106] as:

$$B_{i,n}(t) = \sum_{j=i}^{i+r} \frac{\binom{n}{i}\binom{r}{j-i}}{\binom{n+r}{j}} B_{j,n+r}(t), \quad i = 0, 1, \cdots, n . \tag{1.23}$$

Figure 1.2 shows the Bernstein polynomials of degree 3 and 4. The derivative of a Bernstein polynomial is

$$\frac{dB_{i,n}(t)}{dt} = n[B_{i-1,n-1}(t) - B_{i,n-1}(t)] , \tag{1.24}$$

where $B_{-1,n-1}(t) = B_{n,n-1}(t) = 0$.

1.3.2 Arithmetic operations of polynomials in Bernstein form

Arithmetic operations between polynomials are often required for shape interrogation (see for example Chaps. 4, 5, etc.). Farouki and Rajan [106] provide

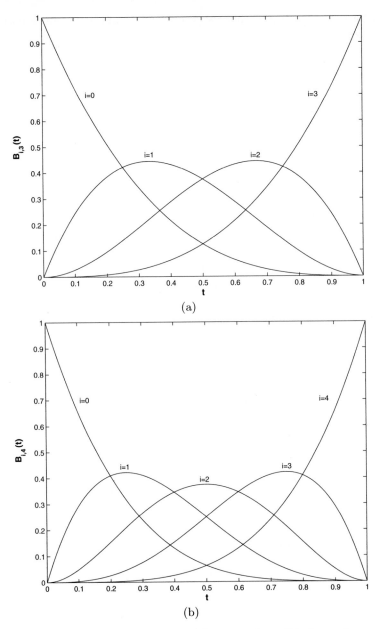

Fig. 1.2. Bernstein polynomials: (a) degree three, (b) degree four

formulae for such arithmetic operations of polynomials in Bernstein form. Let the two polynomials $f(t)$ and $g(t)$ of degree m and n with Bernstein coefficients f_i^m and g_i^n be as follows:

$$f(t) = \sum_{i=0}^{m} f_i^m B_{i,m}(t), \qquad g(t) = \sum_{i=0}^{n} g_i^n B_{i,n}(t), \qquad 0 \le t \le 1. \quad (1.25)$$

- *Addition and subtraction*
  If the degrees of the two polynomials are the same, i.e. $m = n$, we simply add or subtract the coefficients

$$f(t) + g(t) = \sum_{i=0}^{m} (f_i^m \pm g_i^m) B_{i,m}(t). \quad (1.26)$$

If $m > n$, we need to first degree elevate $g(t)$ $m - n$ times using (1.23) and then add or subtract the coefficients

$$f(t) + g(t) = \sum_{i=0}^{m} \left(f_i^m \pm \sum_{j=max(0,i-m+n)}^{min(n,i)} \frac{\binom{n}{j}\binom{m-n}{i-j}}{\binom{m}{i}} g_j^n \right) B_{i,m}(t). \quad (1.27)$$

- *Multiplication*
  Multiplication of two polynomials of degree m and n yields a degree $m+n$ polynomial

$$f(t)g(t) = \sum_{i=0}^{m+n} \left(\sum_{j=max(0,i-n)}^{min(m,i)} \frac{\binom{m}{j}\binom{n}{i-j}}{\binom{m+n}{i}} f_j^m g_{i-j}^n \right) B_{i,m+n}(t). \quad (1.28)$$

1.3.3 Numerical condition of polynomials in Bernstein form

Polynomials in the Bernstein basis have better numerical stability under perturbation of their coefficients than in the power basis. We will introduce the concept of condition numbers for polynomial roots investigated by Farouki and Rajan [105].

Let us consider a polynomial $f(t)$ in the basis $\phi_i(t)$ with coefficients f_i:

$$f(t) = \sum_{i=0}^{n} f_i \phi_i(t). \quad (1.29)$$

If we perturb a single coefficient f_j by δf_j, we have

$$\tilde{f}(t) = f_0\phi_0(t) + f_1\phi_1(t) + \ldots + (f_j + \delta f_j)\phi_j(t) + \ldots + f_n\phi_n(t) \,, \quad (1.30)$$

or using (1.29)

$$\tilde{f}(t) = f(t) + \delta f_j \phi_j(t) \,. \quad (1.31)$$

If $t + \delta t$ is a root of the perturbed polynomial $\tilde{f}(t)$, then

$$\tilde{f}(t + \delta t) = f(t + \delta t) + \delta f_j \phi_j(t + \delta t) = 0 \,, \quad (1.32)$$

or

$$f(t + \delta t) = -\delta f_j \phi_j(t + \delta t) \,. \quad (1.33)$$

Now let us Taylor expand (1.33) about t_0, where t_0 is a root of $f(t)$, i.e. $f(t_0) = 0$,

$$\sum_{i=1}^{n} \frac{(\delta t)^i}{i!} \frac{d^i f}{dt^i}(t_0) = -\delta f_j \sum_{i=0}^{n} \frac{(\delta t)^i}{i!} \frac{d^i \phi_j}{dt^i}(t_0) \,. \quad (1.34)$$

If t_0 is a simple root of $f(t)$, then $\dot{f}(t_0) \neq 0$, and in the limit of infinitesimal perturbations the above equation gives:

$$\lim_{\delta f_j \to 0} \frac{\delta t}{\frac{\delta f_j}{f_j}} = -\frac{f_j \phi_j(t_0)}{\dot{f}(t_0)} \,. \quad (1.35)$$

The absolute value of the right hand side of the above equation

$$C = |f_j \phi_j(t_o)/\dot{f}(t_o)| \,, \quad (1.36)$$

is called the *condition number* of the root t_0 with respect to the single coefficient f_j. For perturbations in each coefficient, $f_j, j = 0, 1, \ldots, n$, the condition number of the root t_0 becomes:

$$C = \frac{\sum_{j=0}^{n} |f_j \phi_j(t_o)|}{|\dot{f}(t_o)|} \,. \quad (1.37)$$

If t_0 is an m-fold root, $m \geq 2$, then a multiple-root condition number $C^{(m)}$ for perturbations in each coefficient $f_j, j = 0, 1, \ldots, n$ is defined as

$$C^{(m)} = \left(\frac{m!}{\left|\frac{d^m f(t_0)}{dt^m}\right|} \sum_{j=0}^{n} |f_j \phi_j(t_0)| \right)^{1/m} \,. \quad (1.38)$$

The following theorem is due to Farouki and Rajan [105].

Table 1.3. Condition numbers for Wilkinson polynomial (adapted from [105])

i	$C_p(x_0)$	$C_b(x_0)$
1	2.100×10^1	3.413×10^0
2	4.389×10^3	1.453×10^2
3	3.028×10^5	2.335×10^3
4	1.030×10^7	2.030×10^4
5	2.059×10^8	1.111×10^5
6	2.667×10^9	4.153×10^5
7	2.409×10^{10}	1.115×10^6
8	1.566×10^{11}	2.215×10^6
9	7.570×10^{11}	3.321×10^6
10	2.775×10^{12}	3.797×10^6
11	7.822×10^{12}	3.321×10^6
12	1.707×10^{13}	2.215×10^6
13	2.888×10^{13}	1.115×10^6
14	3.777×10^{13}	4.153×10^5
15	3.777×10^{13}	1.111×10^5
16	2.833×10^{13}	2.030×10^4
17	1.541×10^{13}	2.335×10^3
18	5.742×10^{12}	1.453×10^2
19	1.310×10^{12}	3.413×10^0
20	1.378×10^{11}	0

Theorem 1.3.1. *For an arbitrary polynomial $f(t)$ with a simple root $t_0 \in [0,1]$, let $C_p(t_0)$ and $C_b(t_0)$ denote the condition numbers (1.37) of the root in the power and Bernstein bases on $[0,1]$, respectively. Then $C_b(t_0) \le C_p(t_0)$ for all $t_0 \in [0,1]$. In particular $C_b(0) = C_p(0) = 0$, while for $t_0 \in (0,1]$ we have the strict inequality $C_b(t_0) < C_p(t_0)$.*

As an illustration of the above theorem, let us consider Wilkinson's polynomial in which twenty real roots are equally distributed on $[0,1]$:

$$f(t) = \prod_{i=1}^{20} (t - i/20) . \qquad (1.39)$$

The condition numbers for each root with respect to a perturbation in the single coefficient of t^{19} are shown in Table 1.3 [105]. We can clearly observe that the condition numbers of the root in the Bernstein basis are several orders of magnitude smaller than in the power basis. This serves to illustrate the attractiveness of using the Bernstein basis in computations in CAD/CAM systems. Although not a panacea, Bernstein basis when used properly in a floating point environment increases reliability of computations (see also detailed discussions in Chaps. 4 and 5).

1.3.4 Definition of Bézier curve and its properties

A *Bézier curve* is a parametric curve that uses the Bernstein polynomials as a basis. A Bézier curve of degree n (order $n+1$) is represented by

$$\mathbf{r}(t) = \sum_{i=0}^{n} \mathbf{b}_i B_{i,n}(t), \qquad 0 \le t \le 1. \qquad (1.40)$$

The coefficients, $\mathbf{b}_i$, are the *control points* or *Bézier points* and together with the basis function $B_{i,n}(t)$ determine the shape of the curve. Lines drawn between consecutive control points of the curve form the *control polygon*. A cubic Bézier curve together with its control polygon is shown in Fig. 1.3 (a). Bézier curves have the following properties:

- *Geometry invariance property:* Partition of unity property of the Bernstein polynomial assures the invariance of the shape of the Bézier curve under translation and rotation of its control points.
- *End points geometric property:*
  - The first and last control points are the endpoints of the curve. In other words, $\mathbf{b}_0 = \mathbf{r}(0)$ and $\mathbf{b}_n = \mathbf{r}(1)$.

  - The curve is tangent to the control polygon at the endpoints. This can be easily observed by taking the first derivative of a Bézier curve

$$\dot{\mathbf{r}}(t) = \frac{d\mathbf{r}(t)}{dt} = n \sum_{i=0}^{n-1} (\mathbf{b}_{i+1} - \mathbf{b}_i) B_{i,n-1}(t), \qquad 0 \le t \le 1. \qquad (1.41)$$

In particular we have $\dot{\mathbf{r}}(0) = n(\mathbf{b}_1 - \mathbf{b}_0)$ and $\dot{\mathbf{r}}(1) = n(\mathbf{b}_n - \mathbf{b}_{n-1})$. Equation (1.41) can be simplified by setting $\Delta\mathbf{b}_i = \mathbf{b}_{i+1} - \mathbf{b}_i$:

$$\dot{\mathbf{r}}(t) = n \sum_{i=0}^{n-1} \Delta\mathbf{b}_i B_{i,n-1}(t), \qquad 0 \le t \le 1. \qquad (1.42)$$

The first derivative of a Bézier curve, which is called *hodograph*, is another Bézier curve whose degree is lower than the original curve by one and has control points $n\Delta\mathbf{b}_i$, $i = 0, \cdots, n-1$. Hodographs are useful in the study of intersection (see Sect. 5.6.2) and other interrogation problems such as singularities and inflection points.
- *Convex hull property:* A domain D is convex if for any two points P_1 and P_2 in the domain, the segment $\overline{P_1 P_2}$ is entirely contained in the domain D [335]. It can be shown that the intersection of convex domains is a

convex domain. The convex hull of a set of points P is the boundary of the smallest convex domain containing P. There are several efficient algorithms for computing the convex hull of a set of points [335, 66, 292].

Using the above definitions and facts, the convex hull of a Bézier curve is the boundary of the intersection of all the convex sets containing all vertices or the intersection of the half spaces generated by taking three vertices at a time to construct a plane and having all other vertices on one side. The convex hull can also be conceptualized at the shape of a rubber band in 2-D or a sheet in 3-D stretched taut over the polygon vertices [75]. The entire curve is contained within the convex hull of the control points as shown in Fig. 1.3 (b). The convex hull property is useful in intersection problems (see Fig. 1.4), in detection of absence of interference and in providing estimates of the position of the curve through simple and efficiently computable bounds.

- *Variation diminishing property*:
  - 2-D: The number of intersections of a straight line with a planar Bézier curve is no greater than the number of intersections of the line with the control polygon. A line intersecting the convex hull of a planar Bézier curve may intersect the curve transversally, be tangent to the curve, or not intersect the curve at all. It may not, however, intersect the curve more times than it intersects the control polygon. This property is illustrated in Fig. 1.5.
  - 3-D: The same relation holds true for a plane with a space Bézier curve.

  From this property, we can roughly say that a Bézier curve oscillates less than its control polygon, or in other words, the control polygon's segments exaggerate the oscillation of the curve. This property is important in intersection algorithms and in detecting the *fairness* of Bézier curves.

- *Symmetry property*:   If we renumber the control points as $\mathbf{b}^*_{n-i} = \mathbf{b}_i$, or in other words relabel from $\mathbf{b}_0, \mathbf{b}_1, \dots, \mathbf{b}_n$ to $\mathbf{b}_n, \mathbf{b}_{n-1}, \dots, \mathbf{b}_0$ and using the symmetry property of the Bernstein polynomial (1.20) the following identity holds:

$$\sum_{i=0}^{n} \mathbf{b}_i B_{i,n}(t) = \sum_{i=0}^{n} \mathbf{b}^*_i B_{i,n}(1-t) . \qquad (1.43)$$

1.3.5 Algorithms for Bézier curves

- *Evaluation and subdivision algorithm*:   A Bézier curve can be evaluated at a specific parameter value t_0 and the curve can be split at that value using the *de Casteljau algorithm* [175], where the following equation

$$\mathbf{b}^k_i(t_0) = (1 - t_0)\mathbf{b}^{k-1}_{i-1} + t_0\mathbf{b}^{k-1}_i, \quad k = 1, 2, \dots, n, \quad i = k, \dots, n ,$$
$$(1.44)$$

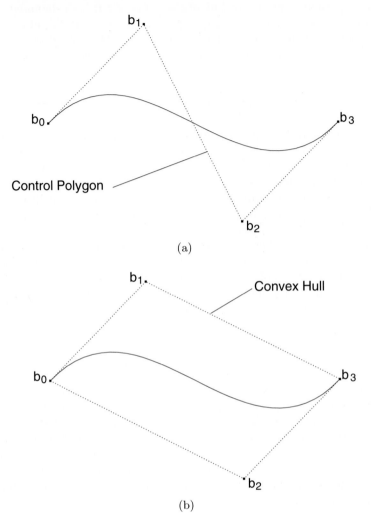

Fig. 1.3. A cubic Bézier curve: **(a)** with control polygon, **(b)** with convex hull

is applied recursively to obtain the new control points. The algorithm is illustrated in Fig. 1.6, and has the following properties:

- The values $\mathbf{b}_i^0$ are the original control points of the curve.
- The value of the curve at parameter value t_0 is $\mathbf{b}_n^n$.
- The curve is split at parameter value t_o and can be represented as two curves, with control points $(\mathbf{b}_0^0, \mathbf{b}_1^1, \ldots, \mathbf{b}_n^n)$ and $(\mathbf{b}_n^n, \mathbf{b}_n^{n-1}, \ldots, \mathbf{b}_n^0)$.
- *Continuity algorithm*:   Bézier curves can represent complex curves by increasing the degree and thus the number of control points. Alternatively, complex curves can be represented using composite curves, which can be

Fig. 1.4. Comparison of convex hulls of Bézier curves as means of detecting intersection

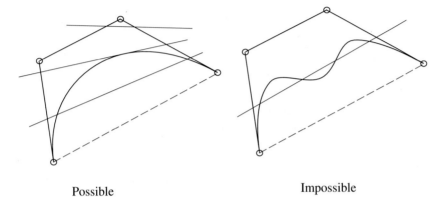

Possible   Impossible

Fig. 1.5. Variation diminishing property of a cubic Bézier curve

formed by joining several Bézier curves end to end. If this method is adopted, the continuity between consecutive curves must be addressed.

One set of continuity conditions are the *geometric* continuity conditions, designated by the letter G with an integer exponent. *Position continuity*, or G^0 continuity, requires the endpoints of the two curves to coincide,

$$\mathbf{r}^a(1) = \mathbf{r}^b(0) . \tag{1.45}$$

The superscripts denote the first and second curves. *Tangent continuity*, or G^1 continuity, requires G^0 continuity and in addition the tangents of the curves to be in the same direction,

$$\dot{\mathbf{r}}^a(1) = \alpha_1 \mathbf{t} , \tag{1.46}$$

$$\dot{\mathbf{r}}^b(0) = \alpha_2 \mathbf{t} , \tag{1.47}$$

where $\mathbf{t}$ is the common unit tangent vector and α_1, α_2 are the magnitude of $\dot{\mathbf{r}}^a(1)$ and $\dot{\mathbf{r}}^b(0)$. G^1 continuity is important in minimizing stress concentrations in physical solids loaded with external forces and in helping prevent flow separation in fluids.

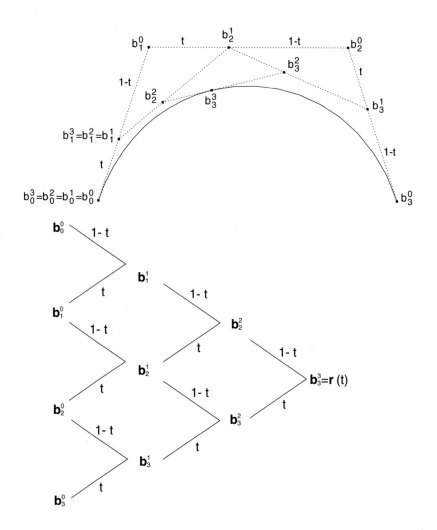

Fig. 1.6. The de Casteljau algorithm

Curvature continuity, or G^2 continuity, requires G^1 continuity and in addition the center of curvature to move continuously past the connection point [116],

$$\ddot{\mathbf{r}}^b(0) = \left(\frac{\alpha_2}{\alpha_1}\right)^2 \ddot{\mathbf{r}}^a(1) + \mu \dot{\mathbf{r}}^a(1) \,, \tag{1.48}$$

where μ is an arbitrary constant. G^2 continuity is important for aesthetic reasons and also for helping prevent fluid flow separation.

More stringent continuity conditions are the *parametric* continuity conditions, where C^k continuity requires the kth derivative (and all lower derivatives) of each curve to be equal at the joining point. In other words,

$$\frac{d^k \mathbf{r}^a(1)}{dt^k} = \frac{d^k \mathbf{r}^b(0)}{dt^k} \ . \tag{1.49}$$

Let us assume that the global parameter t, associated with the i-th segment of a composite degree n Bézier curve with local parameter u_i ($0 \le u_i \le 1$), runs over the interval $[t_i, t_{i+1}]$. Then the i-th segment of a composite Bézier curve is given by:

$$\mathbf{r}_i(t) = \sum_{j=0}^{n} \mathbf{b}_{ni+j} B_{j,n}(u_i) \ , \tag{1.50}$$

where the global parameter t and the local parameter u_i are related by,

$$0 \le u_i = \frac{t - t_i}{t_{i+1} - t_i} \le 1 \ . \tag{1.51}$$

If we denote $h_i = t_{i+1} - t_i$, the C^1 and C^2 continuity conditions for the i-th and $i+1$-th segments of the composite Bézier curve can be stated as [455, 175]:

$$h_{i+1} (\mathbf{b}_{ni} - \mathbf{b}_{ni-1}) = h_i (\mathbf{b}_{ni+1} - \mathbf{b}_{ni}) \ , \tag{1.52}$$

and

$$\mathbf{b}_{ni-1} + \frac{h_{i+1}}{h_i}(\mathbf{b}_{ni-1} - \mathbf{b}_{ni-2}) = \mathbf{b}_{ni+1} + \frac{h_i}{h_{i+1}}(\mathbf{b}_{ni+1} - \mathbf{b}_{ni+2}) \ . \tag{1.53}$$

Figure 1.7 illustrates the connection of two cubic Bézier curve segments at $t = t_{i+1}$.
- *Degree elevation:*   The degree elevation algorithm permits us to increase the degree of a Bézier curve from n to $n+1$ and the number of control points from $n+1$ to $n+2$ without changing the shape of the curve. The new control points $\mathbf{b}_i^{n+1}$ of the degree $n+1$ curve are given by

$$\mathbf{b}_i^{n+1} = \frac{i}{n+1}\mathbf{b}_{i-1}^n + \left(1 - \frac{i}{n+1}\right)\mathbf{b}_i^n, \quad i = 0,\dots,n+1 \ , \tag{1.54}$$

where $\mathbf{b}_{-1}^n = \mathbf{b}_{n+1}^n = \mathbf{0}$. The degree elevation algorithm for a Bézier curve from degree n to $n+r$ is given by [106]:

$$\mathbf{b}_i^{n+r} = \sum_{j=max(0,i-r)}^{min(n,i)} \frac{\binom{n}{j}\binom{r}{i-j}}{\binom{n+r}{i}}\mathbf{b}_j^n, \quad i = 0,1,\cdots,n+r \ . \tag{1.55}$$

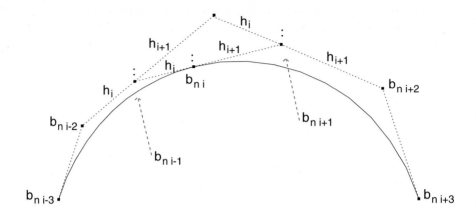

Fig. 1.7. Continuity conditions

1.3.6 Bézier surfaces

A *tensor product surface patch* is formed by moving a curve through space while allowing deformations in that curve. This can be thought of as allowing each control point $\mathbf{b}_i$ to sweep a curve in space. If this surface is represented using Bernstein polynomials, a Bézier surface patch is formed, with the following formula:

$$\mathbf{r}(u,v) = \sum_{i=0}^{m} \sum_{j=0}^{n} \mathbf{b}_{ij} B_{i,m}(u) B_{j,n}(v), \qquad 0 \le u, v \le 1 . \qquad (1.56)$$

Here, the set of straight lines drawn between consecutive control points $\mathbf{b}_{ij}$ is referred to as the *control net*. It is easy to see that boundary iso-parametric curves ($u = 0$, $u = 1$, $v = 0$ and $v = 1$) have the same control points as the corresponding boundary points on the net. An example of a bi-quadratic Bézier surface with its control net can be seen in Fig. 1.8. Since a Bézier surface is a direct extension of univariate Bézier curve to its bivariate form, it inherits many of the properties of the Bézier curve described in Sect. 1.3.4 such as:

- Geometry invariance property.
- End points geometric property.
- Convex hull property.

However, no variation diminishing property is known for Bézier surface patches.

The surface patches treated in this book are mostly topologically quadrilateral. However we sometimes need to use topologically triangular patches. In such cases, we may collapse one boundary curve of a quadrilateral patch into a single point to form a three-sided patch as shown in Fig. 1.9. Such a triangular patch is said to be *degenerate* [116, 92]. Alternatively one could arrange for two partial derivatives $\mathbf{r}_u$ and $\mathbf{r}_v$ at one of the corners of a quadrilateral patch (1.56) to be collinear to create degenerate patches [92]. The differential geometry of degenerated patches is studied in [116, 453, 457].

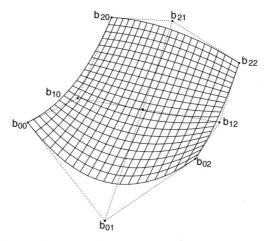

Fig. 1.8. A bi-quadratic Bézier surface with control net

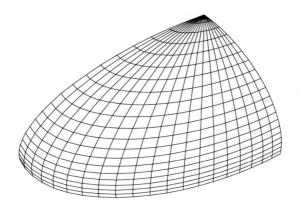

Fig. 1.9. Octant of ellipsoid, represented by a degenerate patch

1.4 B-spline curves and surfaces

The Bézier representation has two main disadvantages. First, the number of control points is directly related to the degree. Therefore, to increase the complexity of the shape of the curve by adding control points requires increasing the degree of the curve or satisfying the continuity conditions between consecutive segments of a composite curve. Second, changing any control point affects the entire curve or surface, making design of specific sections very difficult. These disadvantages are remedied with the introduction of the *B-spline (basis-spline)* representation.

Early fundamental work on the B-spline basis functions was performed almost 50 years ago by Schoenberg [368], and this was followed by development of fundamental algorithms by Cox [67] and de Boor [72, 73]. B-splines in the context of Computer Aided Geometric Design were proven to be a viable and attractive representation method by many pioneers of this field, such as Riesenfeld [345, 130], Boehm [33], Schumaker [369] and many subsequent researchers.

In this section, we provide definitions and the basic properties and algorithms of B-splines. However, we do not deal with fitting, approximation and fairing methods using B-splines which are very important in their own right. For these topics, there are specialized books, monographs and proceedings and a large variety of papers [365, 175, 92, 314, 45].

1.4.1 B-splines

An order k B-spline is formed by joining several pieces of polynomials of degree $k - 1$ with at most C^{k-2} continuity at the breakpoints. A set of nondescending breaking points $t_0 \leq t_1 \leq \ldots \leq t_m$ defines a *knot vector*

$$\mathbf{T} = (t_0, t_1, \ldots, t_m) , \tag{1.57}$$

which determines the parametrization of the basis functions.

Given a knot vector $\mathbf{T}$, the associated B-spline basis functions, $N_{i,k}(t)$, are defined as:

$$N_{i,1}(t) = \begin{cases} 1 \text{ for } t_i \leq t < t_{i+1} \\ 0 \text{ otherwise} , \end{cases} \tag{1.58}$$

for $k = 1$, and

$$N_{i,k}(t) = \frac{t - t_i}{t_{i+k-1} - t_i} N_{i,k-1}(t) + \frac{t_{i+k} - t}{t_{i+k} - t_{i+1}} N_{i+1,k-1}(t) , \tag{1.59}$$

for $k > 1$ and $i = 0, 1, \ldots, n$. These equations have the following properties [175]:

- *Positivity*: $N_{i,k}(t) > 0$, for $t_i < t < t_{i+k}$.

- *Local support:* $N_{i,k}(t) = 0$, for $t_0 \leq t \leq t_i$, and $t_{i+k} \leq t \leq t_{n+k}$.
- *Partition of unity:* $\sum_{i=0}^{n} N_{i,k}(t) = 1$, for $t \in [t_0, t_m]$.
- *Recursion:*   Given by (1.59).
- *Continuity:* $N_{i,k}(t)$ has C^{k-2} continuity at each simple knot.

The concept of *nodes* or *Greville abscissae* [130, 92], which are the averages of the knots, are important in B-spline approximations [130, 452] and defined as follows:

$$\xi_i = \frac{1}{k-1}(t_{i+1} + t_{i+2} + \cdots + t_{i+k-1}) \,. \tag{1.60}$$

The node ξ_i generally lies near the parameter value which corresponds to a maximum of the basis function $N_{i,k}(t)$ [345, 314].

The *derivative* of the B-spline basis function is given by [314]

$$\frac{dN_{i,k}(t)}{dt} = \frac{k-1}{t_{i+k-1} - t_i}N_{i,k-1}(t) - \frac{k-1}{t_{i+k} - t_{i+1}}N_{i+1,k-1}(t) \,. \tag{1.61}$$

1.4.2 B-spline curve

A B-spline curve is defined as a linear combination of control points $\mathbf{p}_i$ and B-spline basis functions $N_{i,k}(t)$ given by

$$\mathbf{r}(t) = \sum_{i=0}^{n} \mathbf{p}_i N_{i,k}(t), \quad n \geq k - 1, \quad t \in [t_{k-1}, t_{n+1}] \,. \tag{1.62}$$

In this context the control points are called *de Boor points*. The basis function $N_{i,k}(t)$ is defined on a *knot vector*

$$\mathbf{T} = (t_0, t_1, \ldots, t_{k-1}, t_k, t_{k+1}, \ldots, t_{n-1}, t_n, t_{n+1}, \ldots, t_{n+k}) \,, \tag{1.63}$$

where there are $n+k+1$ elements, i.e. the number of control points $n+1$ plus the order of the curve k. Each knot *span* $t_i \leq t \leq t_{i+1}$ is mapped onto a poly-nomial curve between two successive joints $\mathbf{r}(t_i)$ and $\mathbf{r}(t_{i+1})$. Normalization of the knot vector, so it covers the interval [0,1], is helpful in improving nu-merical accuracy in floating point arithmetic computation due to the higher density of floating point numbers in this interval [133, 300].

A B-spline curve has the following properties:

- *Geometry invariance property:*   Partition of unity property of the B-spline assures the invariance of the shape of the B-spline curve under translation and rotation.
- *End points geometric property:*
  - Unlike Bézier curves, B-spline curves do not in general pass through the two end control points. Increasing the multiplicity of a knot reduces the continuity of the curve at that knot. Specifically, the curve is $(k - p - 1)$

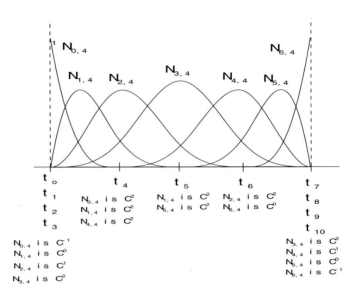

Fig. 1.10. An order four B-spline basis functions with uniform knot vector

times continuously differentiable at a knot with multiplicity p ($\leq k$), and thus has $C^{(k-p-1)}$ continuity. Therefore, the control polygon will coincide with the curve at a knot of multiplicity $k-1$, and a knot with multiplicity k indicates C^{-1} continuity, or a discontinuous curve. Repeating the knots at the end k times will force the endpoints to coincide with the control polygon. Thus the first and the last control points of a curve with a knot vector described by

$$\mathbf{T} = (\underbrace{t_0, t_1, \ldots, t_{k-1}}_{k \text{ equal knots}}, \underbrace{t_k, t_{k+1}, \ldots, t_{n-1}, t_n}_{n\text{-}k+1 \text{ internal knots}}, \underbrace{t_{n+1}, \ldots, t_{n+k}}_{k \text{ equal knots}}) , \quad (1.64)$$

coincide with the endpoints of the curve. Such knot vectors and curves are known as *clamped* [314]. In other words, *clamped/unclamped* refers to whether both ends of the knot vector have multiplicity equal to k or not. Figure 1.10 shows cubic B-spline basis functions defined on a knot vector $\mathbf{T} = (t_0 = t_1 = t_2 = t_3, \ t_4, \ t_5, \ t_6, \ t_7 = t_8 = t_9 = t_{10})$. A clamped cubic B-spline curve based on this knot vector is illustrated in Fig. 1.11 with its control polygon.

- B-spline curves with a knot vector (1.64) are tangent to the control polygon at their endpoints. This is derived from the fact that the first derivative of a B-spline curve is given by [175]

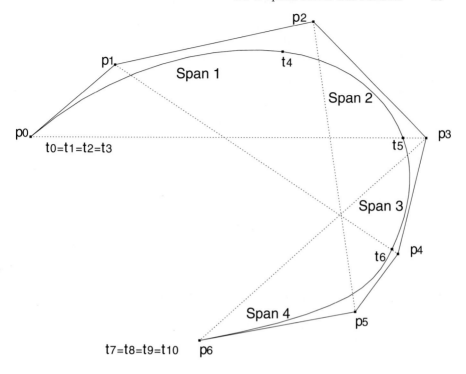

Fig. 1.11. A clamped cubic B-spline curve

$$\dot{\mathbf{r}}(t) = \sum_{i=1}^{n} (k-1) \left(\frac{\mathbf{p}_i - \mathbf{p}_{i-1}}{t_{i+k-1} - t_i} \right) N_{i,k-1}(t) , \qquad (1.65)$$

where the knot vector is obtained by dropping the first and last knots
from (1.64), i.e.

$$\mathbf{T}' = (\;\underbrace{t_1, \ldots, t_{k-1},}_{\text{k-1 equal knots}} \;\underbrace{t_k, t_{k+1}, \ldots, t_{n-1}, t_n,}_{\text{n-k+1 internal knots}} \underbrace{t_{n+1}, \ldots, t_{n+k-1}}_{\text{k-1 equal knots}}) , (1.66)$$

and

$$\dot{\mathbf{r}}(0) = \frac{k-1}{t_k - t_1} (\mathbf{p}_1 - \mathbf{p}_0) , \qquad (1.67)$$

$$\dot{\mathbf{r}}(1) = \frac{k-1}{t_{n+k-1} - t_n} (\mathbf{p}_n - \mathbf{p}_{n-1}) . \qquad (1.68)$$

- *Convex hull property:*   The convex hull property for B-splines applies
  locally, so that a span lies within the convex hull of the control points that
  affect it. This provides a tighter convex hull property than that of a Bézier
  curve, as can be seen in Fig. 1.11. The i-th span of the cubic B-spline curve

in Fig. 1.11 lies within the convex hull formed by control points $\mathbf{p}_{i-1}$, $\mathbf{p}_i$, $\mathbf{p}_{i+1}$, $\mathbf{p}_{i+2}$. In other words, a B-spline curve must lie within the union of all such convex hulls formed by k successive control points [130].

- *Local support property:* A single span of a B-spline curve is controlled only by k control points, and any control point affects k spans. Specifically, changing $\mathbf{p}_i$ affects the curve in the parameter range $t_i < t < t_{i+k}$ and the curve at a point t where $t_r < t < t_{r+1}$ is determined completely by the control points $\mathbf{p}_{r-(k-1)}, \dots, \mathbf{p}_r$ as shown in Fig. 1.11.
- *Variation diminishing property:*
  - 2-D: The number of intersections of a straight line with a planar B-spline curve is no greater than the number of intersections of the line with the control polygon. A line intersecting the convex hull of a planar B-spline curve may intersect the curve transversally, be tangent to the curve, or not intersect the curve at all. It may not, however, intersect the curve more times than it intersects the control polygon.
  - 3-D: The same relation holds true for a plane with a 3-D space B-spline curve.
- *B-spline to Bézier property:* From the discussion of end points geometric property, it can be seen that a Bézier curve of order k (degree $k-1$) is a B-spline curve with no internal knots and the end knots repeated k times. The knot vector is thus

$$\mathbf{T} = (\underbrace{t_0, t_1, \dots, t_{k-1}}_{k \text{ equal knots}}, \underbrace{t_{n+1}, \dots, t_{n+k}}_{k \text{ equal knots}}) , \tag{1.69}$$

where $n + k + 1 = 2k$ or $n = k - 1$.

1.4.3 Algorithms for B-spline curves

- *Evaluation and subdivision algorithm:* A B-spline curve can be evaluated at a specific parameter value $\bar{t}$ using the de Boor algorithm, which is a generalization of the de Casteljau algorithm introduced in Sect. 1.3.5. The repeated substitution of the recursive definition of the B-spline basis function (1.59) into (1.62) and re-indexing leads to the following de Boor algorithm [175]

$$\mathbf{r}(t) = \sum_{i=0}^{n+j} \mathbf{p}_i^j N_{i,k-j}(t), \qquad j = 0, 1, \dots, k-1 , \tag{1.70}$$

where

$$\mathbf{p}_i^j = (1 - \alpha_i^j)\mathbf{p}_{i-1}^{j-1} + \alpha_i^j \mathbf{p}_i^{j-1}, \qquad j > 0 , \tag{1.71}$$

with

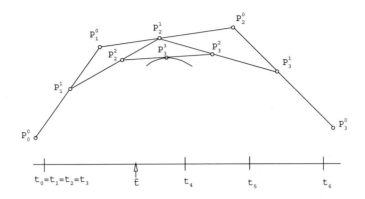

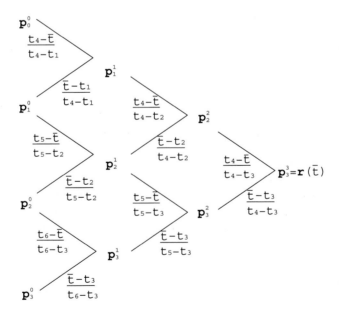

Fig. 1.12. The de Boor algorithm

$$\alpha_i^j = \frac{\bar{t} - t_i}{t_{i+k-j} - t_i} \quad \text{and} \quad \mathbf{p}_j^0 = \mathbf{p}_j . \tag{1.72}$$

For $j = k - 1$, the B-spline basis function reduces to $N_{l,1}$ for $t \in [t_l, t_{l+1}]$, and $\mathbf{p}_l^{k-1}$ coincides with the curve

$$\mathbf{r}(\bar{t}) = \mathbf{p}_l^{k-1} . \tag{1.73}$$

The de Boor algorithm is shown graphically in Fig. 1.12 for a cubic B-spline curve ($\bar{t} \in [t_3, t_4]$). If we compare Figs. 1.6 and 1.12, it is obvious that the de Boor algorithm is a generalization of the de Casteljau algorithm. The de Boor algorithm also permits the subdivision of the B-spline curve into two segments of the same order. In Fig. 1.12, the two new polygons are $\mathbf{p}_0^0 \, \mathbf{p}_1^1 \, \mathbf{p}_2^2 \, \mathbf{p}_3^3$ and $\mathbf{p}_3^3 \, \mathbf{p}_3^2 \, \mathbf{p}_3^1 \, \mathbf{p}_3^0$.

• *Knot insertion*: A knot can be inserted into a B-spline curve without changing the geometry of the curve [34, 314]. The new curve is identical to the old one, with a new basis where

$$\sum_{i=0}^{n} \mathbf{p}_i N_{i,k}(t) \qquad \text{becomes} \qquad \sum_{i=0}^{n+1} \bar{\mathbf{p}}_i \bar{N}_{i,k}(t) \tag{1.74}$$

over $\mathbf{T} = [t_0, t_1, \ldots, t_l, t_{l+1}, \ldots]$ over $\mathbf{T} = [t_0, t_1, \ldots, t_l, \bar{t}, t_{l+1}, \ldots]$,

when a new knot $\bar{t}$ is inserted between knots t_l and t_{l+1}. The new de Boor points are given by

$$\bar{\mathbf{p}}_i = (1 - \alpha_i)\mathbf{p}_{i-1} + \alpha_i \mathbf{p}_i , \tag{1.75}$$

where

$$\alpha_i = \begin{cases} 1 & i \le l - k + 1 \\ 0 & i \ge l + 1 \\ \frac{\bar{t} - t_i}{t_{l+k-1} - t_i} & l - k + 2 \le i \le l . \end{cases} \tag{1.76}$$

The above algorithm is also known as *Boehm's algorithm* [34, 35]. A more general (but also more complex) insertion algorithm permitting insertion of several (possibly multiple) knots into a B-spline knot vector, known as the Oslo algorithm, was developed by Cohen et al. [63]. Both algorithms due to Boehm and Cohen et al. have found wide application in CAD/CAM systems since the early 1980's.

A B-spline curve is C^∞ continuous in the interior of a span. Within exact arithmetic, inserting a knot does not change the curve, so it does not change the continuity. However, if any of the control points are moved after knot insertion, the continuity at the knot will become C^{k-p-1}, where p is the multiplicity of the knot. Figure 1.13 illustrates a single insertion of a knot at parameter value $\bar{t}$, resulting in a knot with multiplicity one.

The B-spline curve can be subdivided into Bézier segments by knot insertion at each internal knot until the multiplicity of each internal knot is equal to k.

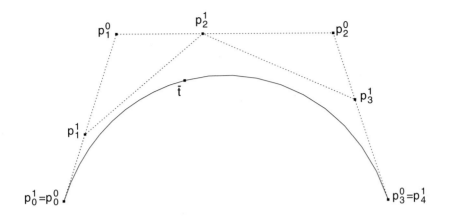

Fig. 1.13. Boehm's algorithm

- *Knot removal:*   Knot removal is the reverse process of knot insertion. It is used for approximation and data reduction [243], and for data reduction in cases the curve or surface does not change neither geometrically nor parametrically [420].

  We briefly review the latter knot removal algorithm developed by Tiller [420]. To demonstrate the process, this example uses a cubic B-spline curve $\mathbf{r}(\bar{t})$ given by control points $(\mathbf{p}_0^0, \ldots, \mathbf{p}_6^0)$ and knot vector $(t_0, \ldots t_{10})$ where $t_0 = \ldots = t_3 = 0$, $t_4 = t_5 = t_6 = 1$ and $t_7 = \ldots = t_{10} = 2$ as shown in Fig. 1.14. As the basis functions only guarantee C^0 continuity at $\bar{t} = 1$, the first derivative may or may not be continuous there. Using the C^1 continuity condition (1.52), the first derivative will be continuous if and only if

  $$(t_7 - t_4)(\mathbf{p}_3^0 - \mathbf{p}_2^0) = (t_6 - t_3)(\mathbf{p}_4^0 - \mathbf{p}_3^0) . \tag{1.77}$$

  Since $t_4 = t_6 = \bar{t}$,

  $$\mathbf{p}_3^0 = \frac{\bar{t} - t_3}{t_7 - t_3}\mathbf{p}_4^0 + \frac{t_7 - \bar{t}}{t_7 - t_3}\mathbf{p}_2^0 . \tag{1.78}$$

  Since $\mathbf{p}_2^0 = \mathbf{p}_2^1$ and $\mathbf{p}_4^0 = \mathbf{p}_3^1$,

  $$\mathbf{p}_3^0 = \alpha_3 \mathbf{p}_3^1 + (1 - \alpha_3)\mathbf{p}_2^1 \qquad \alpha_3 = \frac{\bar{t} - t_3}{t_7 - t_3} . \tag{1.79}$$

  A similar reasoning yields the fact that a knot $\bar{t} = 1$ can be removed a second time, if and only if the second derivative is continuous, yielding

$$\mathbf{p}_2^1 = \alpha_2 \mathbf{p}_2^2 + (1 - \alpha_2)\mathbf{p}_1^2 \,,$$
$$\mathbf{p}_3^1 = \alpha_3 \mathbf{p}_3^2 + (1 - \alpha_3)\mathbf{p}_2^2 \,, \tag{1.80}$$
$$\alpha_i = \frac{\bar{t} - t_i}{t_{i+p+2} - t_i} \qquad i = 2, 3 \,,$$

and a knot $\bar{t} = 1$ can be removed a third time if and only if the third derivative is continuous, yielding

$$\mathbf{p}_1^2 = \alpha_1 \mathbf{p}_1^3 + (1 - \alpha_1)\mathbf{p}_0^3 \,,$$
$$\mathbf{p}_2^2 = \alpha_2 \mathbf{p}_2^3 + (1 - \alpha_2)\mathbf{p}_1^3 \,,$$
$$\mathbf{p}_3^2 = \alpha_3 \mathbf{p}_3^3 + (1 - \alpha_3)\mathbf{p}_2^3 \,, \tag{1.81}$$
$$\alpha_i = \frac{\bar{t} - t_i}{t_{i+p+3} - t_i} \qquad i = 1, 2, 3 \,.$$

Note that there are no unknowns in (1.79), one unknown, $\mathbf{p}_2^2$, in (1.80) and two unknowns, $\mathbf{p}_1^3$ and $\mathbf{p}_2^3$, in (1.81).

For the knot removal process, first the right hand side of (1.79) is computed and compared to $\mathbf{p}_3^0$. If they are equal within a given tolerance, the knot and $\mathbf{p}_3^0$ are removed.

If the first knot removal is successful, equations (1.80) are solved for $\mathbf{p}_2^2$ and compared:

$$\mathbf{p}_2^2 = \frac{\mathbf{p}_2^1 - (1 - \alpha_2)\mathbf{p}_1^2}{\alpha_2} \,, \tag{1.82}$$

$$\mathbf{p}_2^2 = \frac{\mathbf{p}_3^1 - \alpha_3 \mathbf{p}_3^2}{1 - \alpha_3} \,. \tag{1.83}$$

If the two values for $\mathbf{p}_2^2$ are the same, then the knot and control points $\mathbf{p}_2^1$ and $\mathbf{p}_3^1$ are removed and control point $\mathbf{p}_2^2$ is inserted.

If the second knot removal is successful, the third step is to solve the first and third equations of (1.81) for

$$\mathbf{p}_1^3 = \frac{\mathbf{p}_1^2 - (1 - \alpha_1)\mathbf{p}_0^3}{\alpha_1} \,, \tag{1.84}$$

$$\mathbf{p}_2^3 = \frac{\mathbf{p}_3^2 - \alpha_3 \mathbf{p}_3^3}{1 - \alpha_3} \,. \tag{1.85}$$

The two values are then substituted into the second equation of (1.81). If the result is within tolerance of $\mathbf{p}_2^2$, then the knot is removed and control points $\mathbf{p}_1^2$, $\mathbf{p}_2^2$ and $\mathbf{p}_3^2$ are replaced by $\mathbf{p}_1^3$ and $\mathbf{p}_2^3$.

This can be generalized to apply to any number of removals of any particular knot. For the nth removal, there will be a system of n equations with $n - 1$ unknowns. If n is even, two values of the final unknown control point will be calculated and compared. If they are within tolerance, the knot removal is successful. If n is odd, all new control points are computed

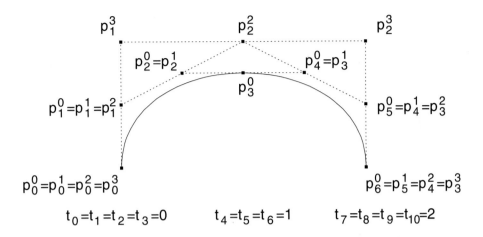

Fig. 1.14. Knot removal (adapted from [420])

and the final two are substituted into the middle equation. If the result is within the tolerance, the knot removal is successful. If the nth removal is successful, n control points will be replaced by $n-1$ control points.

Knot removal from a surface is performed on the $m+1$ rows or $n+1$ columns of control points. However, the knot removal is successful only if the knot can be successfully removed from each row or column. Therefore, the result must be checked for each row or column before any control points are removed.

1.4.4 B-spline surface

The surface analogue of the B-spline curve is the B-spline surface (patch). This is a tensor product surface defined by a topologically rectangular set of control points $\mathbf{p}_{ij}$, $0 \le i \le m$, $0 \le j \le n$ and two knot vectors $\mathbf{U} = (u_0, u_1, \ldots, u_{m+k})$ and $\mathbf{V} = (v_0, v_1, \ldots, v_{n+l})$ associated with each parameter u, v. The corresponding integral B-spline is given by

$$\mathbf{r}(u, v) = \sum_{i=0}^{m} \sum_{j=0}^{n} \mathbf{p}_{ij} N_{i,k}(u) N_{j,l}(v) . \tag{1.86}$$

Parametric lines on a B-spline surface are obtained by letting $u = const$, or $v = const$. A parametric line with $u = u_0$ is a B-spline curve in v with $\mathbf{V}$ as its knot vector and vertices $\mathbf{q}_j$, $0 \le j \le n$ given by $\mathbf{q}_j = \sum_{i=0}^{m} \mathbf{p}_{ij} N_{i,m}(u_0)$.

Some of the properties of the B-spline curves can be easily extended to surfaces such as:

- Geometry invariance property.
- End points geometric property.
- Convex hull property.
- B-spline to Bézier property.

However, no variation diminishing property is known for B-spline surface patches.

1.5 Generalization of B-spline to NURBS

Non-Uniform Rational B-Spline (NURBS) curves and surface patches [433, 314] are the most popular representation method in CAD/CAM due to their generality, excellent properties and incorporation in international standards such as IGES (Initial Graphics Exchange Specification) [182] and STEP (Standard for the Exchange of Product Model Data) [429]. The NURBS functions have the same properties as integral B-splines, and are capable of representing a wider class of geometries. The NURBS curve is represented in a rational form

$$\mathbf{r}(t) = \frac{\sum_{i=0}^{n} w_i \mathbf{p}_i N_{i,k}(t)}{\sum_{i=0}^{n} w_i N_{i,k}(t)} , \qquad (1.87)$$

where $w_i > 0$ is a weighting factor and $N_{i,k}(u)$ is the B-spline basis function. If all the weights are equal to one, the integral B-spline is recovered. If the number of control points equals the order of the NURBS curve, then the curve reduces to a rational Bézier curve

$$\mathbf{r}(t) = \frac{\sum_{i=0}^{n} w_i \mathbf{b}_i B_{i,n}(t)}{\sum_{i=0}^{n} w_i B_{i,n}(t)} . \qquad (1.88)$$

The NURBS formulation permits exact representation of conics, such as circle, ellipse and hyperbola.

Example 1.5.1. Let us express the first quadrant of an ellipse as a rational Bézier curve as shown in Fig. 1.15. A parametric representation of such ellipse segment is given by

$$x = a \cos \theta, \qquad y = b \sin \theta, \qquad 0 \le \theta \le \frac{\pi}{2} ,$$

where θ is an angle parameter. If we set

$$t = \tan \frac{\theta}{2} = \sqrt{\frac{1 - \cos \theta}{1 + \cos \theta}} ,$$

then

$$\cos\theta = \frac{1-t^2}{t^2+1}, \qquad \sin\theta = \frac{2t}{t^2+1}.$$

Therefore the first quadrant of the ellipse can be described by

$$x(t) = a\frac{1-t^2}{t^2+1}, \qquad y(t) = b\frac{2t}{t^2+1}, \qquad 0 \le t \le 1. \qquad (1.89)$$

On the other hand a third order rational Bézier curve is given by

$$\mathbf{r}(t) = \frac{w_0(1-t)^2\mathbf{b}_0 + w_1 2t(1-t)\mathbf{b}_1 + w_2 t^2\mathbf{b}_2}{w_0(1-t)^2 + w_1 2t(1-t) + w_2 t^2}, \qquad 0 \le t \le 1. \qquad (1.90)$$

By equating the denominators of (1.89) and (1.90), we find the weights to be $w_0 = 1$, $w_1 = 1$ and $w_2 = 2$. The three control points $\mathbf{b}_0$, $\mathbf{b}_1$, $\mathbf{b}_2$ can then be easily obtained by the end points geometric property of the Bézier curve as $\mathbf{b}_0 = (a,0)^T$, $\mathbf{b}_1 = (a,b)^T$, and $\mathbf{b}_2 = (0,b)^T$.

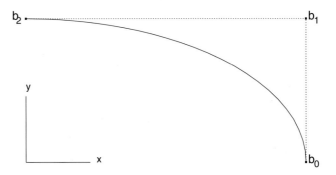

Fig. 1.15. The first quadrant of an ellipse described by a rational Bézier curve

A NURBS surface patch can be represented as

$$\mathbf{r}(u,v) = \frac{\sum_{i=0}^{m}\sum_{j=0}^{n} w_{ij}\mathbf{P}_{ij}N_{i,k}(u)N_{j,l}(v)}{\sum_{i=0}^{m}\sum_{j=0}^{n} w_{ij}N_{i,k}(u)N_{j,l}(v)}, \qquad (1.91)$$

where $w_{ij} > 0$ is a weighting factor. This formulation allows for exact representation of quadrics, tori, surfaces of revolution and very general free-form surfaces. If all $w_{ij} = 1$, the integral B-spline surface is recovered. If the number of control points are equal to the order of the B-spline basis function in both parameters u and v, then the NURBS surface reduces to a rational Bézier surface patch:

$$\mathbf{r}(u,v) = \frac{\sum_{i=0}^{m}\sum_{j=0}^{n} w_{ij}\mathbf{b}_{ij}B_{i,m}(u)B_{j,n}(v)}{\sum_{i=0}^{m}\sum_{j=0}^{n} w_{ij}B_{i,m}(u)B_{j,n}(v)} \ . \qquad (1.92)$$

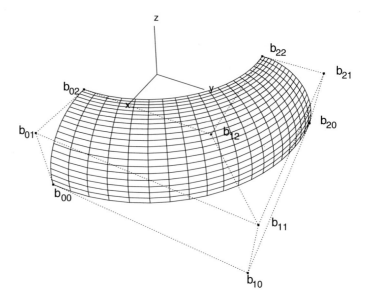

Fig. 1.16. 1/16 of a torus represented by a rational Bézier surface patch

Example 1.5.2. Let us express 1/16th of a torus (in the first octant of a coordinate frame) as a rational Bézier surface as shown in Fig. 1.16. A parametric representation of such a toroidal surface patch is given by

$$\mathbf{r}(\theta,\phi) = (R + a\cos\phi)\cos\theta\,\mathbf{i} + (R + a\cos\phi)\sin\theta\,\mathbf{j} + a\sin\phi\,\mathbf{k}, \quad 0 \le \theta, \phi \le \frac{\pi}{2}\,,$$

where θ and ϕ are angle parameters, $R > a$, and $\mathbf{i}, \mathbf{j}, \mathbf{k}$ are unit vectors having the directions of the positive x, y and z axes, respectively. If we set

$$u = \tan\frac{\theta}{2} = \sqrt{\frac{1-\cos\theta}{1+\cos\theta}}, \qquad v = \tan\frac{\phi}{2} = \sqrt{\frac{1-\cos\phi}{1+\cos\phi}}, \qquad 0 \le u,v \le 1\,,$$

then

$$\cos\theta = \frac{1-u^2}{u^2+1}, \qquad \sin\theta = \frac{2u}{u^2+1}, \qquad \cos\phi = \frac{1-v^2}{v^2+1}, \qquad \sin\phi = \frac{2v}{v^2+1}\ .$$

Thus the toroidal surface patch under consideration can be described by

$$\mathbf{r}(u,v) = \left(R + a\frac{1-v^2}{v^2+1}\right)\frac{1-u^2}{u^2+1}\mathbf{i} + \left(R + a\frac{1-v^2}{v^2+1}\right)\frac{2u}{u^2+1}\mathbf{j} + \frac{2va}{v^2+1}\mathbf{k},$$
$$0 \le u, v \le 1. \tag{1.93}$$

Now we will convert this rational polynomial surface patch into a rational Bézier surface patch. A biquadratic Bézier surface is given by

$$\mathbf{r}(u,v) = \frac{\sum_{i=0}^{2}\sum_{j=0}^{2}w_{ij}\mathbf{b}_{ij}B_{i,2}(u)B_{j,2}(v)}{\sum_{i=0}^{2}\sum_{j=0}^{2}w_{ij}B_{i,2}(u)B_{j,2}(v)}. \tag{1.94}$$

By equating the denominators of (1.93) and (1.94), we find the weights to be

$$w_{00} = 1, \quad w_{01} = 1, \quad w_{02} = 2,$$
$$w_{10} = 1, \quad w_{11} = 1, \quad w_{12} = 2,$$
$$w_{20} = 2, \quad w_{21} = 2, \quad w_{22} = 4.$$

The nine control points $\mathbf{b}_{ij}$, $0 \le i, j \le 2$ can then be easily obtained by the end points geometric property of the Bézier surface as:

$$\mathbf{b}_{00} = (R+a,0,0)^T, \quad \mathbf{b}_{01} = (R+a,0,a)^T, \quad \mathbf{b}_{02} = (R,0,a)^T,$$
$$\mathbf{b}_{10} = (R+a,R+a,0)^T, \quad \mathbf{b}_{11} = (R+a,R+a,a)^T, \quad \mathbf{b}_{12} = (R,R,a)^T,$$
$$\mathbf{b}_{20} = (0,R+a,0)^T, \quad \mathbf{b}_{21} = (0,R+a,a)^T, \quad \mathbf{b}_{22} = (0,R,a)^T.$$

2. Differential Geometry of Curves

The differential geometry of curves and surfaces is fundamental in Computer Aided Geometric Design (CAGD). The curves and surfaces treated in differential geometry are defined by functions which can be differentiated a certain number of times. Books by Hilbert and Cohn-Vossen [165], Koenderink [205] provide intuitive introductions to the extensive mathematical literature on three-dimensional shape analysis. The books by Struik [412], Willmore [444], Kreyszig [206], Lipschutz [235], do Carmo [76] offer firm theoretical basis to the differential geometry aspects of three-dimensional shape description. A book by Gray [136] combines the traditional textbook style and a symbolic manipulation program MATHEMATICA. In a recent textbook, Gallier [122] provides a thorough introduction to differential geometry as well as a comprehensive treatment of affine and projective geometry and their applications to rational curves and surfaces in addition to basic topics of computational geometry (eg. convex hulls, Voronoi diagrams and Delaunay triangulations). We briefly review elementary differential geometry of curves in this chapter and surfaces in Chap. 3.

2.1 Arc length and tangent vector

Let us consider a segment of a parametric curve $\mathbf{r} = \mathbf{r}(t)$ between two points P $(\mathbf{r}(t))$ and Q $(\mathbf{r}(t + \Delta t))$ as shown in Fig. 2.1. Its length Δs can be approximated by a chord length $|\Delta \mathbf{r}| = |\mathbf{r}(t + \Delta t) - \mathbf{r}(t)|$, and by means of a Taylor expansion we have

$$\Delta s \simeq |\Delta \mathbf{r}| = |\mathbf{r}(t + \Delta t) - \mathbf{r}(t)| = \left| \frac{d\mathbf{r}}{dt} \Delta t + \frac{d^2\mathbf{r}}{dt^2}(\Delta t)^2 \right| \simeq \left| \frac{d\mathbf{r}}{dt} \right| \Delta t , \quad (2.1)$$

to the first order approximation.

Thus as point Q approaches P or in other words $\Delta t \to 0$, the length Δs becomes the differential arc length of the curve:

$$ds = \left| \frac{d\mathbf{r}}{dt} \right| dt = |\dot{\mathbf{r}}| dt = \sqrt{\dot{\mathbf{r}} \cdot \dot{\mathbf{r}}} dt . \quad (2.2)$$

Here the dot ˙ denotes differentiation with respect to the parameter t. Therefore the arc length of a segment of the curve between points $\mathbf{r}(t_o)$ and $\mathbf{r}(t)$ can

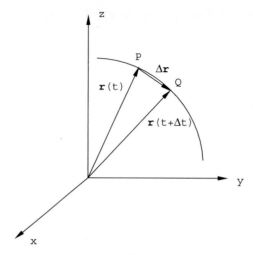

Fig. 2.1. A segment $\Delta\mathbf{r}$ connecting two point P and Q on a parametric curve $\mathbf{r}(t)$

be obtained as follows (provided the function $t \in [t_0, t] \to \mathbf{r}(t)$ is one-to-one almost everywhere):

$$s(t) = \int_{t_o}^{t} ds = \int_{t_o}^{t} \sqrt{\dot{\mathbf{r}} \cdot \dot{\mathbf{r}}}\, dt = \int_{t_o}^{t} \sqrt{\dot{x}^2(t) + \dot{y}^2(t) + \dot{z}^2(t)}\, dt \ . \tag{2.3}$$

The vector $\frac{d\mathbf{r}}{dt}$ is called the *tangent vector* at point P. This tangent vector has a simple geometrical interpretation. The vector $\mathbf{r}(t + \Delta t) - \mathbf{r}(t)$ indicates the direction from $\mathbf{r}(t)$ to $\mathbf{r}(t + \Delta t)$. If we divide the vector by Δt and take the limit as $\Delta t \to 0$, then the vector will converge to the finite magnitude vector $\dot{\mathbf{r}}(t)$, i.e. the tangent vector. The magnitude of the tangent vector is derived from (2.2) as

$$|\dot{\mathbf{r}}| = \frac{ds}{dt} \ , \tag{2.4}$$

hence the unit tangent vector becomes

$$\mathbf{t} = \frac{\dot{\mathbf{r}}}{|\dot{\mathbf{r}}|} = \frac{\frac{d\mathbf{r}}{dt}}{\frac{ds}{dt}} = \frac{d\mathbf{r}}{ds} \equiv \mathbf{r}' \ . \tag{2.5}$$

Here the prime $'$ denotes differentiation with respect to the arc length. We will keep these notations, i.e. dot $\dot{\ }$ is for differentiation with respect to non-arc-length parameter t and prime $'$ with respect to arc length parameter s throughout the book. We list some useful formulae of the derivatives of arc length s with respect to parameter t and vice versa:

$$\dot{s} = \frac{ds}{dt} = |\dot{\mathbf{r}}| = \sqrt{\dot{\mathbf{r}} \cdot \dot{\mathbf{r}}} \ , \tag{2.6}$$

$$\ddot{s} = \frac{d\dot{s}}{dt} = \frac{\dot{\mathbf{r}} \cdot \ddot{\mathbf{r}}}{\sqrt{\dot{\mathbf{r}} \cdot \dot{\mathbf{r}}}} , \tag{2.7}$$

$$\dddot{s} = \frac{d\ddot{s}}{dt} = \frac{(\dot{\mathbf{r}} \cdot \dot{\mathbf{r}})(\dot{\mathbf{r}} \cdot \dddot{\mathbf{r}} + \ddot{\mathbf{r}} \cdot \ddot{\mathbf{r}}) - (\dot{\mathbf{r}} \cdot \ddot{\mathbf{r}})^2}{(\dot{\mathbf{r}} \cdot \dot{\mathbf{r}})^{\frac{3}{2}}} , \tag{2.8}$$

$$t' = \frac{dt}{ds} = \frac{1}{|\dot{\mathbf{r}}|} = \frac{1}{\sqrt{\dot{\mathbf{r}} \cdot \dot{\mathbf{r}}}} , \tag{2.9}$$

$$t'' = \frac{dt'}{ds} = -\frac{\dot{\mathbf{r}} \cdot \ddot{\mathbf{r}}}{(\dot{\mathbf{r}} \cdot \dot{\mathbf{r}})^2} , \tag{2.10}$$

$$t''' = \frac{dt''}{ds} = -\frac{(\ddot{\mathbf{r}} \cdot \ddot{\mathbf{r}} + \dot{\mathbf{r}} \cdot \dddot{\mathbf{r}})(\dot{\mathbf{r}} \cdot \dot{\mathbf{r}}) - 4(\dot{\mathbf{r}} \cdot \ddot{\mathbf{r}})^2}{(\dot{\mathbf{r}} \cdot \dot{\mathbf{r}})^{\frac{7}{2}}} . \tag{2.11}$$

Definition 2.1.1. *A regular (ordinary) point P on a parametric curve $\mathbf{r} = \mathbf{r}(t) = (x(t), y(t), z(t))^T$ is defined as a point where $|\dot{\mathbf{r}}(t)| \neq 0$. A point which is not a regular point is called a singular point.*

Definition 2.1.2. *A parametrization $\mathbf{r} = \mathbf{r}(t) = (x(t), y(t), z(t))^T$ of a curve defined in the interval I is called an allowable representation of class r [207], if it satisfies the following:*

1. *the mapping $\mathbf{r} : I \to \mathbf{R}^3$, $t \mapsto \mathbf{r}(t) = (x(t), y(t), z(t))^T$ is one-to-one,*
2. *the vector function $\mathbf{r} = \mathbf{r}(t)$ is of class $r \geq 1$ in the interval I,*
3. *$|\dot{\mathbf{r}}(t)| \neq 0$ for all $t \in I$.*

A parametric curve satisfying Definition 2.1.2 is also referred to as a *regular curve*. The magnitude of the tangent vector $\frac{ds}{dt}$ can be interpreted as a rate of change of the arc length s with respect to the parameter t and is called the *parametric speed*. If we assume the curve $\dot{\mathbf{r}}(t)$ to be regular, then by definition $|\dot{\mathbf{r}}(t)|$ is never zero and hence $\frac{ds}{dt}$ is always positive. When $\frac{ds}{dt} = 1$, the curve is said to be *arc length parametrized* or to have *unit speed*. If the parametric speed does not vary significantly, points of the curve obtained at parameter values $t_0, t_1, \cdots, t_N$ corresponding to a uniform increment $\Delta t = t_k - t_{k-1}$, will be nearly evenly distributed along the curve, as illustrated in Fig. 2.2. It is well known that every regular curve has an arc length parametrization [109], however, in practice it is very difficult to find it analytically, due to the fact that (2.3) is hard to integrate analytically. *Pythagorean hodograph (PH) curves*, introduced by Farouki and Sakkalis [108, 110], form a class of special planar polynomial curves whose parametric speed is a polynomial. Accordingly, its arc length is a polynomial function $s(t)$ of the parameter t. We provide a further review of Pythagorean hodograph curves and surfaces in Sect. 11.4.

Definition 2.1.3. *A point (x_0, y_0) of a planar irreducible implicit curve $f(x, y) = 0$ is said to be singular if $f(x_0, y_0) = f_x(x_0, y_0) = f_y(x_0, y_0) = 0$.*

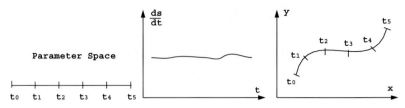

Fig. 2.2. When parametric speed does not vary significantly, points with uniformly spaced parameter values are nearly uniformly spaced along a parametric curve

The unit tangent vector for implicit curves can also be derived as follows. First we start with the planar curve $f(x, y) = 0$. The differential df of the implicit form $f = 0$ is zero, thus by letting $f_x = \frac{\partial f}{\partial x}$ and $f_y = \frac{\partial f}{\partial y}$ we have

$$df = f_x dx + f_y dy = 0 , \qquad (2.12)$$

or assuming $f_y \neq 0$,

$$\frac{dy}{dx} = -\frac{f_x}{f_y} . \qquad (2.13)$$

Therefore the tangent vector on the implicit curve is given by $\pm(f_y, -f_x)^T$, and hence the unit tangent vector is

$$\mathbf{t} = \pm \frac{(f_y, -f_x)^T}{\sqrt{f_x^2 + f_y^2}} . \qquad (2.14)$$

The sign depends on the sense in which s increases.

As shown in Table 1.1, an implicit space curve is defined as the intersection of two implicit surfaces, $f(x, y, z) = 0$ and $g(x, y, z) = 0$. As we will see in Sect. 3.1, the normal vectors of these two implicit surfaces are ∇f and ∇g, respectively, where the symbol ∇ represents the gradient vector operator which is of the form $\nabla = \left(\frac{\partial}{\partial x}, \frac{\partial}{\partial y}, \frac{\partial}{\partial z} \right)^T$.

Since the tangent vector to the intersection curve is orthogonal to the normals of the two implicit surfaces, the unit tangent vector is given by

$$\mathbf{t} = \pm \frac{\nabla f \times \nabla g}{|\nabla f \times \nabla g|} , \qquad (2.15)$$

provided that the denominator is nonzero ($\nabla f \neq \mathbf{0}$ and $\nabla g \neq \mathbf{0}$ or in other words the two surfaces are nonsingular and the surfaces are not tangent to each other at their common point under consideration). The unit tangent vector of the intersection of two implicit surfaces, when the two surfaces intersect tangentially is given in Sect. 6.4. Also here the sign depends on the sense in which s increases. A more detailed treatment of the tangent vector of implicit curves resulting from intersection of various types of surfaces can be found in Chap.6.

Example 2.1.1. The semi-cubical parabola, which is illustrated in Fig. 2.3, can be represented in parametric form as the curve $\mathbf{r}(t) = (t^2, t^3)^T$ [227]. The parametric speed is evaluated as $|\dot{\mathbf{r}}(t)| = \sqrt{t^2(4 + 9t^2)}$. It becomes zero when $t = 0$, hence it is singular at the origin and forms a cusp, which is illustrated in Fig. 2.3. The curve can be also represented implicitly $f(x, y) = x^3 - y^2 = 0$. We can also observe that $f(0, 0) = f_x(0, 0) = f_y(0, 0) = 0$.

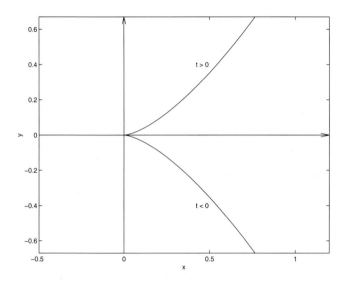

Fig. 2.3. A singular point occurs on a semi-cubical parabola in the form of a cusp

2.2 Principal normal and curvature

If $\mathbf{r}(s)$ is an arc length parametrized curve, then $\mathbf{r}'(s)$ is a unit vector (see (2.5)), and hence $\mathbf{r}' \cdot \mathbf{r}' = 1$. Differentiating this relation, we obtain

$$\mathbf{r}' \cdot \mathbf{r}'' = 0 , \tag{2.16}$$

which states that $\mathbf{r}''$ is orthogonal to the tangent vector, provided it is not a null vector. This fact can be also interpreted from the definition of the second derivative $\mathbf{r}''(s)$

$$\mathbf{r}''(s) = \lim_{\Delta s \to 0} \frac{\mathbf{r}'(s + \Delta s) - \mathbf{r}'(s)}{\Delta s} . \tag{2.17}$$

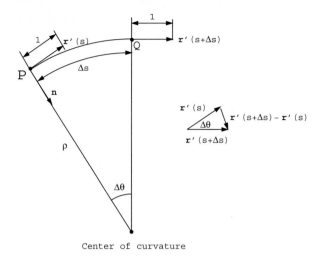

Fig. 2.4. Derivation of the normal vector of a curve (adapted from [455])

As shown in Fig. 2.4, the direction of $\mathbf{r}'(s + \Delta s) - \mathbf{r}'(s)$ becomes perpendicular to the tangent vector as $\Delta s \to 0$. The unit vector

$$\mathbf{n} = \frac{\mathbf{r}''(s)}{|\mathbf{r}''(s)|} = \frac{\mathbf{t}'(s)}{|\mathbf{t}'(s)|} , \tag{2.18}$$

which has the direction and sense of $\mathbf{t}'(s)$ is called the *unit principal normal vector* at s. The plane determined by the unit tangent and normal vectors $\mathbf{t}(s)$ and $\mathbf{n}(s)$ is called the *osculating plane* at s. It is also well known that the plane through three consecutive points of the curve approaching a single point defines the osculating plane at that point [412].

When $\mathbf{r}'(s + \Delta s)$ is moved from Q to P, then $\mathbf{r}'(s)$, $\mathbf{r}'(s + \Delta s)$ and $\mathbf{r}'(s + \Delta s) - \mathbf{r}'(s)$ form an isosceles triangle (see Fig. 2.4), since $\mathbf{r}'(s + \Delta s)$ and $\mathbf{r}'(s)$ are unit tangent vectors. Thus we have $|\mathbf{r}'(s + \Delta s) - \mathbf{r}'(s)| = \Delta \theta \cdot 1 = \Delta \theta = |\mathbf{r}''(s) \Delta s|$ as $\Delta s \to 0$ and hence

$$|\mathbf{r}''(s)| = \lim_{\Delta s \to 0} \frac{\Delta \theta}{\Delta s} = \lim_{\Delta s \to 0} \frac{\Delta \theta}{\varrho \Delta \theta} = \frac{1}{\varrho} \equiv \kappa . \tag{2.19}$$

κ is called the *curvature* , and its reciprocal ϱ is called the *radius of curvature* at s. It follows that

$$\mathbf{r}'' = \mathbf{t}' = \kappa \mathbf{n} . \tag{2.20}$$

The vector $\mathbf{k} = \mathbf{r}'' = \mathbf{t}'$ is called the *curvature vector*, and measures the rate of change of the tangent along the curve. By definition κ is nonnegative, thus the sense of the normal vector is the same as that of $\mathbf{r}''(s)$.

The curvature for arbitrary speed (non-arc-length parametrized) curve can be obtained as follows. First we evaluate $\dot{\mathbf{r}}$ and $\ddot{\mathbf{r}}$ by the chain rule

$$\dot{\mathbf{r}} = \frac{d\mathbf{r}}{ds}\frac{ds}{dt} = \mathbf{t}v , \tag{2.21}$$

$$\ddot{\mathbf{r}} = \frac{d}{dt}[\mathbf{t}v] = \frac{d\mathbf{t}}{ds}v^2 + \mathbf{t}\frac{dv}{dt} = \kappa\mathbf{n}v^2 + \mathbf{t}\frac{dv}{dt} , \tag{2.22}$$

where $v = \frac{ds}{dt}$ is the parametric speed. Taking the cross product of $\dot{\mathbf{r}}$ and $\ddot{\mathbf{r}}$ we obtain

$$\dot{\mathbf{r}} \times \ddot{\mathbf{r}} = \kappa v^3 \mathbf{t} \times \mathbf{n} . \tag{2.23}$$

For the planar curve, we can give the curvature κ a sign by defining the normal vector such that $(\mathbf{t}, \mathbf{n}, \mathbf{e}_z)$ form a right-handed screw, where $\mathbf{e}_z = (0, 0, 1)^T$ as shown in Fig. 2.5. The point where the curvature changes sign is called an *inflection point* (see also Fig. 8.3).

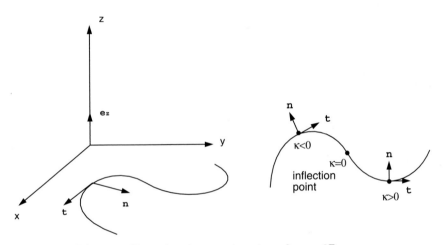

Fig. 2.5. Normal and tangent vectors along a 2D curve

According to this definition the unit normal vector of the plane curve is given by

$$\mathbf{n} = \mathbf{e}_z \times \mathbf{t} = \frac{(-\dot{y}, \dot{x})^T}{\sqrt{\dot{x}^2 + \dot{y}^2}} , \tag{2.24}$$

and hence from (2.23) we have

$$\kappa = \frac{(\dot{\mathbf{r}} \times \ddot{\mathbf{r}}) \cdot \mathbf{e}_z}{v^3} = \frac{\dot{x}\ddot{y} - \dot{y}\ddot{x}}{(\dot{x}^2 + \dot{y}^2)^{\frac{3}{2}}} . \tag{2.25}$$

For a space curve, by taking the norm of (2.23) and using (2.4), we obtain

$$\kappa = \frac{|\dot{\mathbf{r}} \times \ddot{\mathbf{r}}|}{|\dot{\mathbf{r}}|^3} \ . \tag{2.26}$$

The normal vector for the arbitrary speed curve can be obtained from $\mathbf{n} = \mathbf{b} \times \mathbf{t}$, where $\mathbf{b}$ is the unit *binormal vector* which will be introduced in Sect. 2.3 (see (2.41)).

The unit principal normal vector and curvature for implicit curves can be obtained as follows. For the planar curve the normal vector can be deduced by combining (2.14) and (2.24) yielding

$$\mathbf{n} = \mathbf{e}_z \times \mathbf{t} = \frac{(f_x, f_y)^T}{\sqrt{f_x^2 + f_y^2}} = \frac{\nabla f}{|\nabla f|} \ , \tag{2.27}$$

where only the $+$ sign of $\mathbf{t}$ was used (although it is not necessary).

We will introduce a derivative operator with respect to arc length so that the derivation becomes simple. If we rewrite the plane implicit curve as $f(x(s), y(s)) = 0$ where s is arc length along the implicit curve, the total derivative with respect to the arc length becomes

$$\frac{df}{ds} = \frac{\partial f}{\partial x}\frac{dx}{ds} + \frac{\partial f}{\partial y}\frac{dy}{ds} \ . \tag{2.28}$$

Now if we replace $\frac{dx}{ds}$ and $\frac{dy}{ds}$ by using (2.5) and (2.14) ($+$ sign), we obtain the derivative operator with respect to arc length

$$\frac{d}{ds} = \frac{1}{|\nabla f|}\left(f_y\frac{\partial}{\partial x} - f_x\frac{\partial}{\partial y}\right) \ . \tag{2.29}$$

By applying the operator (2.29) to (2.14) ($+$ sign) and equating with $\kappa\mathbf{n}$ (using (2.20) and (2.27)), we obtain

$$\kappa = -\frac{f_{xx}f_y^2 - 2f_{xy}f_xf_y + f_x^2f_{yy}}{(f_x^2 + f_y^2)^{\frac{3}{2}}} \ . \tag{2.30}$$

For a 3-D implicit curve, we can deduce a derivative operator [444] similar to (2.29),

$$\frac{d}{ds} = \frac{1}{|\boldsymbol{\alpha}|}\left(\alpha_1\frac{\partial}{\partial x} + \alpha_2\frac{\partial}{\partial y} + \alpha_3\frac{\partial}{\partial z}\right) \ , \tag{2.31}$$

where $\boldsymbol{\alpha}$ is the tangent vector of the 3-D implicit curve (see (2.15)) given by

$$\boldsymbol{\alpha} = (\alpha_1, \alpha_2, \alpha_3) = \nabla f \times \nabla g \ , \tag{2.32}$$

and

$$\alpha_1 = \frac{\partial f}{\partial y}\frac{\partial g}{\partial z} - \frac{\partial g}{\partial y}\frac{\partial f}{\partial z} \ , \tag{2.33}$$

$$\alpha_2 = \frac{\partial g}{\partial x}\frac{\partial f}{\partial z} - \frac{\partial f}{\partial x}\frac{\partial g}{\partial z} \ , \tag{2.34}$$

$$\alpha_3 = \frac{\partial f}{\partial x}\frac{\partial g}{\partial y} - \frac{\partial g}{\partial x}\frac{\partial f}{\partial y} \ . \tag{2.35}$$

By applying the derivative operator (2.31) to $|\boldsymbol{\alpha}|\mathbf{t} = \boldsymbol{\alpha}$ we obtain

$$\frac{d|\boldsymbol{\alpha}|\mathbf{t}}{ds} = \frac{1}{|\boldsymbol{\alpha}|}\left(\alpha_1\frac{\partial \boldsymbol{\alpha}}{\partial x} + \alpha_2\frac{\partial \boldsymbol{\alpha}}{\partial y} + \alpha_3\frac{\partial \boldsymbol{\alpha}}{\partial z} \right) , \tag{2.36}$$

which gives

$$|\boldsymbol{\alpha}|^2\kappa\mathbf{n} + |\boldsymbol{\alpha}||\boldsymbol{\alpha}|'\mathbf{t} = \left(\alpha_1\frac{\partial \boldsymbol{\alpha}}{\partial x} + \alpha_2\frac{\partial \boldsymbol{\alpha}}{\partial y} + \alpha_3\frac{\partial \boldsymbol{\alpha}}{\partial z} \right) . \tag{2.37}$$

Taking the cross product of $|\boldsymbol{\alpha}|\mathbf{t} = \boldsymbol{\alpha}$ and (2.37) yields

$$|\boldsymbol{\alpha}|^3\kappa\mathbf{b} = \boldsymbol{\alpha} \times \left(\alpha_1\frac{\partial \boldsymbol{\alpha}}{\partial x} + \alpha_2\frac{\partial \boldsymbol{\alpha}}{\partial y} + \alpha_3\frac{\partial \boldsymbol{\alpha}}{\partial z} \right) . \tag{2.38}$$

Thus,

$$\kappa = \frac{\left| \boldsymbol{\alpha} \times \left(\alpha_1\frac{\partial \boldsymbol{\alpha}}{\partial x} + \alpha_2\frac{\partial \boldsymbol{\alpha}}{\partial y} + \alpha_3\frac{\partial \boldsymbol{\alpha}}{\partial z} \right) \right|}{|\boldsymbol{\alpha}|^3} . \tag{2.39}$$

A different derivation of the curvature of a 3-D implicit curve is given in Sect. 6.3.2.

2.3 Binormal vector and torsion

In Sects. 2.1 and 2.2, we have introduced the tangent and normal vectors, which are orthogonal to each other and lie in the osculating plane. Let us define a unit binormal vector $\mathbf{b}$ such that $(\mathbf{t}, \mathbf{n}, \mathbf{b})$ form a right-handed screw, i.e.

$$\mathbf{b} = \mathbf{t} \times \mathbf{n}, \qquad \mathbf{t} = \mathbf{n} \times \mathbf{b}, \qquad \mathbf{n} = \mathbf{b} \times \mathbf{t} , \tag{2.40}$$

which is shown in Fig. 2.6. The plane defined by normal and binormal vectors is called the *normal plane* and the plane defined by binormal and tangent vectors is called the *rectifying plane* (see Fig. 2.6). As mentioned before, the plane defined by tangent and normal vectors is called the *osculating plane*. The binormal vector for the arbitrary speed curve with nonzero curvature can be obtained by using (2.23) and the first equation of (2.40) as follows:

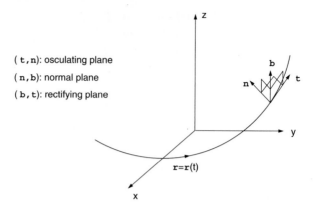

(t , n): osculating plane
(n , b): normal plane
(b , t): rectifying plane

Fig. 2.6. The tangent, normal, and binormal vectors define an orthogonal coordinate system along a space curve

$$\mathbf{b} = \frac{\dot{\mathbf{r}} \times \ddot{\mathbf{r}}}{|\dot{\mathbf{r}} \times \ddot{\mathbf{r}}|} \ . \tag{2.41}$$

The binormal vector is perpendicular to the osculating plane and its rate of change is expressed by the vector

$$\mathbf{b}' = \frac{d}{ds}(\mathbf{t} \times \mathbf{n}) = \frac{d\mathbf{t}}{ds} \times \mathbf{n} + \mathbf{t} \times \frac{d\mathbf{n}}{ds} = \mathbf{t} \times \mathbf{n}' \ , \tag{2.42}$$

where we used the fact that $\frac{d\mathbf{t}}{ds} = \mathbf{r}'' = \kappa\mathbf{n}$.

Since $\mathbf{n}$ is a unit vector $\mathbf{n} \cdot \mathbf{n} = 1$, we have $\mathbf{n} \cdot \mathbf{n}' = 0$. Therefore $\mathbf{n}'$ is parallel to the rectifying plane $(\mathbf{b}, \mathbf{t})$, and hence $\mathbf{n}'$ can be expressed as a linear combination of $\mathbf{b}$ and $\mathbf{t}$:

$$\mathbf{n}' = \mu\mathbf{t} + \tau\mathbf{b} \ . \tag{2.43}$$

Thus, using (2.42) and (2.43), we obtain

$$\mathbf{b}' = \mathbf{t} \times (\mu\mathbf{t} + \tau\mathbf{b}) = \tau\mathbf{t} \times \mathbf{b} = -\tau\mathbf{b} \times \mathbf{t} = -\tau\mathbf{n} \ . \tag{2.44}$$

The coefficient τ is called the *torsion* and measures how much the curve deviates from the osculating plane. By taking the dot product with $-\mathbf{n}$, we obtain the torsion of the curve at a nonzero curvature point

$$\tau = -\mathbf{n} \cdot \mathbf{b}' = -\frac{\mathbf{r}''}{\kappa} \cdot \left(\mathbf{r}' \times \frac{\mathbf{r}''}{\kappa}\right)' = -\frac{\mathbf{r}''}{\kappa} \cdot \left(\mathbf{r}' \times \frac{\mathbf{r}'''}{\kappa}\right) = \frac{(\mathbf{r}'\mathbf{r}''\mathbf{r}''')}{\mathbf{r}'' \cdot \mathbf{r}''} \ , \tag{2.45}$$

where (2.20) is used and $(\mathbf{r}'\mathbf{r}''\mathbf{r}''')$ is a *triple scalar product*. [1]

---

[1] A triple scalar product $(\mathbf{a} \ \mathbf{b} \ \mathbf{c})$ is numerically equal to the volume of the parallelepiped having the edge vectors $\mathbf{a}$, $\mathbf{b}$ and $\mathbf{c}$, and is given by

The torsion for an arbitrary speed curve is given by

$$\tau = \frac{(\dot{\mathbf{r}} \ddot{\mathbf{r}} \dddot{\mathbf{r}})}{(\dot{\mathbf{r}} \times \ddot{\mathbf{r}}) \cdot (\dot{\mathbf{r}} \times \ddot{\mathbf{r}})} \ . \tag{2.48}$$

The evaluation of torsion when curvature vanishes is discussed in Sect. 6.2.

While the curvature is determined only in magnitude, except for plane curves, torsion is determined both in magnitude and sign. Torsion is positive when the rotation of the osculating plane is in the direction of a right-handed screw moving in the direction of $\mathbf{t}$ as s increases. If the torsion is zero at all points, the curve is planar.

The binormal vector of a 3-D implicit curve can be obtained from (2.38) as follows:

$$\mathbf{b} = \frac{\boldsymbol{\alpha} \times \left(\alpha_1 \frac{\partial \boldsymbol{\alpha}}{\partial x} + \alpha_2 \frac{\partial \boldsymbol{\alpha}}{\partial y} + \alpha_3 \frac{\partial \boldsymbol{\alpha}}{\partial z} \right)}{\left| \boldsymbol{\alpha} \times \left(\alpha_1 \frac{\partial \boldsymbol{\alpha}}{\partial x} + \alpha_2 \frac{\partial \boldsymbol{\alpha}}{\partial y} + \alpha_3 \frac{\partial \boldsymbol{\alpha}}{\partial z} \right) \right|} \ . \tag{2.49}$$

The torsion for a 3-D implicit curve can be derived by applying the derivative operator (2.31) to (2.38) [444], which gives

$$\frac{d}{ds}(|\boldsymbol{\alpha}|^3 \kappa \mathbf{b}) = \tag{2.50}$$
$$\frac{1}{|\boldsymbol{\alpha}|} \left(\alpha_1 \frac{\partial}{\partial x} + \alpha_2 \frac{\partial}{\partial y} + \alpha_3 \frac{\partial}{\partial z} \right) \left(\boldsymbol{\alpha} \times \left(\alpha_1 \frac{\partial \boldsymbol{\alpha}}{\partial x} + \alpha_2 \frac{\partial \boldsymbol{\alpha}}{\partial y} + \alpha_3 \frac{\partial \boldsymbol{\alpha}}{\partial z} \right) \right) ,$$

and therefore

$$|\boldsymbol{\alpha}|(|\boldsymbol{\alpha}|^3 \kappa)' \mathbf{b} - |\boldsymbol{\alpha}|^4 \kappa \tau \mathbf{n} = \tag{2.51}$$
$$\left(\alpha_1 \frac{\partial}{\partial x} + \alpha_2 \frac{\partial}{\partial y} + \alpha_3 \frac{\partial}{\partial z} \right) \left(\boldsymbol{\alpha} \times \left(\alpha_1 \frac{\partial \boldsymbol{\alpha}}{\partial x} + \alpha_2 \frac{\partial \boldsymbol{\alpha}}{\partial y} + \alpha_3 \frac{\partial \boldsymbol{\alpha}}{\partial z} \right) \right) .$$

Taking the dot product with (2.37) we obtain

$$-|\boldsymbol{\alpha}|^6 \kappa^2 \tau = \left(\alpha_1 \frac{\partial \boldsymbol{\alpha}}{\partial x} + \alpha_2 \frac{\partial \boldsymbol{\alpha}}{\partial y} + \alpha_3 \frac{\partial \boldsymbol{\alpha}}{\partial z} \right) \tag{2.52}$$
$$\cdot \left(\alpha_1 \frac{\partial}{\partial x} + \alpha_2 \frac{\partial}{\partial y} + \alpha_3 \frac{\partial}{\partial z} \right) \left(\boldsymbol{\alpha} \times \left(\alpha_1 \frac{\partial \boldsymbol{\alpha}}{\partial x} + \alpha_2 \frac{\partial \boldsymbol{\alpha}}{\partial y} + \alpha_3 \frac{\partial \boldsymbol{\alpha}}{\partial z} \right) \right) ,$$

from which we calculate τ. An alternative approach for evaluating the torsion of 3-D implicit curves is presented in Sect. 6.3.3.

$$(\mathbf{a}\,\mathbf{b}\,\mathbf{c}) = \begin{vmatrix} a_x & b_x & c_x \\ a_y & b_y & c_y \\ a_z & b_z & c_z \end{vmatrix} = \begin{vmatrix} a_x & a_y & a_z \\ b_x & b_y & b_z \\ c_x & c_y & c_z \end{vmatrix} = (\mathbf{a} \times \mathbf{b}) \cdot \mathbf{c} = \mathbf{a} \cdot (\mathbf{b} \times \mathbf{c}) \ . \tag{2.46}$$

Also a cyclic permutation maintains the value of the triple scalar product:

$$(\mathbf{a}\,\mathbf{b}\,\mathbf{c}) = (\mathbf{b}\,\mathbf{c}\,\mathbf{a}) = (\mathbf{c}\,\mathbf{a}\,\mathbf{b}) \ . \tag{2.47}$$

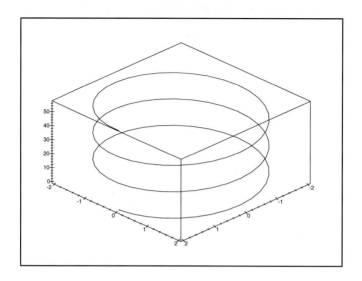

Fig. 2.7. Circular helix with $a = 2$, $b = 3$ for $0 \le t \le 6\pi$

Example 2.3.1. A circular helix in parametric representation is given by $\mathbf{r}(t) = (a \cos t, a \sin t, bt)^T$. Figure 2.7 shows a circular helix with $a = 2$, $b = 3$ for $0 \le t \le 6\pi$. The parametric speed is easily computed as $|\dot{\mathbf{r}}(t)| = \sqrt{a^2 + b^2} \equiv c$, which is a constant. Therefore the curve is regular and its arc length is

$$s(t) = \int_0^t |\dot{\mathbf{r}}| dt = \int_0^t \sqrt{a^2 + b^2} dt = ct \; .$$

We can easily reparametrize the curve with arc length by replacing t by $\frac{s}{c}$ yielding $\mathbf{r} = (a \cos \frac{s}{c}, a \sin \frac{s}{c}, \frac{bs}{c})^T$. The first three derivatives are evaluated as

$$\mathbf{r}'(s) = \left(-\frac{a}{c} \sin \frac{s}{c}, \frac{a}{c} \cos \frac{s}{c}, \frac{b}{c} \right)^T, \quad \mathbf{r}''(s) = \left(-\frac{a}{c^2} \cos \frac{s}{c}, -\frac{a}{c^2} \sin \frac{s}{c}, 0 \right)^T,$$

$$\mathbf{r}'''(s) = \left(\frac{a}{c^3} \sin \frac{s}{c}, -\frac{a}{c^3} \cos \frac{s}{c}, 0 \right)^T.$$

The curvature and torsion are evaluated as follows:

$$\kappa^2 = \mathbf{r}'' \cdot \mathbf{r}'' = \frac{a^2}{c^4} \left(\cos^2 \frac{s}{c} + \sin^2 \frac{s}{c} \right) = \frac{a^2}{c^4} = constant \; ,$$

$$\tau = \frac{(\mathbf{r}'\mathbf{r}''\mathbf{r}''')}{\mathbf{r}'' \cdot \mathbf{r}''} = \frac{(\mathbf{r}'\mathbf{r}''\mathbf{r}''')}{\kappa^2} = \frac{c^4}{a^2} \begin{vmatrix} -\frac{a}{c} \sin \frac{s}{c} & \frac{a}{c} \cos \frac{s}{c} & \frac{b}{c} \\ -\frac{a}{c^2} \cos \frac{s}{c} & -\frac{a}{c^2} \sin \frac{s}{c} & 0 \\ \frac{a}{c^3} \sin \frac{s}{c} & -\frac{a}{c^3} \cos \frac{s}{c} & 0 \end{vmatrix}$$

$$= \frac{c^4}{a^2} \frac{b}{c} \frac{a^2}{c^5} \left(\cos^2 \frac{s}{c} + \sin^2 \frac{s}{c} \right) = \frac{b}{c^2} = constant .$$

Note that the circular helix has constant curvature and torsion and when $b > 0$, it is a right-handed helix while when $b < 0$, it is a left-handed helix.

2.4 Frenet-Serret formulae

From (2.20) and (2.44), we found that

$$\mathbf{t}' = \kappa \mathbf{n} , \qquad (2.53)$$
$$\mathbf{b}' = -\tau \mathbf{n} . \qquad (2.54)$$

From these equations we deduce

$$\mathbf{n}' = (\mathbf{b} \times \mathbf{t})' = \mathbf{b}' \times \mathbf{t} + \mathbf{b} \times \mathbf{t}' = -\tau \mathbf{n} \times \mathbf{t} + \mathbf{b} \times (\kappa \mathbf{n}) = -\kappa \mathbf{t} + \tau \mathbf{b} . \qquad (2.55)$$

In matrix form we can express the differential equations as

$$\begin{pmatrix} \mathbf{t}' \\ \mathbf{n}' \\ \mathbf{b}' \end{pmatrix} = \begin{pmatrix} 0 & \kappa & 0 \\ -\kappa & 0 & \tau \\ 0 & -\tau & 0 \end{pmatrix} \begin{pmatrix} \mathbf{t} \\ \mathbf{n} \\ \mathbf{b} \end{pmatrix} . \qquad (2.56)$$

Thus, $\mathbf{t}$, $\mathbf{n}$, $\mathbf{b}$ are completely determined by the curvature and torsion of the curve as a function of parameter s. The equations $\kappa = \kappa(s)$, $\tau = \tau(s)$ are called *intrinsic equations* of the curve. The formulae (2.56) are known as the Frenet-Serret formulae and describe the motion of a moving trihedron ($\mathbf{t}$, $\mathbf{n}$, $\mathbf{b}$) along the curve. From these $\mathbf{t}$, $\mathbf{n}$, $\mathbf{b}$ the shape of the curve can be determined apart for a translation and rotation. For arbitrary speed curve the Frenet-Serret formulae are given by

$$\begin{pmatrix} \dot{\mathbf{t}} \\ \dot{\mathbf{n}} \\ \dot{\mathbf{b}} \end{pmatrix} = \begin{pmatrix} 0 & v\kappa & 0 \\ -v\kappa & 0 & v\tau \\ 0 & -v\tau & 0 \end{pmatrix} \begin{pmatrix} \mathbf{t} \\ \mathbf{n} \\ \mathbf{b} \end{pmatrix} , \qquad (2.57)$$

where $v = \frac{ds}{dt}$ is the parametric speed.

Example 2.4.1. As shown in Example 2.3.1 the intrinsic equations of circular helix are given by $\kappa(s) = \frac{a}{c^2}$, $\tau(s) = \frac{b}{c^2}$, where $c = \sqrt{a^2 + b^2}$. In this example we derive the parametric equations of circular helix from these intrinsic equations. Substituting the intrinsic equations into the Frenet-Serret equations we obtain

$$\frac{d\mathbf{t}}{ds} = \frac{a}{c^2} \mathbf{n}, \qquad \frac{d\mathbf{n}}{ds} = -\frac{a}{c^2} \mathbf{t} + \frac{b}{c^2} \mathbf{b}, \qquad \frac{d\mathbf{b}}{ds} = -\frac{b}{c^2} \mathbf{n} .$$

We first differentiate the first equation twice and the second equation once with respect to s, which yield

$$\frac{d^2\mathbf{t}}{ds^2} = \frac{a}{c^2}\frac{d\mathbf{n}}{ds}, \qquad \frac{d^3\mathbf{t}}{ds^3} = \frac{a}{c^2}\frac{d^2\mathbf{n}}{ds^2}, \qquad \frac{d^2\mathbf{n}}{ds^2} = -\frac{a}{c^2}\frac{d\mathbf{t}}{ds} - \frac{b^2}{c^4}\mathbf{n},$$

where the third equation is used to replace $\frac{d\mathbf{b}}{ds}$. Eliminating $\mathbf{n}$, $\frac{d\mathbf{n}}{ds}$, $\frac{d^2\mathbf{n}}{ds^2}$ and recognizing that $\mathbf{t} = \frac{d\mathbf{r}}{ds}$, we obtain the fourth order differential equation

$$\frac{d^4\mathbf{r}}{ds^4} + \frac{1}{c^2}\frac{d^2\mathbf{r}}{ds^2} = 0 .$$

The general solution to this differential equation is given by

$$\mathbf{r}(s) = \mathbf{C}_1 + \mathbf{C}_2 s + \mathbf{C}_3 \cos\frac{s}{c} + \mathbf{C}_4 \sin\frac{s}{c} ,$$

where $\mathbf{C}_1$, $\mathbf{C}_2$, $\mathbf{C}_3$ and $\mathbf{C}_4$ are the vector constants determined by the initial conditions. In this case we assume the following initial conditions

$$\mathbf{r}(0) = (a,0,0)^T, \qquad \mathbf{r}'(0) = \left(0, \frac{a}{c}, \frac{b}{c}\right)^T, \qquad \mathbf{r}''(0) = \left(-\frac{a}{c^2},0,0\right)^T,$$

$$\mathbf{r}'''(0) = \left(0, -\frac{a}{c^3}, 0\right)^T,$$

which yield

$$\mathbf{C}_1 = (0,0,0)^T, \qquad \mathbf{C}_2 = \left(0,0,\frac{b}{c}\right)^T, \quad \mathbf{C}_3 = (a,0,0)^T, \qquad \mathbf{C}_4 = (0,a,0)^T ,$$

thus, we have $\mathbf{r}(s) = \left(a\cos\frac{s}{c}, a\sin\frac{s}{c}, \frac{bs}{c}\right)^T$.

3. Differential Geometry of Surfaces

3.1 Tangent plane and surface normal

Let us consider a curve $u = u(t)$, $v = v(t)$ in the parametric domain of a parametric surface $\mathbf{r} = \mathbf{r}(u, v)$ as shown in Fig. 3.1. Then $\mathbf{r} = \mathbf{r}(t) = \mathbf{r}(u(t), v(t))$ is a parametric curve lying on the surface $\mathbf{r} = \mathbf{r}(u, v)$. The tangent vector to the curve on the surface is evaluated by differentiating $\mathbf{r}(t)$ with respect to the parameter t using the chain rule and is given by

$$\dot{\mathbf{r}}(t) = \mathbf{r}_u \dot{u} + \mathbf{r}_v \dot{v} , \tag{3.1}$$

where subscripts u and v denote partial differentiation with respect to u and v, respectively. The *tangent plane* at point P can be considered as a union

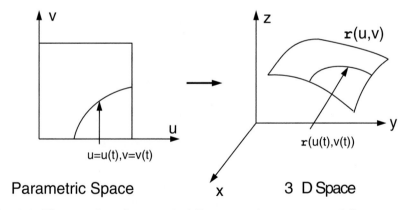

Fig. 3.1. The mapping of a curve in 2-D parametric space onto a 3-D parametric surface

of the tangent vectors of the form (3.1) for all $\mathbf{r}(t)$ through P as illustrated in Fig. 3.2. Point P corresponds to parameters u_p, v_p. Since the tangent vector (3.1) consists of a linear combination of two surface tangents along isoparametric curves $\mathbf{r}_u$ and $\mathbf{r}_v$, the equation of the tangent plane at $\mathbf{r}(u_p, v_p)$ in parametric form with parameters μ, ν is given by

$$\mathbf{T}_p(\mu, \nu) = \mathbf{r}(u_p, v_p) + \mu \mathbf{r}_u(u_p, v_p) + \nu \mathbf{r}_v(u_p, v_p) \ . \tag{3.2}$$

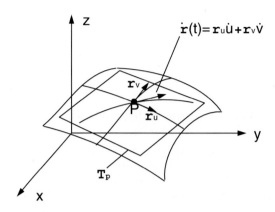

Fig. 3.2. The tangent plane at a point on a surface

The *surface normal vector* is perpendicular to the tangent plane (see Fig. 3.3) and hence the unit normal vector is given by

$$\mathbf{N} = \frac{\mathbf{r}_u \times \mathbf{r}_v}{|\mathbf{r}_u \times \mathbf{r}_v|} \ . \tag{3.3}$$

By using (3.3), the equation of the tangent plane at $\mathbf{r}(u_p, v_p)$ can be written in the implicit form as

$$(\mathbf{r} - \mathbf{r}(u_p, v_p)) \cdot \mathbf{N}(u_p, v_p) = 0 \ , \tag{3.4}$$

where $\mathbf{r}$ is a point on the tangent plane.

Definition 3.1.1. *A regular (ordinary) point P on a parametric surface is defined as a point where* $\mathbf{r}_u \times \mathbf{r}_v \neq \mathbf{0}$. *A point which is not a regular point is called a singular point.*

The condition $\mathbf{r}_u \times \mathbf{r}_v \neq \mathbf{0}$ requires that at point P the vectors $\mathbf{r}_u$ and $\mathbf{r}_v$ do not vanish and have different directions, i.e. $\mathbf{r}_u$ and $\mathbf{r}_v$ are linearly independent. As we discussed in Sect. 1.3.6, in some design problems we need to employ triangular patches defined by parametrization over a rectangular domain. Such a degenerated patch can be generated by collapsing one boundary curve into a single point or by arranging for two partial derivatives $\mathbf{r}_u$ and $\mathbf{r}_v$ at one of the corners of a quadrilateral patch to be collinear. In both cases $\mathbf{r}_u \times \mathbf{r}_v$ has zero magnitude at the degenerate corner point and (3.3) cannot be used. Conditions for the existence of surface normals at these degenerate

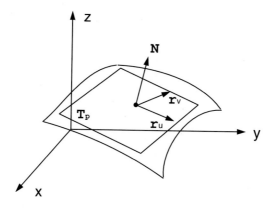

Fig. 3.3. The normal to the point on a surface

corner points have been discussed in [116, 92, 453, 457]. The concept of a regular surface requires additional conditions beyond the existence of a tangent plane everywhere on the surface, such as absence of self-intersections. This concept is presented fully in do Carmo [76].

There are *essential* and *artificial* singularities [444]. The essential singularities arise from specific features of the surface geometry such as the apex of a cone. The artificial singularities arise from the choice of parametrization.

Example 3.1.1. The elliptic cone can be described in a parametric form $\mathbf{r} = (at\cos\theta, bt\sin\theta, ct)^T$, where $0 \le \theta \le 2\pi$, $0 \le t \le l$ and a, b, c are constants. We have

$$\mathbf{r}_\theta = (-at\sin\theta, bt\cos\theta, 0)^T, \quad \mathbf{r}_t = (a\cos\theta, b\sin\theta, c)^T ,$$

thus

$$|\mathbf{r}_\theta \times \mathbf{r}_t| = |bct\cos\theta\mathbf{e}_x + act\sin\theta\mathbf{e}_y - abt\mathbf{e}_z|$$
$$= \sqrt{t^2(b^2c^2\cos^2\theta + a^2c^2\sin^2\theta + a^2b^2)} .$$

We can easily observe that the surface becomes singular only at $t = 0$, which corresponds to the apex of the cone.

The unit normal vector for an implicit surface can be derived by considering two parametric curves $\mathbf{r}_1 = (x_1(t_1), y_1(t_1), z_1(t_1))^T$, $\mathbf{r}_2 = (x_2(t_2), y_2(t_2), z_2(t_2))^T$ lying on an implicit surface $f(x, y, z) = 0$, and intersecting at point P on the surface with different tangent directions. Thus we have the relations:

$$f(x_1(t_1), y_1(t_1), z(t_1)) = 0, \quad f(x_2(t_2), y_2(t_2), z(t_2)) = 0 . \tag{3.5}$$

Total differentiation of (3.5) with respect to t_1 and t_2, respectively, yields

$$f_x \frac{dx_1}{dt_1} + f_y \frac{dy_1}{dt_1} + f_z \frac{dz_1}{dt_1} = 0 , \tag{3.6}$$

$$f_x \frac{dx_2}{dt_2} + f_y \frac{dy_2}{dt_2} + f_z \frac{dz_2}{dt_2} = 0 . \tag{3.7}$$

Now if we multiply (3.6) by $\frac{dx_2}{dt_2}$ and subtract (3.7) multiplied by $\frac{dx_1}{dt_1}$, and if we multiply (3.6) by $\frac{dy_2}{dt_2}$ and subtract (3.7) multiplied by $\frac{dy_1}{dt_1}$ we can deduce the following relation

$$f_x : f_y : f_z = \tag{3.8}$$

$$\frac{dz_2}{dt_2}\frac{dy_1}{dt_1} - \frac{dz_1}{dt_1}\frac{dy_2}{dt_2} : \frac{dz_1}{dt_1}\frac{dx_2}{dt_2} - \frac{dz_2}{dt_2}\frac{dx_1}{dt_1} : \frac{dx_1}{dt_1}\frac{dy_2}{dt_2} - \frac{dx_2}{dt_2}\frac{dy_1}{dt_1} ,$$

which indicates that vector $\nabla f = (f_x, f_y, f_z)^T$ (also known as gradient of f) is in the direction of the cross product of the two tangent vectors at P, i.e. in the normal direction. Thus the unit normal vector of the implicit surface is given by

$$\mathbf{N} = \frac{(f_x, f_y, f_z)^T}{\sqrt{f_x^2 + f_y^2 + f_z^2}} = \frac{\nabla f}{|\nabla f|} , \tag{3.9}$$

provided that $|\nabla f| \neq 0$.

Alternatively, we can derive (3.9) by considering an arbitrary parametric curve $\mathbf{r} = \mathbf{r}(t)$ on an implicit surface $f(x, y, z) = 0$, leading to the relation $\nabla f \cdot \dot{\mathbf{r}} = 0$. Since $\mathbf{r} = \mathbf{r}(t)$ is arbitrary, ∇f must be perpendicular to the tangent plane, and hence it is a normal vector.

The tangent plane of an implicit surface $f(x, y, z) = 0$ at point P with coordinates (x_p, y_p, z_p) can be obtained by replacing the normal vector of parametric surface in (3.4) with (3.9), which leads to

$$f_x(x - x_p) + f_y(y - y_p) + f_z(z - z_p) = 0 , \tag{3.10}$$

where $f(x_p, y_p, z_p) = 0$ and f_x, f_y f_z in (3.10) are evaluated at (x_p, y_p, z_p).

Example 3.1.2. The elliptic cone of Example 3.1.1 has also the following implicit representation $f(x, y, z) = (\frac{x}{a})^2 + (\frac{y}{b})^2 - (\frac{z}{c})^2 = 0$. The magnitude of the normal vector $\nabla f = (\frac{2x}{a^2}, \frac{2y}{b^2}, -\frac{2z}{c^2})^T$, where $(x, y, z) \in f(x, y, z) = 0$, becomes 0 only when $x=y=z=0$ corresponding to the apex of the cone as also derived in Example 3.1.1.

3.2 First fundamental form I (metric)

The differential arc length of a parametric curve is given by (2.2). Now if we replace the parametric curve by a curve $u = u(t)$, $v = v(t)$ which lies on the parametric surface $\mathbf{r} = \mathbf{r}(u, v)$, then

$$ds = \left| \frac{d\mathbf{r}}{dt} \right| dt = \left| \mathbf{r}_u \frac{du}{dt} + \mathbf{r}_v \frac{dv}{dt} \right| dt = \sqrt{(\mathbf{r}_u \dot{u} + \mathbf{r}_v \dot{v}) \cdot (\mathbf{r}_u \dot{u} + \mathbf{r}_v \dot{v})} dt$$

$$= \sqrt{E du^2 + 2F du\, dv + G dv^2} \ , \quad (3.11)$$

where

$$E = \mathbf{r}_u \cdot \mathbf{r}_u, \quad F = \mathbf{r}_u \cdot \mathbf{r}_v, \quad G = \mathbf{r}_v \cdot \mathbf{r}_v \ . \quad (3.12)$$

The *first fundamental form* is defined as

$$I = ds^2 = d\mathbf{r} \cdot d\mathbf{r} = E du^2 + 2F du\, dv + G dv^2 \ , \quad (3.13)$$

and E, F, G are called the first fundamental form coefficients and play important roles in many intrinsic properties of a surface. The first fundamental form I can be rewritten as

$$I = \frac{1}{E}(E\, du + F\, dv)^2 + \frac{EG - F^2}{E} dv^2 \ . \quad (3.14)$$

Since $(\mathbf{r}_u \times \mathbf{r}_v)^2 = (\mathbf{r}_u \times \mathbf{r}_v) \cdot (\mathbf{r}_u \times \mathbf{r}_v) = (\mathbf{r}_u \cdot \mathbf{r}_u)(\mathbf{r}_v \cdot \mathbf{r}_v) - (\mathbf{r}_u \cdot \mathbf{r}_v)^2 = EG - F^2 > 0$[1] and $E = \mathbf{r}_u \cdot \mathbf{r}_u > 0$, I is positive definite, provided that the surface is regular. That is $I \geq 0$ and $I = 0$ if and only if $du = 0$ and $dv = 0$.

Example 3.2.1. Let us compute the arc length of a curve $u = t$, $v = t$ for $0 \leq t \leq 1$ on a hyperbolic paraboloid $\mathbf{r}(u, v) = (u, v, uv)^T$ where $0 \leq u, v \leq 1$ as shown in Fig. 3.4 (a). We have

$$\mathbf{r}_u = (1, 0, v)^T, \quad \mathbf{r}_v = (0, 1, u)^T \ ,$$

$$E = \mathbf{r}_u \cdot \mathbf{r}_u = 1 + v^2, \quad F = \mathbf{r}_u \cdot \mathbf{r}_v = uv, \quad G = \mathbf{r}_v \cdot \mathbf{r}_v = 1 + u^2 \ ,$$

and along the curve the first fundamental form coefficients are

$$E = 1 + t^2, \quad F = t^2, \quad G = 1 + t^2 \ ,$$

thus,

$$ds = \sqrt{E \dot{u}^2 + 2F \dot{u}\dot{v} + G \dot{v}^2}\, dt = 2\sqrt{t^2 + \frac{1}{2}}\, dt \ .$$

Finally the arc length for $0 \leq t \leq 1$ is given by

---

[1] Here the vector identity

$$(\mathbf{a} \times \mathbf{b}) \cdot (\mathbf{c} \times \mathbf{d}) = (\mathbf{a} \cdot \mathbf{c})(\mathbf{b} \cdot \mathbf{d}) - (\mathbf{a} \cdot \mathbf{d})(\mathbf{b} \cdot \mathbf{c}) \ , \quad (3.15)$$

with the special case

$$(\mathbf{a} \times \mathbf{b}) \cdot (\mathbf{a} \times \mathbf{b}) = (\mathbf{a} \cdot \mathbf{a})(\mathbf{b} \cdot \mathbf{b}) - (\mathbf{a} \cdot \mathbf{b})^2 \ , \quad (3.16)$$

is used.

$$s = 2 \int_0^1 \sqrt{t^2 + \frac{1}{2}} dt = \left[t\sqrt{t^2 + \frac{1}{2}} + \frac{1}{2} log \left(t + \sqrt{t^2 + \frac{1}{2}} \right) \right]_0^1$$

$$= \sqrt{\frac{3}{2}} + \frac{1}{2} log(\sqrt{2} + \sqrt{3}) .$$

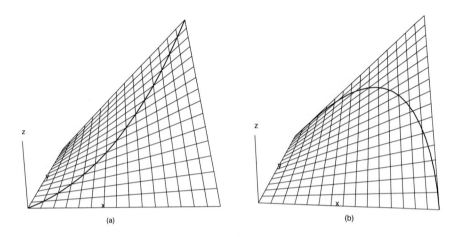

Fig. 3.4. Hyperbolic paraboloid: (a) arc length along $u = t$, $v = t$, (b) area bounded by positive u and v axes and a quarter circle

The angle between two curves on a parametric surface $\mathbf{r}_1 = \mathbf{r}(u_1(t), v_1(t))$ and $\mathbf{r}_2 = \mathbf{r}(u_2(t), v_2(t))$ can be evaluated by taking the inner product of the tangent vectors of $\mathbf{r}_1$ and $\mathbf{r}_2$, yielding

$$\begin{aligned}
\cos \omega &= \frac{E du_1 du_2 + F(du_1 dv_2 + dv_1 du_2) + G dv_1 dv_2}{\sqrt{E du_1^2 + 2F du_1 dv_1 + G dv_1^2} \sqrt{E du_2^2 + 2F du_2 dv_2 + G dv_2^2}} \\
&= E\frac{du_1}{ds_1}\frac{du_2}{ds_2} + F \left(\frac{du_1}{ds_1}\frac{dv_2}{ds_2} + \frac{dv_1}{ds_1}\frac{du_2}{ds_2} \right) + G\frac{dv_1}{ds_1}\frac{dv_2}{ds_2} .
\end{aligned} \tag{3.17}$$

As a result of the above equation, the orthogonality condition for the two tangent vectors $\dot{\mathbf{r}}_1$ and $\dot{\mathbf{r}}_2$ is:

$$E du_1 du_2 + F(du_1 dv_2 + dv_1 du_2) + G dv_1 dv_2 = 0 . \tag{3.18}$$

In particular when the two curves are the u and v iso-parametric curves, (3.17) reduces to

$$\cos \omega = \frac{\mathbf{r}_u \cdot \mathbf{r}_v}{|\mathbf{r}_u||\mathbf{r}_v|} = \frac{\mathbf{r}_u \cdot \mathbf{r}_v}{\sqrt{\mathbf{r}_u \cdot \mathbf{r}_u}\sqrt{\mathbf{r}_v \cdot \mathbf{r}_v}} = \frac{F}{\sqrt{EG}} . \tag{3.19}$$

Thus the iso-parametric curves are orthogonal if $F = 0$.

The area of a small parallelogram with vertices $\mathbf{r}(u, v)$, $\mathbf{r}(u+\delta u, v)$, $\mathbf{r}(u, v+\delta v)$ and $\mathbf{r}(u + \delta u, v + \delta v)$ as illustrated in Fig. 3.5, is approximated by

$$\delta A = |\mathbf{r}_u \delta u \times \mathbf{r}_v \delta v| = \sqrt{EG - F^2} \delta u \delta v , \qquad (3.20)$$

or in differential form

$$dA = \sqrt{EG - F^2} du dv . \qquad (3.21)$$

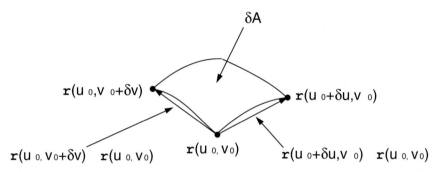

Fig. 3.5. Area of small surface patch

Example 3.2.2. Let us compute the area of a region of the hyperbolic paraboloid that is used in Example 3.2.1. The region is bounded by positive u and v axes and a quarter circle $u^2 + v^2 = 1$ as shown in Fig. 3.4 (b). Substituting $EG - F^2 = (1 + v^2)(1 + u^2) - u^2 v^2 = 1 + u^2 + v^2$ into (3.21), we obtain

$$A = \int_D \sqrt{1 + u^2 + v^2} du dv .$$

To perform the integration it is easier to change variables, $u = r \cos \theta$, $v = r \sin \theta$, so that

$$A = \int_0^{\frac{\pi}{2}} \int_0^1 \sqrt{1 + r^2}\, r\, d\theta\, dr = \frac{\pi}{6}(\sqrt{8} - 1) .$$

3.3 Second fundamental form II (curvature)

In order to quantify the curvatures of a surface S, we consider a curve C on S which passes through point P as shown in Fig. 3.6. The unit tangent vector $\mathbf{t}$ and the unit normal vector $\mathbf{n}$ of the curve C at point P are related by (2.20) as follows:

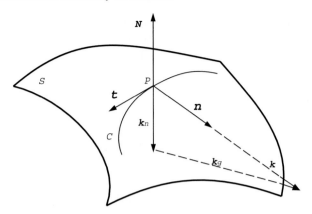

Fig. 3.6. Definition of normal curvature

$$\mathbf{k} = \frac{d\mathbf{t}}{ds} = \kappa\mathbf{n} = \mathbf{k}_n + \mathbf{k}_g \ , \tag{3.22}$$

where $\mathbf{k}_n$ is the *normal curvature vector* and $\mathbf{k}_g$ is the *geodesic curvature vector* which are the components of the curvature vector $\mathbf{k}$ of C in the surface normal direction and in the direction perpendicular to $\mathbf{t}$ in the surface tangent plane. Thus, the normal curvature vector can be expressed as

$$\mathbf{k}_n = \kappa_n\mathbf{N} \ , \tag{3.23}$$

where κ_n is called the normal curvature of the surface at P in the direction $\mathbf{t}$. In other words, κ_n is the magnitude of the projection of $\mathbf{k}$ onto the surface normal at P, with a sign determined by the orientation of the surface normal at P.

By differentiating $\mathbf{N} \cdot \mathbf{t} = 0$ along the curve with respect to s we obtain

$$\frac{d\mathbf{t}}{ds} \cdot \mathbf{N} + \mathbf{t} \cdot \frac{d\mathbf{N}}{ds} = 0 \ , \tag{3.24}$$

thus

$$\kappa_n = \frac{d\mathbf{t}}{ds} \cdot \mathbf{N} = -\mathbf{t} \cdot \frac{d\mathbf{N}}{ds} = -\frac{d\mathbf{r}}{ds} \cdot \frac{d\mathbf{N}}{ds} = -\frac{d\mathbf{r} \cdot d\mathbf{N}}{d\mathbf{r} \cdot d\mathbf{r}} \tag{3.25}$$

$$= \frac{L du^2 + 2M dudv + N dv^2}{E du^2 + 2F dudv + G dv^2} \ , \tag{3.26}$$

where

$$L = -\mathbf{r}_u \cdot \mathbf{N}_u, \quad M = -\frac{1}{2}(\mathbf{r}_u \cdot \mathbf{N}_v + \mathbf{r}_v \cdot \mathbf{N}_u) = -\mathbf{r}_u \cdot \mathbf{N}_v = -\mathbf{r}_v \cdot \mathbf{N}_u \ ,$$
$$N = -\mathbf{r}_v \cdot \mathbf{N}_v \ . \tag{3.27}$$

Since $\mathbf{r}_u$ and $\mathbf{r}_v$ are perpendicular to $\mathbf{N}$, we have $\mathbf{r}_u \cdot \mathbf{N} = 0$ and $\mathbf{r}_v \cdot \mathbf{N} = 0$, and hence we have an alternative expression for L, M and N

$$L = \mathbf{r}_{uu} \cdot \mathbf{N}, \quad M = \mathbf{r}_{uv} \cdot \mathbf{N}, \quad N = \mathbf{r}_{vv} \cdot \mathbf{N} . \tag{3.28}$$

Computation of curvatures at points where the surface representation is degenerate (see Sect. 1.3.6) is given in [453].

The numerator of (3.26) is the *second fundamental form II*, i.e.

$$II = Ldu^2 + 2Mdudv + Ndv^2 , \tag{3.29}$$

and L, M, N are called second fundamental form coefficients. Therefore the normal curvature is given by

$$\kappa_n = \frac{II}{I} = \frac{L + 2M\lambda + N\lambda^2}{E + 2F\lambda + G\lambda^2} , \tag{3.30}$$

where $\lambda = \frac{dv}{du}$ is the direction of the tangent line to C at P. We can observe that κ_n at a given point P on the surface depends only on λ which leads to the following theorem due to Meusnier.

Theorem 3.3.1. *All curves lying on a surface S passing through a given point $p \in S$ with the same tangent line have the same normal curvature at this point.*

Using this theorem we can say that the normal curvature is positive when the center of the curvature of the normal section curve, which is a curve through P cut out by a plane that contains $\mathbf{t}$ and $\mathbf{N}$, is on the same side of the surface normal (see Fig. 3.7 (a)). Sometimes the positive normal curvature is defined in the opposite direction, i.e. the center of curvature of the normal section curve is on the opposite side of the surface normal as illustrated in Fig. 3.7 (b). In such cases (3.23) (3.30) become

$$\mathbf{k}_n = -\kappa_n \mathbf{N}, \quad \kappa_n = -\frac{II}{I} = -\frac{L + 2M\lambda + N\lambda^2}{E + 2F\lambda + G\lambda^2} . \tag{3.31}$$

The latter convention is often used in the area of offset curves and surfaces in the context of NC machining. Throughout this book we refer to the first convention as convention (a) and to the second one as convention (b). We have listed all the equations, which involve changes due to this convention in the last page of this chapter.

Suppose P is a point on a surface and Q is a point in the neighborhood of P and $\mathbf{r} = \mathbf{r}(u, v)$ is the surface containing P and Q, as in Fig. 3.8. Now suppose P and Q are the points $\mathbf{r}(u, v)$ and $\mathbf{r}(u + du, v + dv)$, then Taylor's expansion gives

$$\begin{aligned} \mathbf{r}(u + du, v + dv) =\ & \mathbf{r}(u, v) + \mathbf{r}_u du + \mathbf{r}_v dv \\ & + \frac{1}{2}(\mathbf{r}_{uu} du^2 + 2\mathbf{r}_{uv} dudv + \mathbf{r}_{vv} dv^2) + \cdots . \end{aligned} \tag{3.32}$$

Therefore

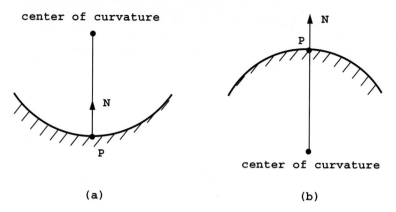

Fig. 3.7. Definition of positive normal curvature: (a) $\kappa \mathbf{n} \cdot \mathbf{N} = \kappa_n$, (b) $\kappa \mathbf{n} \cdot \mathbf{N} = -\kappa_n$

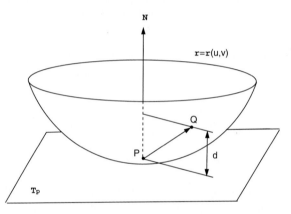

Fig. 3.8. Geometrical illustration of the second fundamental form

$$\mathbf{PQ} = \mathbf{r}(u + du, v + dv) - \mathbf{r}(u, v) = \mathbf{r}_u du + \mathbf{r}_v dv \qquad (3.33)$$
$$+ \frac{1}{2}(\mathbf{r}_{uu} du^2 + 2\mathbf{r}_{uv} du dv + \mathbf{r}_{vv} dv^2) + \dots .$$

Thus using (3.28), (3.29), the projection of $\mathbf{PQ}$ onto $\mathbf{N}$ is

$$d = \mathbf{PQ} \cdot \mathbf{N} = (\mathbf{r}_u du + \mathbf{r}_v dv) \cdot \mathbf{N} + \frac{1}{2} II , \qquad (3.34)$$

where the higher order terms are neglected and since $\mathbf{r}_u \cdot \mathbf{N} = \mathbf{r}_v \cdot \mathbf{N} = 0$, we get

$$d = \frac{1}{2} II = \frac{1}{2}(L du^2 + 2M du dv + N dv^2) . \qquad (3.35)$$

Thus $|II|$ is equal to twice the distance from Q to the tangent plane of the surface at P within second order terms. We want to observe in which situation d is positive and negative or in other words we want to examine in which side of the tangent plane Q lies. When $d = 0$, (3.35) becomes $L du^2 + 2M du dv + N dv^2 = 0$, which can be considered as a quadratic equation in terms of du or dv. If we solve for du, assuming $L \neq 0$, we obtain

$$du = \frac{-M \pm \sqrt{M^2 - LN}}{L} dv , \qquad (3.36)$$

which leads us to the following four cases:

- If $M^2 - LN < 0$, there is no real root. This means there is no intersection between the surface and its tangent plane except at point P. Point P is called *elliptic point* (Fig. 3.9(a)). For example, an ellipsoid consists entirely of elliptic points.
- If $M^2 - LN = 0$ and $L^2 + M^2 + N^2 \neq 0$, there are double roots. The surface intersects its tangent plane with one line $du = -\frac{M}{L} dv$, which passes through point P. Point P is called *parabolic point* (Fig. 3.9(b)). For example, a circular cylinder consists entirely of parabolic points.
- If $M^2 - LN > 0$, there are two roots. The surface intersects its tangent plane with two lines $du = \frac{-M \pm \sqrt{M^2 - LN}}{L} dv$, which intersect at point P. Point P is called *hyperbolic point* (Fig. 3.9(c)). For example, a hyperboloid of revolution consists entirely of hyperbolic points.
- If $L = M = N = 0$, the surface and the tangent plane have a contact of higher order than in the preceding cases. Point P is called a *flat* or *planar point*.

If $L = 0$ and $N \neq 0$, we can solve for dv instead of du. If $L = N = 0$ and $M \neq 0$, we have $2M du dv = 0$, thus the iso-parametric lines $u = constant$, $v = constant$ will be the two intersection lines.

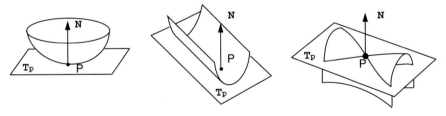

Fig. 3.9. (a) Elliptic point; (b) parabolic point; (c) hyperbolic point

3.4 Principal curvatures

As we can see from (3.30) the normal curvature at a point P depends on the direction of $\lambda = \frac{dv}{du}$. Now we will seek the directions in which the extrema of principal curvature occur following Struik [412]. The extreme values of κ_n can be obtained by evaluating $\frac{d\kappa_n}{d\lambda} = 0$ of (3.30), which gives:

$$(E + 2F\lambda + G\lambda^2)(N\lambda + M) - (L + 2M\lambda + N\lambda^2)(G\lambda + F) = 0 , \quad (3.37)$$

and hence

$$\kappa_n = \frac{L + 2M\lambda + N\lambda^2}{E + 2F\lambda + G\lambda^2} = \frac{M + N\lambda}{F + G\lambda} . \quad (3.38)$$

Furthermore since

$$E + 2F\lambda + G\lambda^2 = (E + F\lambda) + \lambda(F + G\lambda) ,$$
$$L + 2M\lambda + N\lambda^2 = (L + M\lambda) + \lambda(M + N\lambda) ,$$

(3.37) can be reduced to

$$(E + F\lambda)(M + N\lambda) = (L + M\lambda)(F + G\lambda) , \quad (3.39)$$

and hence

$$\kappa_n = \frac{L + 2M\lambda + N\lambda^2}{E + 2F\lambda + G\lambda^2} = \frac{M + N\lambda}{F + G\lambda} = \frac{L + M\lambda}{E + F\lambda} . \quad (3.40)$$

Therefore, the extreme values of κ_n satisfy the two simultaneous equations

$$(L - \kappa_n E)du + (M - \kappa_n F)dv = 0 ,$$
$$(M - \kappa_n F)du + (N - \kappa_n G)dv = 0 . \quad (3.41)$$

These equations form a homogeneous linear system of equations for du, dv, which will have a nontrivial solution if and only if

$$\begin{vmatrix} L - \kappa_n E & M - \kappa_n F \\ M - \kappa_n F & N - \kappa_n G \end{vmatrix} = 0 , \quad (3.42)$$

where $|\ |$ denotes the determinant of a matrix, or expanding

$$(EG - F^2)\kappa_n^2 - (EN + GL - 2FM)\kappa_n + (LN - M^2) = 0 . \quad (3.43)$$

The discriminant D of this quadratic equation in κ_n can be re-formulated as

$$D = 4\left(\frac{EG - F^2}{E^2}\right)(EM - FL)^2 + \left(EN - GL - \frac{2F}{E}(EM - FL)\right)^2 , \quad (3.44)$$

after some algebraic manipulations. Thus the discriminant D is always greater than or equal to zero and (3.43) has real roots. The discriminant D becomes zero if and only if $EM - FL = 0$ and $EN - GL = 0$ or if and only if there is a constant k such that

$$L = kE, \quad M = kF, \quad N = kG \,. \tag{3.45}$$

Such a point is called an *umbilic* and the normal curvature is the same in all directions. Therefore (3.43) has either two distinct real roots, or a double root. If we set

$$K = \frac{LN - M^2}{EG - F^2} \,, \tag{3.46}$$

$$H = \frac{EN + GL - 2FM}{2(EG - F^2)} \,, \tag{3.47}$$

the quadratic equation for κ_n (3.43) simplifies to:

$$\kappa_n^2 - 2H\kappa_n + K = 0 \,. \tag{3.48}$$

The quantities K and H are called *Gaussian (Gauss) curvature* and *mean curvature*, respectively. Upon solving (3.48) for the extreme values of curvature, we have

$$\kappa_{max} = H + \sqrt{H^2 - K} \,, \tag{3.49}$$

$$\kappa_{min} = H - \sqrt{H^2 - K} \,, \tag{3.50}$$

where κ_{max} is the *maximum principal curvature* and κ_{min} is the *minimum principal curvature*. The directions in the tangent plane for which κ_n takes maximum and minimum values are called *principal directions*. The corresponding directions in the uv-plane can be determined by using (3.40), which leads to

$$\lambda = -\frac{M - \kappa_n F}{N - \kappa_n G} \,, \tag{3.51}$$

or

$$\lambda = -\frac{L - \kappa_n E}{M - \kappa_n F} \,, \tag{3.52}$$

where κ_n is replaced by either κ_{max} or κ_{min}.

When the discriminant is zero or $H^2 = K$, κ_n is a double root with value equal to H and the corresponding point of the surface is an *umbilical point*. At an umbilical point a surface is locally a part of sphere with radius of curvature $\frac{1}{|H|}$. In the special case where both K and H vanish, the point is a *flat* or *planar point*.

Alternatively we can derive the principal directions by solving a quadratic equation in λ

$$(FN - GM)\lambda^2 + (EN - GL)\lambda + (EM - FL) = 0 , \qquad (3.53)$$

which is deduced from (3.37). The discriminant of this equation is easily shown to be the same as that of (3.43), and hence it is greater than or equal to zero. At an umbilical point the discriminant vanishes and (3.45) hold, thus we have $FN = GM$, $EN = GL$ and $EM = FL$. Therefore, the coefficients of the quadratic equation become all zero and thus the principal directions are not defined. When a point P on the surface is a non-umbilical point, there are always two principal directions determined by the quadratic equations. Let λ_{max} and λ_{min} be the directions of maximum and minimum principal curvature in the uv-plane. Then, λ_{max} and λ_{min} satisfy the quadratic equation (3.53):

$$(FN - GM)\lambda_{max}^2 + (EN - GL)\lambda_{max} + (EM - FL) = 0 , \qquad (3.54)$$
$$(FN - GM)\lambda_{min}^2 + (EN - GL)\lambda_{min} + (EM - FL) = 0 . \qquad (3.55)$$

From these equations we can deduce

$$\lambda_{max} + \lambda_{min} = -\frac{EN - GL}{FN - GM} , \qquad (3.56)$$

$$\lambda_{max}\lambda_{min} = \frac{EM - FL}{FN - GM} , \qquad (3.57)$$

thus,

$$E + F(\lambda_{max} + \lambda_{min}) + G\lambda_{max}\lambda_{min} \qquad (3.58)$$
$$= \frac{1}{FN - GM}[E(FN - GM) - F(EN - GL) + G(EM - FL)] = 0 .$$

Consequently, it is evident from (3.18) that the two tangent vectors in the principal directions are orthogonal.

A curve on a surface whose tangent at each point is in a principal direction at that point is called a *line of curvature*. Since at each (non-umbilical) point there are two principal directions that are orthogonal, the lines of curvatures form an orthogonal net of lines. Figure 3.10 shows an example of the lines of curvature on a saddle-shaped surface where all points are hyperbolic. The solid lines correspond to the maximum principal curvature direction, while the dashed lines correspond to the minimum principal curvature direction (convention (a) is used). Since there is no umbilical point on the surface, we do not encounter any singularity on the net of lines of curvature. The lines of curvature in the presence of umbilical points are discussed in Chap. 9.

This orthogonal net of lines can be used as a parametrization of a surface. In such cases, we have $F = 0$ (see (3.19)), and (3.41) reduce to

$$(L - \kappa_n E)du + Mdv = 0, \quad Mdu + (N - \kappa_n G)dv = 0 . \qquad (3.59)$$

If these equations are satisfied by $du = 0$ and by $dv = 0$, this implies $M = 0$, and the two principal curvatures are $\kappa_1 = \frac{L}{E}$ and $\kappa_2 = \frac{N}{G}$, in the absence of

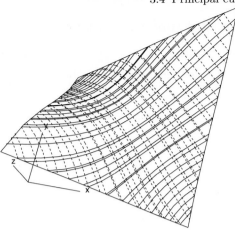

Fig. 3.10. Lines of curvature

umbilical points. Therefore the necessary condition for the parametric lines
to be lines of curvature is

$$F = M = 0 .\tag{3.60}$$

The converse is also true and the condition is also sufficient.

Example 3.4.1. As a curve C in the xz-plane $x = f(t)$, $z = g(t)$ revolves
about the z-axis, it generates a *surface of revolution* S. The curves C in
different rotated positions are called the *meridians* of S, while the circles
generated by each point on C are called the *parallels* of S. If we denote
the rotation angle in the xy-plane as θ, the surface of revolution can be
parametrized as

$$\mathbf{r} = (f(t) \cos \theta, f(t) \sin \theta, g(t))^T .$$

Thus,

$$\mathbf{r}_t = (\dot{f}(t) \cos \theta, \dot{f}(t) \sin \theta, \dot{g}(t))^T, \qquad \mathbf{r}_\theta = (-f(t) \sin \theta, f(t) \cos \theta, 0)^T,$$

and hence

$$E = \dot{f}^2(t) + \dot{g}^2(t), \qquad F = 0, \qquad G = f^2(t) .$$

Since $F = 0$, (3.19) shows that the meridians and parallels are orthogonal.
Furthermore we have

$$L = \frac{-\ddot{f}\dot{g} + \dot{f}\ddot{g}}{\sqrt{\dot{f}^2(t) + \dot{g}^2(t)}}, \qquad M = 0, \qquad N = \frac{f\dot{g}}{\sqrt{\dot{f}^2(t) + \dot{g}^2(t)}} ,$$

which lead us to the conclusion that the meridians and parallels of a surface
of revolution are the lines of curvature.

3.5 Gaussian and mean curvatures

From (3.49), (3.50), it is readily seen that the Gaussian and mean curvatures are the product and the average of the two principal curvatures, respectively:

$$K = \kappa_{max}\kappa_{min} , \tag{3.61}$$

$$H = \frac{\kappa_{max} + \kappa_{min}}{2} . \tag{3.62}$$

The sign of the Gaussian curvature coincides with sign of $LN - M^2$, since $K = \frac{LN-M^2}{EG-F^2}$ (see (3.46)) and $EG - F^2 > 0$. Consequently a point on a surface is elliptic if $K > 0$ (κ_{max} and κ_{min} are of the same sign), hyperbolic if $K < 0$ (κ_{max} and κ_{min} have different signs) and parabolic if $K = 0$ and $H \neq 0$ (either κ_{max} or κ_{min} is zero), flat or planar point if $K = H = 0$ ($\kappa_{max} = \kappa_{min} = 0$).

3.5.1 Explicit surfaces

Very often a surface is given by an explicit form $z = h(x, y)$. It is, therefore, convenient to have analytic equations for the Gaussian and mean curvatures expressed in terms of the derivatives of the height function $h(x, y)$. As we mentioned in Sect. 1.1 the explicit form can be converted into a parametric form $\mathbf{r} = (u, v, h(u, v))^T$ where $u = x$ and $v = y$. This form is often referred to as *Monge form* , and the surface is called a Monge patch. It is straightforward to evaluate

$$E = 1 + h_x^2, \quad F = h_x h_y, \quad G = 1 + h_y^2 , \tag{3.63}$$

$$\mathbf{N} = \frac{(-h_x, -h_y, 1)^T}{\sqrt{1 + h_x^2 + h_y^2}} , \tag{3.64}$$

$$L = \frac{h_{xx}}{\sqrt{1 + h_x^2 + h_y^2}}, \quad M = \frac{h_{xy}}{\sqrt{1 + h_x^2 + h_y^2}}, \quad N = \frac{h_{yy}}{\sqrt{1 + h_x^2 + h_y^2}} , \tag{3.65}$$

and hence

$$K = \frac{LN - M^2}{EG - F^2} = \frac{h_{xx}h_{yy} - h_{xy}^2}{(1 + h_x^2 + h_y^2)^2} , \tag{3.66}$$

$$H = \frac{EN + GL - 2FM}{2(EG - F^2)} = \frac{(1 + h_x^2)h_{yy} - 2h_x h_y h_{xy} + (1 + h_y^2)h_{xx}}{2(1 + h_x^2 + h_y^2)^{3/2}} . \tag{3.67}$$

Example 3.5.1. Let us compute the Gaussian and mean curvatures of the hyperbolic paraboloid $z = xy$ (in Example 3.2.1 we used its parametric form) using the explicit formulae (3.63) to (3.67). Since

$$h_x = y, \quad h_y = x, \quad h_{xx} = 0, \quad h_{xy} = 1, \quad h_{yy} = 0, \quad \mathbf{N} = \frac{(-y, -x, 1)^T}{\sqrt{x^2 + y^2 + 1}},$$

we have

$$E = 1 + y^2, \quad F = xy, \quad G = 1 + x^2, \quad L = 0, \quad M = \frac{1}{\sqrt{x^2 + y^2 + 1}}, \quad N = 0,$$

and hence

$$K = -\frac{1}{(x^2 + y^2 + 1)^2}, \qquad H = -\frac{xy}{(x^2 + y^2 + 1)^{\frac{3}{2}}}.$$

Here we can observe that the Gaussian curvature is always negative and thus all the points on a hyperbolic paraboloid are hyperbolic points. Furthermore, since $L = N = 0$ and $M \neq 0$, the surface intersects its tangent plane at the iso-parametric lines (see Sect. 3.3 last paragraph). Also from (3.49) and (3.50) we obtain

$$\kappa_{max} = \frac{-xy + \sqrt{(x^2 + 1)(y^2 + 1)}}{(x^2 + y^2 + 1)^{\frac{3}{2}}}, \qquad \kappa_{min} = \frac{-xy - \sqrt{(x^2 + 1)(y^2 + 1)}}{(x^2 + y^2 + 1)^{\frac{3}{2}}},$$

where it is very easy to show that $\kappa_{max} > 0$ and $\kappa_{min} < 0$ for all (x, y).

3.5.2 Implicit surfaces

Using (3.66), (3.67) for an explicit surface, we can derive equations for the Gaussian and mean curvature of an implicit surface $f(x, y, z) = 0$. At a point where $f_z \neq 0$, z can be expressed as a function of x and y, say $z = h(x, y)$ [166]. In such cases variables x and y are independent but z is a function of both x and y. Since f constantly satisfies the equation $f(x, y, z) = 0$, the partial differentiation of f with respect to the independent variable x (by holding y fixed) must vanish [166]. Thus,

$$\left(\frac{\partial f}{\partial x}\right)_y = \frac{\partial f}{\partial x} + \frac{\partial f}{\partial z}\frac{\partial z}{\partial x} = 0, \tag{3.68}$$

where $\left(\frac{\partial f}{\partial x}\right)_y$ on the left-hand side is considered as f being expressed in terms of x and y only and y is held constant in the differentiation with respect to x, while $\frac{\partial f}{\partial x}$ on the right-hand side is considered as f being expressed in terms of x, y, z and y, z are held constant in the x differentiation. Similarly we have

$$\left(\frac{\partial f}{\partial y}\right)_x = \frac{\partial f}{\partial y} + \frac{\partial f}{\partial z}\frac{\partial z}{\partial y} = 0 \,. \tag{3.69}$$

Consequently we have

$$h_x = -\frac{f_x}{f_z}, \qquad h_y = -\frac{f_y}{f_z} \,. \tag{3.70}$$

The second order partial derivatives h_{xx}, h_{xy}, h_{yy} are provided by differentiating (3.70). For example,

$$h_{xx} = -\frac{\left(\frac{\partial f_x}{\partial x}\right)_y f_z - \left(\frac{\partial f_z}{\partial x}\right)_y f_x}{f_z^2} = \frac{2 f_x f_z f_{xz} - f_x^2 f_{zz} - f_z^2 f_{xx}}{f_z^3} \,. \tag{3.71}$$

Similarly, we have

$$h_{xy} = \frac{f_x f_z f_{yz} + f_y f_z f_{xz} - f_x f_y f_{zz} - f_z^2 f_{xy}}{f_z^3} \,, \tag{3.72}$$

$$h_{yy} = \frac{2 f_y f_z f_{yz} - f_y^2 f_{zz} - f_z^2 f_{yy}}{f_z^3} \,. \tag{3.73}$$

Equations (3.70) to (3.73) may be substituted into (3.63) and (3.65) to obtain the first and second fundamental form coefficients, and into (3.66) and (3.67) to compute the Gaussian and mean curvature of an implicit surface. If $f_z = 0$, alternate formulae may be found by cyclic permutation of x, y, z.

For every quadric surface, it is possible to find a suitable 3-D rotation such that the cross terms dxy, eyz and fxz cancel out in (1.15). If a quadric surface has a center[2], its axes can be translated to the center as origin so that the equation of the quadric surface does not have any first degree terms [79]. Therefore after these transformations the implicit quadrics, ellipsoids, hyperboloids of one and two sheets, elliptic cones, elliptic cylinders and hyperbolic cylinders can be expressed in a *standard form*

$$f(x, y, z) = \zeta\frac{x^2}{a^2} + \eta\frac{y^2}{b^2} + \xi\frac{z^2}{c^2} - \delta = 0 \,, \tag{3.74}$$

where ζ, η and ξ take values either -1, 0 or 1 and δ takes values either 0 or 1, depending on the classification of quadrics (see Table 3.1).

By evaluating (3.70), (3.71), (3.72) and (3.73) for f given in (3.74), and substituting into (3.66) and (3.67), we obtain

---

[2] A center of a quadric surface is defined as a point bisecting every chord passing through it [79]. Here chord is a line which joins two points on a surface. Ellipsoids and hyperboloids have centers, while paraboloids do not have centers. The elliptic/hyperbolic cylinder is a limiting case of the ellipsoid/hyperboloid and the elliptic cone is asymptotic to hyperboloids of one and two sheets.

Table 3.1. Classification of implicit quadrics

Implicit Quadrics	ζ	η	ξ	δ
Ellipsoid	1	1	1	1
Hyperboloid of One Sheet	1	1	-1	1
	1	-1	1	1
	-1	1	1	1
Hyperboloid of Two Sheets	1	-1	-1	1
	-1	1	-1	1
	-1	-1	1	1
Elliptic Cone	1	1	-1	0
	1	-1	1	0
	-1	1	1	0
Elliptic Cylinder	1	1	0	1
	1	0	1	1
	0	1	1	1
Hyperbolic Cylinder	1	-1	0	1
	-1	1	0	1
	1	0	-1	1
	-1	0	1	1
	0	1	-1	1
	0	-1	1	1

$$K(x,y,z) = \frac{\zeta\eta\xi\delta}{a^2b^2c^2(\zeta^2\frac{x^2}{a^4} + \eta^2\frac{y^2}{b^4} + \xi^2\frac{z^2}{c^4})^2} , \tag{3.75}$$

$$H(x,y,z) = \tag{3.76}$$
$$-\frac{\zeta^2b^2c^2(\xi b^2 + \eta c^2)x^2 + \eta^2a^2c^2(\xi a^2 + \zeta c^2)y^2 + \xi^2a^2b^2(\eta a^2 + \zeta b^2)z^2}{2a^4b^4c^4(\zeta^2\frac{x^2}{a^4} + \eta^2\frac{y^2}{b^4} + \xi^2\frac{z^2}{c^4})^{\frac{3}{2}}} ,$$

where (x,y,z) satisfy $f(x,y,z) = 0$. The principal curvatures can be obtained by substituting (3.75) and (3.77) into

$$\kappa(x,y,z) = H \pm \sqrt{H^2 - K} , \tag{3.77}$$

where we will not show the substituted expression because it is too cumbersome.

The curvatures of a hyperbolic cylinder ($\zeta = \delta = 1$, $\eta = -1$, $\xi = 0$)

$$f(x,y) = \frac{x^2}{a^2} - \frac{y^2}{b^2} - 1 = 0 , \tag{3.78}$$

can be obtained by evaluating (3.75), (3.77) and (3.77) resulting

$$K = 0, \quad H = \frac{b^2x^2 - a^2y^2}{2a^4b^4(\frac{x^2}{a^4} + \frac{y^2}{b^4})^{\frac{3}{2}}} , \tag{3.79}$$

$$\kappa_{max} = \frac{b^2x^2 - a^2y^2}{a^4b^4(\frac{x^2}{a^4} + \frac{y^2}{b^4})^{\frac{3}{2}}}, \quad \kappa_{min} = 0 , \tag{3.80}$$

where $(x, y) \in f(x, y) = \frac{x^2}{a^2} - \frac{y^2}{b^2} - 1 = 0$.

Similarly, the curvatures of an ellipsoid $(\zeta = \eta = \xi = \delta = 1)$

$$f(x, y, z) = \frac{x^2}{a^2} + \frac{y^2}{b^2} + \frac{z^2}{c^2} - 1 = 0 , \tag{3.81}$$

are evaluated as

$$K = \frac{1}{a^2 b^2 c^2 \left(\frac{x^2}{a^4} + \frac{y^2}{b^4} + \frac{z^2}{c^4} \right)^2}, \quad H = \frac{x^2 + y^2 + z^2 - a^2 - b^2 - c^2}{2a^2 b^2 c^2 \left(\frac{x^2}{a^4} + \frac{y^2}{b^4} + \frac{z^2}{c^4} \right)^{\frac{3}{2}}},$$

$$\tag{3.82}$$

$$\kappa = \frac{x^2 + y^2 + z^2 - a^2 - b^2 - c^2}{2a^2 b^2 c^2 \left(\frac{x^2}{a^4} + \frac{y^2}{b^4} + \frac{z^2}{c^4} \right)^{\frac{3}{2}}} \tag{3.83}$$

$$\pm \frac{\sqrt{(x^2 + y^2 + z^2 - a^2 - b^2 - c^2)^2 - 4a^2 b^2 c^2 \left(\frac{x^2}{a^4} + \frac{y^2}{b^4} + \frac{z^2}{c^4} \right)}}{2a^2 b^2 c^2 \left(\frac{x^2}{a^4} + \frac{y^2}{b^4} + \frac{z^2}{c^4} \right)^{\frac{3}{2}}} ,$$

where $(x, y, z) \in f(x, y, z) = \frac{x^2}{a^2} + \frac{y^2}{b^2} + \frac{z^2}{c^2} - 1 = 0$. Here we note that in the derivation of the mean curvature in (3.82), we used (3.81) to simplify the expression. For the case of a sphere of radius R, (3.81) simplifies to $f(x, y, z) = \frac{1}{R^2}(x^2 + y^2 + z^2) - 1 = 0$, and (3.82) and (3.83) simplify to $K = \frac{1}{R^2}$, $H = \kappa = -\frac{1}{R}$, which shows that a sphere is made of entirely nonflat umbilics (see Sects. 9.1 and 9.2). The negative sign comes from the sign convention of the curvature (see Fig. 3.7 and Table 3.2).

Finally, the curvatures of an elliptic cone $(\zeta = \eta = 1, \xi = -1$ and $\delta = 0)$

$$f(x, y, z) = \frac{x^2}{a^2} + \frac{y^2}{b^2} - \frac{z^2}{c^2} = 0 , \tag{3.84}$$

excluding the apex $(0,0,0)$ are given by

$$K = 0, \quad H = -\frac{x^2 + y^2 + z^2}{2a^2 b^2 c^2 \left(\frac{x^2}{a^4} + \frac{y^2}{b^4} + \frac{z^2}{c^4} \right)^{\frac{3}{2}}} , \tag{3.85}$$

$$\kappa_{max} = 0, \kappa_{min} = -\frac{x^2 + y^2 + z^2}{a^2 b^2 c^2 \left(\frac{x^2}{a^4} + \frac{y^2}{b^4} + \frac{z^2}{c^4} \right)^{\frac{3}{2}}}, \tag{3.86}$$

where $(x, y, z) \in f(x, y, z) = \frac{x^2}{a^2} + \frac{y^2}{b^2} - \frac{z^2}{c^2} = 0$. Here we also used (3.84) to simplify the expression of mean curvature in (3.85).

3.6 Euler's theorem and Dupin's indicatrix

The normal curvatures of a surface in an arbitrary direction (in the tangent plane) at point P can be expressed in terms of principal curvatures κ_1 and κ_2

at point P and the angle Φ between the arbitrary direction and the principal direction corresponding to κ_1, namely,

$$\kappa_n = \kappa_1 \cos^2 \Phi + \kappa_2 \sin^2 \Phi . \tag{3.87}$$

This is known as *Euler's theorem*. For simplicity, we assume that the iso-parametric curves of a surface are lines of curvature, which leads to $F = M = 0$ (see (3.60)). Now (3.26) takes the form

$$\kappa_n = \frac{L du^2 + N dv^2}{E du^2 + G dv^2} . \tag{3.88}$$

For $v = const$ iso-parametric lines $dv = 0$ and for $u = const$ iso-parametric lines $du = 0$, thus the principal curvatures κ_1 and κ_2 are given by:

$$\kappa_1 = \frac{L}{E}, \qquad \kappa_2 = \frac{N}{G} . \tag{3.89}$$

The angle Φ between the direction $\frac{dv}{du}$ and the principal direction corresponding to κ_1 ($dv_1 = 0$, u_1 arbitrary) is evaluated by (3.17) as

$$\cos \Phi = E \frac{du}{ds} \frac{du_1}{ds_1} . \tag{3.90}$$

Since $ds_1 = \sqrt{E du_1^2}$ and $ds = \sqrt{E du^2 + G dv^2}$ we deduce

$$\cos \Phi = \sqrt{E} \frac{du}{ds}, \qquad \sin \Phi = \sqrt{G} \frac{dv}{ds} . \tag{3.91}$$

As a consequence, we have the Euler's theorem (3.87).

Next we explain Euler's theorem in a more simple way. Let us consider a section of the surface cut by a plane parallel to the tangent plane at the point P, and at an infinitesimal distance $h > 0$ from it [441]. We also consider a plane through P containing the normal vector. If we denote the intersection points of the surface and the two planes by Q and Q', the signed radius of curvature of this normal section by ϱ, and the length of QQ' by $2R$ as shown in Fig. 3.11, we have the relation

$$(|\varrho| - h)^2 + R^2 = |\varrho|^2 , \tag{3.92}$$

thus

$$R^2 = 2h|\varrho| , \tag{3.93}$$

to the first order. If Φ is the inclination of this normal section to the principal direction corresponding to κ_1, Euler's theorem provides

$$\kappa_1 \cos^2 \Phi + \kappa_2 \sin^2 \Phi = \frac{1}{\varrho} = \pm \frac{2h}{R^2} . \tag{3.94}$$

If we set

$$\xi = R\cos\Phi, \qquad \eta = R\sin\Phi\,, \tag{3.95}$$

we obtain

$$\frac{\xi^2}{2h\varrho_1} + \frac{\eta^2}{2h\varrho_2} = \pm 1\,, \tag{3.96}$$

where ϱ_1 and ϱ_2 are principal radius of curvatures. Consequently a section of the surface cut by a plane parallel to the tangent plane at the point P, and at an infinitesimal distance is a conic section. If we scale the ξ-η coordinates as follows

$$X = \frac{\xi}{\sqrt{2h}} = \frac{R}{\sqrt{2h}}\cos\Phi = \sqrt{|\varrho|}\cos\Phi, \tag{3.97}$$

$$Y = \frac{\eta}{\sqrt{2h}} = \frac{R}{\sqrt{2h}}\sin\Phi = \sqrt{|\varrho|}\sin\Phi\,, \tag{3.98}$$

we obtain

$$\frac{X^2}{\varrho_1} + \frac{Y^2}{\varrho_2} = \pm 1\,. \tag{3.99}$$

This equation determines a conic section called *Dupin's indicatrix* as shown in Fig. 3.12. If P is an elliptic point, both principal curvatures have the same sign, and the indicatrix is an ellipse, while if it is a hyperbolic point, the principal curvatures have different sign and the indicatrix consists of a pair of hyperbolas with asymptotic lines $Y = \pm\sqrt{\frac{|\varrho_2|}{|\varrho_1|}}\,X$. If one of the principal curvatures vanishes, it is a parabolic point and the indicatrix yields a pair of parallel lines.

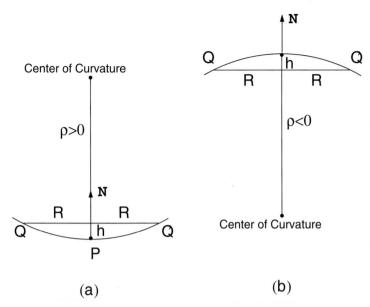

Fig. 3.11. Cross section of the surface cut by a normal plane: (a) normal curvature is positive, (b) normal curvature is negative (Here we followed the curvature convention (a); see Fig. 3.7)

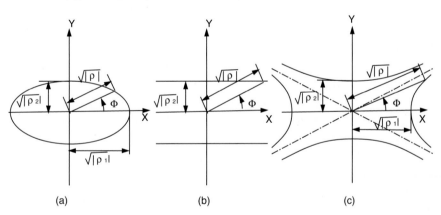

Fig. 3.12. Dupin's indicatrix for (a) elliptic point, (b) parabolic point, (c) hyperbolic point

Table 3.2. A list of equations which involves a sign change due to the sign convention of curvature of the planar curve or the normal curvature of the surface (see Fig. 3.7). In sign convention (a) the center of curvature is on the same side of the normal vector, while in sign convention (b) it is on the opposite direction

Equation	Convention (a)	Convention (b)
(2.20)	$\mathbf{r}'' = \mathbf{t}' = \kappa\mathbf{n}$	$\mathbf{r}'' = \mathbf{t}' = -\kappa\mathbf{n}$
(2.22)	$\ddot{\mathbf{r}} = \kappa\mathbf{n}v^2 + \mathbf{t}\frac{dv}{dt}$	$\ddot{\mathbf{r}} = -\kappa\mathbf{n}v^2 + \mathbf{t}\frac{dv}{dt}$
(2.24)	$\mathbf{n} = \mathbf{e}_z \times \mathbf{t} = \frac{(-\dot{y},\dot{x})^T}{\sqrt{\dot{x}^2+\dot{y}^2}}$	$\mathbf{n} = \mathbf{t} \times \mathbf{e}_z = \frac{(\dot{y},-\dot{x})^T}{\sqrt{\dot{x}^2+\dot{y}^2}}$
(2.27)	$\mathbf{n} = \mathbf{e}_z \times \mathbf{t} = \frac{(f_x,f_y)^T}{\sqrt{f_x^2+f_y^2}}$ $= \frac{\nabla f}{\lvert\nabla f\rvert}$	$\mathbf{n} = \mathbf{t} \times \mathbf{e}_z = \frac{(-f_x,-f_y)^T}{\sqrt{f_x^2+f_y^2}}$ $= -\frac{\nabla f}{\lvert\nabla f\rvert}$
(2.55)	$\mathbf{n}' = -\kappa\mathbf{t}\ (\tau = 0)$	$\mathbf{n}' = \kappa\mathbf{t}\ (\tau = 0)$
(3.23)	$\mathbf{k}_n = \kappa_n\mathbf{N}$	$\mathbf{k}_n = -\kappa_n\mathbf{N}$
(3.25)	$\kappa_n = \frac{d\mathbf{t}}{ds}\cdot\mathbf{N} = -\mathbf{t}\cdot\frac{d\mathbf{N}}{ds}$ $= -\frac{d\mathbf{r}}{ds}\cdot\frac{d\mathbf{N}}{ds} = -\frac{d\mathbf{r}\cdot d\mathbf{N}}{d\mathbf{r}\cdot d\mathbf{r}}$	$\kappa_n = -\frac{d\mathbf{t}}{ds}\cdot\mathbf{N} = \mathbf{t}\cdot\frac{d\mathbf{N}}{ds}$ $= \frac{d\mathbf{r}}{ds}\cdot\frac{d\mathbf{N}}{ds} = \frac{d\mathbf{r}\cdot d\mathbf{N}}{d\mathbf{r}\cdot d\mathbf{r}}$
(3.26)	$\kappa_n = \frac{Ldu^2+2Mdudv+Ndv^2}{Edu^2+2Fdudv+Gdv^2}$	$\kappa_n = -\frac{Ldu^2+2Mdudv+Ndv^2}{Edu^2+2Fdudv+Gdv^2}$
(3.30)	$\kappa_n = \frac{II}{I} = \frac{L+2M\lambda+N\lambda^2}{E+2F\lambda+G\lambda^2}$	$\kappa_n = -\frac{II}{I} = -\frac{L+2M\lambda+N\lambda^2}{E+2F\lambda+G\lambda^2}$
(3.38)	$\kappa_n = \frac{L+2M\lambda+N\lambda^2}{E+2F\lambda+G\lambda^2}$ $= \frac{M+N\lambda}{F+G\lambda}$	$\kappa_n = -\frac{L+2M\lambda+N\lambda^2}{E+2F\lambda+G\lambda^2}$ $= -\frac{M+N\lambda}{F+G\lambda}$
(3.41)	$(L - \kappa_n E)du + (M - \kappa_n F)dv$ $= 0$ $(M - \kappa_n F)du + (N - \kappa_n G)dv$ $= 0$	$(L + \kappa_n E)du + (M + \kappa_n F)dv$ $= 0$ $(M + \kappa_n F)du + (N + \kappa_n G)dv$ $= 0$
(3.42)	$\begin{vmatrix} L - \kappa_n E & M - \kappa_n F \\ M - \kappa_n F & N - \kappa_n G \end{vmatrix} = 0$	$\begin{vmatrix} L + \kappa_n E & M + \kappa_n F \\ M + \kappa_n F & N + \kappa_n G \end{vmatrix} = 0$
(3.43)	$(EG - F^2)\kappa_n^2$ $-(EN + GL - 2FM)\kappa_n$ $+(LN - M^2) = 0$	$(EG - F^2)\kappa_n^2$ $+(EN + GL - 2FM)\kappa_n$ $+(LN - M^2) = 0$
(3.47)	$H = \frac{EN+GL-2FM}{2(EG-F^2)}$	$H = \frac{2FM-EN-GL}{2(EG-F^2)}$
(3.51)	$\lambda = -\frac{M-\kappa_n F}{N-\kappa_n G}$	$\lambda = -\frac{M+\kappa_n F}{N+\kappa_n G}$
(3.52)	$\lambda = -\frac{L-\kappa_n E}{M-\kappa_n F}$	$\lambda = -\frac{L+\kappa_n E}{M+\kappa_n F}$
(3.67)	$H =$ $\frac{(1+h_x^2)h_{yy}-2h_xh_yh_{xy}+(1+h_y^2)h_{xx}}{2(1+h_x^2+h_y^2)^{3/2}}$	$H =$ $\frac{2h_xh_yh_{xy}-(1+h_x^2)h_{yy}-(1+h_y^2)h_{xx}}{2(1+h_x^2+h_y^2)^{3/2}}$
(3.88)	$\kappa_n = \frac{Ldu^2+Ndv^2}{Edu^2+Gdv^2}$	$\kappa_n = -\frac{Ldu^2+Ndv^2}{Edu^2+Gdv^2}$
(3.89)	$\kappa_1 = \frac{L}{E}, \quad \kappa_2 = \frac{N}{G}$	$\kappa_1 = -\frac{L}{E}, \quad \kappa_2 = -\frac{N}{G}$

4. Nonlinear Polynomial Solvers and Robustness Issues

4.1 Introduction

We have seen in Chap. 1 that curves and surfaces in CAD/CAM systems are usually represented by piecewise polynomial equations of various types. As we will see in the remaining chapters of this book, the governing equations for general interrogation problems on such curve and surface representations (intersections, distance functions, curvature extrema, etc.) reduce to solving systems of nonlinear polynomial equations as follows:

$$\mathbf{f}(\mathbf{x}) = 0 , \tag{4.1}$$

where $\mathbf{f}$ consists of n functions $f_1, f_2, \ldots, f_n$, each of which is a polynomial in the l independent variables $x_1, x_2, \ldots, x_l$.

Frequently such systems also include square roots of polynomials, which arise from normalization of the normal vector and analytical expressions of the principal curvatures of a surface (see (2.24), (3.3), (3.49), (3.50)).

Example 4.1.1. As an illustrative example, let us consider a simple intersection problem of two planar implicit polynomial (algebraic) curves. Consider two circles $x^2 + y^2 = \frac{9}{16}$ and $(x - 1)^2 + y^2 = \frac{1}{4}$ intersecting as shown in Fig. 4.1. In this case $n = l = 2$, and if we set $x_1 = x$ and $x_2 = y$, the system of equations becomes

$$f_1(x_1, x_2) = x_1^2 + x_2^2 - \frac{9}{16} = 0 , \tag{4.2}$$

$$f_2(x_1, x_2) = (x_1 - 1)^2 + x_2^2 - \frac{1}{4} = 0 . \tag{4.3}$$

The roots can be obtained by eliminating x_2, and solving for x_1, which gives

$$(x_1, x_2) = \left(\frac{21}{32}, \frac{\pm\sqrt{135}}{32} \right) \simeq (0.65625, \pm 0.36309) .$$

In this example the degree of the polynomials and the number of variables were low, so we could solve the system by elementary analytical (elimination)

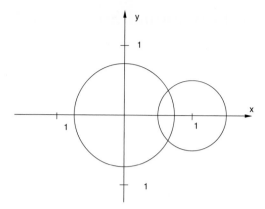

Fig. 4.1. Intersection of two circles

calculations. However, most problems that arise in CAD/CAM interrogation have higher degrees and number of variables. Such systems of equations
have been solved in earlier approaches by *local* numerical techniques such as
Newton-type methods which require good initial approximation to all roots
[69, 126], and hence cannot provide full assurance that all roots will be found.
On the other hand *global* techniques find all the roots without initial approximation. We will briefly introduce Newton's method in Sect. 4.2, and the rest
of Chap. 4 will be spent on global techniques as well as robustness issues.

4.2 Local solution methods

Newton-type methods are based on local linearization and are conceptually
simple. They are designed to compute roots based on initial approximations.
To begin with, we consider Newton's method in one variable [69, 293] where
we want to find roots for $f(x) = 0$. If we denote the initial guess of the
root as x_0, then in the neighborhood of x_0 the function $f(x)$ can be linearly
approximated using Taylor expansion as follows:

$$f(x) \simeq f(x_0) + (x - x_0)\dot{f}(x_0) . \tag{4.4}$$

Provided that $\dot{f}(x_i) \neq 0$, the iteration formula immediately follows:

$$x_{i+1} = x_i - \frac{f(x_i)}{\dot{f}(x_i)}, \quad i = 0, 1, 2 \dots . \tag{4.5}$$

This is illustrated in Fig. 4.2(a). A modified Newton's method [151] for
one variable is shown in Fig. 4.2(b), where we take a fractional step as follows
in order to reduce the possibility of divergence in Fig. 4.2(b) of the full step
method given by (4.5)

$$x_{i+1} = x_i - \mu \frac{f(x_i)}{f'(x_i)} , \tag{4.6}$$

where $\mu = max[1, \frac{1}{2}, \frac{1}{4}, ...]$ such that $|f(x_{i+1})| < |f(x_i)|$.

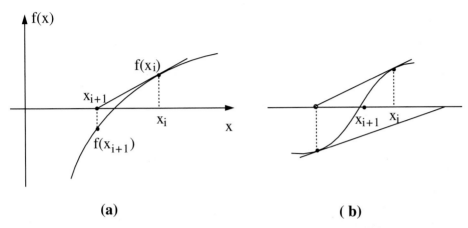

(a) **(b)**

Fig. 4.2. Newton's method for $f(x) = 0$

If $n = l$ in (4.1), we can easily extend Newton's method for a single variable to n variables as follows:

$$\mathbf{x}_{i+1} = \mathbf{x}_i + \Delta\mathbf{x}_i , \tag{4.7}$$

where

$$\mathbf{J}(\mathbf{x}_i) \cdot \Delta\mathbf{x}_i = -\mathbf{f}(\mathbf{x}_i) , \tag{4.8}$$

and $\mathbf{J}(\mathbf{x}_i) = \left[\frac{\partial f_j}{\partial x_k}\right]$ is the Jacobian matrix of (4.1) [69].

Advantages of the Newton's method are its quadratic convergence and simplicity of implementation. Disadvantages are that for each root a good initial approximation is required, otherwise the method may diverge. Also the method cannot by itself provide full assurance that all roots have been found.

Example 4.2.1. Let us solve the intersection of two circles discussed in Example 4.1.1 using Newton's method. The Jacobian matrix is evaluated as follows:

$$[J]_i = \begin{pmatrix} \frac{\partial f_1}{\partial x_1} & \frac{\partial f_1}{\partial x_2} \\ \frac{\partial f_2}{\partial x_1} & \frac{\partial f_2}{\partial x_2} \end{pmatrix}_i = \begin{pmatrix} 2x_1 & 2x_2 \\ 2(x_1 - 1) & 2x_2 \end{pmatrix}_i .$$

Thus the iteration scheme becomes

$$\begin{pmatrix} x_1 \\ x_2 \end{pmatrix}_{i+1} = \begin{pmatrix} x_1 \\ x_2 \end{pmatrix}_i + \begin{pmatrix} \Delta x_1 \\ \Delta x_2 \end{pmatrix}_i ,$$

where Δx_1 and Δx_2 are obtained from the solution of the following linear system

$$\begin{pmatrix} 2x_1 & 2x_2 \\ 2(x_1 - 1) & 2x_2 \end{pmatrix}_i \begin{pmatrix} \Delta x_1 \\ \Delta x_2 \end{pmatrix}_i = - \begin{pmatrix} f_1 \\ f_2 \end{pmatrix}_i .$$

4.3 Classification of global solution methods

Global solution methods are designed to compute all roots in some area of interest. In recent computational algebraic geometry related research, three classes of methods for the computation of solutions of nonlinear polynomial systems can be distinguished [300]: (1) algebraic and hybrid techniques, (2) homotopy (continuation) methods, (3) subdivision methods. We will briefly review these three types of techniques.

4.3.1 Algebraic and Hybrid Techniques

Algebraic techniques for solving a nonlinear polynomial system are based on *elimination theory*. This theory deals with the problem of eliminating one or more variables from a system of polynomial equations, thus reducing the given problem to a problem of higher degree but in *fewer* variables. There are basically two fundamental approaches in elimination theory: (1) Resultants, and (2) Gröbner bases. Both of the above operate ideally in a symbolic algebra environment, and the coefficients of the polynomials involved are either rational or real algebraic numbers. There are several algorithms for solving nonlinear polynomial systems using the above approaches. All the algorithms are based on some fundamental algorithm that "finds" all the roots, real and complex, of a univariate polynomial. The word "finds" means, either the algorithm isolates the roots using intervals and rectangles, or encodes them as algebraic numbers, for further manipulation. Let $f(x)$ be a polynomial with integer coefficients of degree m, d be a bound for the size of the coefficients of $f(x)$, and $L(d)$ be the number of binary digits of d. Then, the (worst) running times of real root finding algorithms are functions of m, $L(d)$ and are given in [64]. On the other hand, bisection methods for finding all roots of f, real and complex with similar running times, can be found in [442, 357]. As it can be seen from the computing times found in [442, 64, 357], there is an enormous coefficient growth of all the quantities involved along the way (requiring significant computer memory). The latter is one of the most serious problems that all the algorithms using these techniques suffer from.

   Resultant type algorithms: A *resultant* is a function of the coefficients of a given system of polynomials and when it is zero it provides an algebraic criterion for determining when this polynomial system has a solution. Resultants can be classified as classical, like the Sylvester, Bezout, Macaulay and

u resultants, and non-classical like the *sparse* resultants. A good introduction to resultants and applications can be found in [89, 409, 413, 310, 381].

Algorithms based on resultant computation have been presented in [48, 258, 187, 414, 49]. They work well on systems with a small number of solutions M. However, on systems with large M, these algorithms suffer from efficiency problems. The main reason for that is that finding roots of high degree univariate polynomials can be a very slow procedure, as discussed above, due to the use of exact arithmetic.

Gröbner bases type algorithms: The theory of Gröbner bases was developed by Buchberger [46]. Gröbner bases are very special and useful bases (generator sets) for a special class of subsets of polynomial rings in l variables, called polynomial ideals. They are named after Gröbner who was Buchberger's thesis advisor. Gröbner bases can be thought of as a generalization of Euclid's algorithm for computing the greatest common divisor of two polynomials and of the Gauss triangularization algorithm for linear systems.

The usefulness of Gröbner basis for solving nonlinear polynomial systems comes from the fact that, whenever the system has a finite number of solutions, Gröbner basis provides an equivalent system of triangular form. Algorithms using Gröbner bases use the above fact, and appear in [47, 218, 228, 115, 445]. Using Gröbner bases, polynomial systems are converted to polynomial triangular systems, which can be solved by backward substitution, much in the manner of the Gauss triangularization algorithm for linear systems.

If the system has a finite number of solutions in the affine plane, as well as in the projective plane, then a Gröbner basis can be computed in $O(m^l)$ time, where m is the highest degree among the polynomials and l is the number of variables. In case, however, that the system is not zero-dimensional at infinity, the time becomes $O(m^{l^2})$. These bounds do not take into account the coefficient growth. Gröbner basis algorithms work well on systems with few roots. This is one reason they have been considered seriously as a practical equation-solving tool. But when their complexity is measured as a function of the number of solutions, their performance is poor. As reported in [259], these algorithms frequently exhaust memory and computer resources even for low number of equations n and variables l (e.g. $n, l \leq 5$) and moderate degrees m. To overcome this difficulty, algorithms that combine resultant and linear algebra techniques are more promising concerning efficiency [15, 288, 260, 259]. These algorithms are generally *hybrid* and are based on algebraic and numerical analysis methods. In particular, this approach based on resultants transforms the problem into a sequence of eigenvalue problems. This method has found extensive application in various types of intersection problems [212].

4.3.2 Homotopy (Continuation) Methods

Homotopy methods [123, 459, 219] are mathematically elegant, but unfortunately, investigation of such methods indicates that they tend to be numerically ill-conditioned. If we try to get around this problem by implementing the algorithm in rational arithmetic, we end up with enormous memory requirements because we have to solve large systems of complex initial value problems (IVP). Interval methods can be applied to the solution of these IVPs but they can be slow in practice [259].

4.3.3 Subdivision Methods

Subdivision methods [221, 333, 286, 392, 402, 133] are generally efficient (in finding simple intersections) and stable. Therefore, they are the most frequently used methods in practice. As we will see, they can be combined with interval methods to numerically guarantee that certain subdomains do not contain solutions. Interval Newton methods [273, 131, 191, 27, 159, 158] are a promising class of subdivision methods. However, the subdivision methods are not as general as algebraic methods, since they are only capable of isolating zero-dimensional solutions. Furthermore, although the chances, that all roots have been found, increase as the resolution tolerance is lowered, there is no certainty that each root has been extracted/isolated. Subdivision methods typically do not provide a guarantee as to how many roots there may be in the remaining subdomains. However, if these subdomains are very small, the existence of a (single) root within these subdomains is a typical assumption. Lastly, subdivision techniques provide no explicit information about root multiplicities without additional computation. Despite these drawbacks, subdivision methods are very useful in practice and are further described below.

4.4 Projected Polyhedron algorithm

In this section we introduce an iterative global root-finding algorithm for an n-dimensional nonlinear polynomial equation system, which belongs to the class of subdivision methods, called *Projected Polyhedron* (PP) algorithm developed by Sherbrooke and Patrikalakis [392]. It is easy to visualize and simple in that it only requires two straightforward algorithms in order to implement it: one for subdividing multivariate polynomials in Bernstein form, and one for finding the convex hull of a two-dimensional set of points. This algorithm is an extension and generalization of earlier adaptive subdivision algorithms: for $n = 1$ used in finding the real roots and extrema of a polynomial within an interval by Lane and Riesenfeld [221], and for $n = 2$ used in shape interrogation by Geisow [124] or in intersecting rays with trimmed rational polynomial surface patches by Nishita et al. [286] (a method known as Bézier clipping). The PP algorithm has found many applications in shape interrogation problems (see also Grandine and Klein [133]) as we will see in subsequent sections and its convergence, rate of convergence and complexity properties are developed in [392].

For illustration, we will enumerate the procedures required by the PP algorithm to find roots of a degree m polynomial equation $f(x) = c_o + c_1 x + c_2 x^2 + \cdots + c_m x^m = 0$ over the interval $a \leq x \leq b$.

1. Make an affine parameter transformation $x = a + t(b - a)$ such that $0 \leq t \leq 1$ as follows:

$$f(t) = \sum_{i=0}^{m} c_i^M t^i, \qquad 0 \leq t \leq 1 . \tag{4.9}$$

   The transition from the interval $a \leq x \leq b$ to the interval $0 \leq t \leq 1$ is an affine map, and the polynomials are invariant under affine parameter transformation [92].
2. Convert the basis from monomial to Bernstein [106]:

$$f(t) = \sum_{i=0}^{m} c_i^B B_{i,m}(t) , \tag{4.10}$$

where

$$c_i^B = \sum_{j=0}^{i} \frac{\binom{i}{j}}{\binom{m}{j}} c_j^M , \tag{4.11}$$

   and $B_{i,m}(t)$ is the ith Bernstein polynomial of degree m.
3. Create a graph of function $f(t)$ using the linear precision property of Bernstein polynomials (see (1.21)). Then the graph will become a Bézier curve

$$\mathbf{f}(t) = \begin{pmatrix} t \\ f(t) \end{pmatrix} = \sum_{i=0}^{m} \begin{pmatrix} \frac{i}{m} \\ c_i^B \end{pmatrix} B_{i,m}(t) , \tag{4.12}$$

   where $(\frac{i}{m}, c_i^B)^T$ are control points. Now the problem of finding roots of the univariate polynomial has been transformed into a geometric problem of finding the intersection of a Bézier curve with the parameter axis, a transformation already used in Geisow [124] for surface interrogation.
4. Construct the convex hull of the Bézier curve.
5. Intersect the convex hull with the parameter axis.
6. Discard the regions which do not contain roots by applying the de Casteljau subdivision algorithm and find a sub-region of [0,1] which may contain the root(s).

7. If the sub-region is sufficiently small, we conclude that there is a root inside and return it. But when there are more than one root in the sub-region, the sub-region will not be reduced. In such case we split the region evenly by applying the de Casteljau subdivision algorithm and we go back to 4.

Example 4.4.1. PP algorithm in one dimension.
Find the roots of $f(x) = -1.1x^2 + 1.4x - 0.2 = 0$ where $0 \le x \le 2$. The roots are approximately, 0.164 and 1.108.

1. Make an affine parameter transformation by plugging $x = 0 + (2-0)t = 2t$ into $f(x)$ yielding $f(t) = -4.4t^2 + 2.8t - 0.2 = 0$ where $0 \le t \le 1$.
2. Convert from monomial to Bernstein basis using (4.11) as below

$$c_i^B = \sum_{j=0}^{i} \frac{\binom{i}{j}}{\binom{2}{j}} c_j^M ,$$

   where $c_0^M = -0.2$, $c_1^M = 2.8$ and $c_2^M = -4.4$, thus leading to $c_0^B = -0.2$, $c_1^B = 1.2$ and $c_2^B = -1.8$.
3. Create a graph of function $f(t)$ using linear precision property of the Bernstein polynomial

$$t = \sum_{i=0}^{2} \frac{i}{2} B_{i,2}(t) ,$$

   yielding a Bézier curve

$$\mathbf{f}(t) = \begin{pmatrix} t \\ f(t) \end{pmatrix} = \sum_{i=0}^{2} \begin{pmatrix} \frac{i}{2} \\ c_i^B \end{pmatrix} B_{i,2}(t) .$$

   with control points of (0, -0.2), (0.5, 1.2) and (1, -1.8).
4. Construct a convex hull of the Bézier curve, which is a triangle as shown in Fig. 4.3.
5. The convex hull intersects the t-axis with $t = 0.0714$ and $t = 0.7$.
6. Discard the regions $0 \le t \le 0.0714$ and $0.7 \le t \le 1$, which do not contain roots, by applying the de Casteljau algorithm. Now we have a smaller convex hull which contains the roots (see shaded triangular in Fig. 4.3).
7. If the sub-region is sufficiently small, we conclude that there is a root inside and return it. In this case there are two roots in the convex hull and the sub-region does not reduce much (even after several iteration steps), thus we split the region evenly by applying the de Casteljau subdivision algorithm and go back to 4.

Example 4.4.2. PP algorithm in two dimensions.
Let us solve the system of polynomial equations (4.2) and (4.3) over the region $-1 \le x_1, x_2 \le 1$.

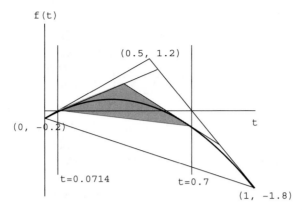

Fig. 4.3. de Casteljau algorithm applied to the quadratic Bézier curve

1. Make an affine parameter transformation by substituting $x_1 = 2u - 1$ and $x_2 = 2v - 1$ into (4.2) and (4.3) so that $0 \le u, v \le 1$, then:

$$f_1(u,v) = 4u^2 - 4u + 4v^2 - 4v + \frac{23}{16} = 0 \;,$$

$$f_2(u,v) = 4u^2 - 8u + 4v^2 - 4v + \frac{19}{4} = 0 \;.$$

2. Convert from monomial to Bernstein basis using

$$c_{ij}^B = \sum_{k=0}^{i} \sum_{l=0}^{j} \frac{\binom{i}{k}\binom{j}{l}}{\binom{m}{k}\binom{n}{l}} c_{kl}^M \;,$$

where in this case $m = n = 2$ leading to

$$c_{100}^B = 1.4375, \quad c_{101}^B = -0.5625, \quad c_{102}^B = 1.4375 \;,$$

$$c_{110}^B = -0.5625, \quad c_{111}^B = -2.5625, \quad c_{112}^B = -0.5625 \;,$$

$$c_{120}^B = 1.4375, \quad c_{121}^B = -0.5625, \quad c_{122}^B = 1.4375 \;,$$

and

$$c_{200}^B = 4.75, \quad c_{201}^B = 2.75, \quad c_{202}^B = 4.75 \;,$$

$$c_{210}^B = 0.75, \quad c_{211}^B = -1.25, \quad c_{212}^B = 0.75 \;,$$

$$c_{220}^B = 0.75, \quad c_{221}^B = -1.25, \quad c_{222}^B = 0.75 \;.$$

3. Create graphs of functions $f_1(u,v)$ and $f_2(u,v)$ using the linear precision property of Bernstein polynomials. Then the graphs will become two Bézier surfaces as follows:

$$\mathbf{f}_1(u,v) = \begin{pmatrix} u \\ v \\ f_1(u,v) \end{pmatrix} = \sum_{i=0}^{2}\sum_{j=0}^{2} \begin{pmatrix} \frac{i}{2} \\ \frac{j}{2} \\ c_{1ij}^B \end{pmatrix} B_{i,2}(u)B_{j,2}(v)\,,$$

$$\mathbf{f}_2(u,v) = \begin{pmatrix} u \\ v \\ f_2(u,v) \end{pmatrix} = \sum_{i=0}^{2}\sum_{j=0}^{2} \begin{pmatrix} \frac{i}{2} \\ \frac{j}{2} \\ c_{2ij}^B \end{pmatrix} B_{i,2}(u)B_{j,2}(v)\,.$$

The two Bézier surfaces are shown in Fig. 4.4. Now the root-finding problem of the bivariate polynomial system has been transformed to find the intersections of three surfaces, $\mathbf{f}_1(u,v)$, $\mathbf{f}_2(u,v)$ and the xy-plane. Figure 4.5 shows the intersection between the plane and both Bézier surfaces. We can easily observe that the two intersection curves are the circles in Fig. 4.1 but trimmed in the resulting (u,v) domain.

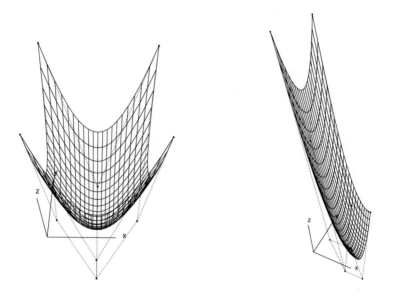

Fig. 4.4. Bézier surfaces and their control points

4. Project the control points of $\mathbf{f}_1(u,v)$ and $\mathbf{f}_2(u,v)$ onto xz and yz planes. Here $x = u$, $y = v$. For each xz and yz plane, construct 2-D convex hulls. Figure 4.6 (a) shows 2-D convex hulls on the xz plane, while Fig. 4.6 (b) shows 2-D convex hulls on the yz plane. The solid line corresponds to

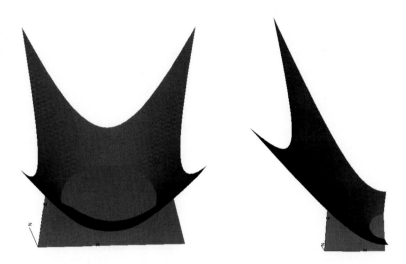

Fig. 4.5. Bézier surfaces intersecting with xy-plane

convex hull of $\mathbf{f}_1(u, v)$ and the dashed line corresponds to that of $\mathbf{f}_2(u, v)$.

5. Intersect each 2-D convex hull on the xz plane with x (u) axis. The parameter interval $[0, 1]$ contains solutions of (4.2), while the interval $[0.34375, 1]$ contains solutions of (4.3). The root is contained in the common interval $[0.34375, 1]$ of u. We will repeat the same procedures for the 2-D convex hulls on the yz plane to find the common interval $[0.1875, 0.8125]$ of v.

6. Discard the reigion $[0, 0.34375]$ of u and regions $[0, 0.1875]$ $[0.8125, 1]$ of v which do not contain the roots by applying the de Casteljau subdivision algorithm to both Bézier surfaces.

7. If the sub-regions of both parameters are sufficiently small, we conclude that there is a root. In this case, the sub-region of the v parameter will not decrease much in size because there are two roots, while the interval decreases more in u parameter, since the two roots have the same u value. We split the sub-region of v evenly by applying the de Casteljau subdivision algorithm for both surfaces and we go back to step 4.

Many applications in shape interrogation result in systems of n nonlinear polynomial equations with n unknowns, referred to as *balanced systems*. However, there exist some problems such as tangential intersection or implicit curve/surface rendering consisting of n nonlinear polynomial equations with l unknowns, where n could be larger or smaller than l. When $n > l$ the sy-

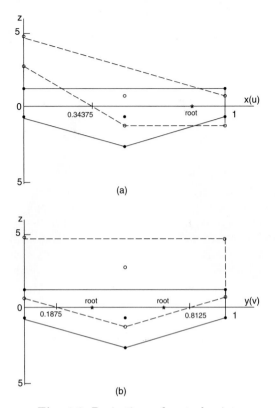

Fig. 4.6. Projections of control points

stem is called *overconstrained* and when $n < l$ it is called *underconstrained*.
Now we will introduce an n-dimensional Projected-Polyhedron algorithm
such that it can effectively handle overconstrained problems [392, 179]. This
algorithm can also be used in underconstrained problems but in such cases it
tends to be slow (especially in the presence of infinite roots); for such cases
more specialized algorithms are necessary (e.g. parametric surface intersections where $n = 3$, and $l = 4$). In such cases the PP algorithm is used to find
characteristic points efficiently and marching methods are used to trace the
intersection curves (see Sect. 5.8.2).

Suppose we are given a set of n nonlinear polynomials $f_1, f_2, \ldots, f_n$, each
of which is polynomial in the independent variables $x_1, x_2, \ldots, x_l$. Let $m_i^{(k)}$
denote the degree in x_i of polynomial f_k, so that the multi-index $M^{(k)} =
(m_1^{(k)}, m_2^{(k)}, \ldots, m_l^{(k)})$ describes all the degree information of f_k. Furthermore,
suppose we are given an l-dimensional rectangular subset of $\mathbf{R}^l$

$$B = [a_1, b_1] \times [a_2, b_2] \times \ldots \times [a_l, b_l] . \qquad (4.13)$$

A priori knowledge of B is one of the main features of geometric modeling and shape interrogation problems. We wish to find all points $\mathbf{x} = (x_1, x_2, \ldots, x_l) \in B$ such that

$$f_1(\mathbf{x}) = f_2(\mathbf{x}) = \ldots = f_n(\mathbf{x}) = 0 . \tag{4.14}$$

By making the *affine parameter transformation* [92] $x_i = a_i + u_i(b_i - a_i)$ for each i between 1 and l inclusive, we simplify the problem to one of determining all $\mathbf{u} \in [0, 1]^l$ such that

$$f_1(\mathbf{u}) = f_2(\mathbf{u}) = \ldots = f_n(\mathbf{u}) = 0 . \tag{4.15}$$

Since all of the f_k are polynomial in each of the l independent variables, a simple *change of basis* [92] allows us to express them in the multivariate Bernstein basis, which has better numerical stability under perturbation of its coefficients than the power basis [105] as we discussed in Sect. 1.3.3 and in addition permits transformation of an algebraic problem to a geometric problem. In other words, for each f_k there exists an l-dimensional array of real coefficients $w^{(k)}_{i_1 i_2 \ldots i_l}$ such that for each $k \in \{1, \ldots, n\}$

$$f_k(\mathbf{u}) = \sum_{i_1=0}^{m_1^{(k)}} \sum_{i_2=0}^{m_2^{(k)}} \cdots \sum_{i_l=0}^{m_l^{(k)}} w^{(k)}_{i_1 i_2 \ldots i_l} B_{i_1,m_1^{(k)}}(u_1) B_{i_2,m_2^{(k)}}(u_2) \ldots B_{i_l,m_l^{(k)}}(u_l) . \tag{4.16}$$

The notation in (4.16) may be simplified by letting $I = (i_1, i_2, \ldots, i_l)$, $M^{(k)} = (m_1^{(k)}, m_2^{(k)}, \ldots, m_l^{(k)})$ and writing (4.16) in the equivalent form

$$f_k(\mathbf{u}) = \sum_I^{M^{(k)}} w_I^{(k)} B_{I,M^{(k)}}(\mathbf{u}) . \tag{4.17}$$

Representation of algebraic and piecewise algebraic surfaces (i.e., for $l = 3$) in terms of tensor products of Bernstein polynomials or B-splines has been studied earlier by Patrikalakis and Kriezis [299]. Equation (4.17) is simply an extension to n dimensions. Provided that conversion of the problem to the Bernstein basis is exact or sufficiently accurate, the use of the Bernstein basis in conjunction with subdivision is known to be numerically stable [105]. The conversion process itself may be numerically ill-conditioned [106]. Therefore, we recommend that the problem be formulated in the Bernstein basis from the very beginning or that the conversion is carried out in exact arithmetic. If necessary, polynomials may be converted from the multivariate power basis to the multivariate Bernstein basis using the following formula:

$$c^B_{i_1 i_2 \ldots i_l} = \sum_{j_1=0}^{i_1} \sum_{j_2=0}^{i_2} \cdots \sum_{j_l=0}^{i_l} \frac{\binom{i_1}{j_1}\binom{i_2}{j_2}\cdots\binom{i_l}{j_l}}{\binom{m_1}{j_1}\binom{m_2}{j_2}\cdots\binom{m_l}{j_l}} c^M_{j_1 j_2 \ldots j_l} . \tag{4.18}$$

We now restate the problem as the intersection of the *graphs* of the f_k (each of which is a hypersurface in $\mathbf{R}^{l+1}$) and the hyperplane $u_{l+1} = 0$ of

$\mathbf{R}^{l+1}$. This idea is designed to impart geometrical significance to the coefficients of the polynomials and to the solution process.

Let us build a graph $\mathbf{f}_k$ for each f_k:

$$\begin{aligned} \mathbf{f}_k(\mathbf{u}) &= (u_1, u_2, \ldots, u_l, f_k(\mathbf{u}))^T \\ &= (\mathbf{u}, f_k(\mathbf{u}))^T \,. \end{aligned} \tag{4.19}$$

Clearly, (4.15) is satisfied by the point $\mathbf{u}$ if and only if

$$\mathbf{f}_1(\mathbf{u}) = \mathbf{f}_2(\mathbf{u}) = \ldots = \mathbf{f}_n(\mathbf{u}) = (\mathbf{u}, 0) \,. \tag{4.20}$$

Using the linear precision property of the Bernstein basis (1.21), we obtain an equivalent expression for each of the u_j in equation (4.19):

$$u_j = \sum_I^{M^{(k)}} \frac{i_j}{m_j^{(k)}} B_{I, M^{(k)}}(\mathbf{u}) \,. \tag{4.21}$$

Substituting (4.21) into (4.19) gives a more useful representation for the $\mathbf{f}_k$:

$$\mathbf{f}_k(\mathbf{u}) = \sum_I^{M^{(k)}} \mathbf{v}_I^{(k)} B_{I, M^{(k)}}(\mathbf{u}) \,, \tag{4.22}$$

where

$$\mathbf{v}_I^{(k)} = \left(\frac{i_1}{m_1^{(k)}}, \frac{i_2}{m_2^{(k)}}, \ldots, \frac{i_l}{m_l^{(k)}}, w_I^{(k)} \right)^T \,. \tag{4.23}$$

The $\mathbf{v}_I^{(k)}$ are called the *control points* of $\mathbf{f}_k$. Using the parametric hypersurfaces $\mathbf{f}_k$ instead of the real-valued f_k permits use of the powerful *convex-hull property* of the multivariate Bernstein basis.

We assume we are given n nonlinear polynomial equations with l variables in the power basis, where $n \geq l$, and a box $B = [a_1, b_1] \times [a_2, b_2] \times \ldots \times [a_l, b_l]$, in which we need to determine the roots of the given system. In this case we first scale the box by performing an appropriate affine parameter transformation described above to the functions f_k, so that the box becomes $[0, 1]^l$. Next we express the transformed nonlinear polynomial equations in the multivariate Bernstein basis using (4.18). Now we summarize the PP algorithm.

1. Using the convex hull property, find a sub-box of $[0, 1]^l$ which contains all the roots. The essential idea behind the box generation scheme in this algorithm is to transform a complicated $l + 1$-dimensional problem into a series of l two-dimensional problems. Suppose $\mathbf{R}^{l+1}$ can be coordinatized with the $u_1, u_2, \ldots, u_{l+1}$ axes; we can then employ these steps:

   a) Project the $\mathbf{v}_I^{(k)}$ of all of the $\mathbf{f}_k$ into l different coordinate planes; specifically, the (u_1, u_{l+1})-plane, the (u_2, u_{l+1})-plane, and so on, up to the (u_l, u_{l+1}) plane.

b) In each one of these planes,

    i. Construct n two-dimensional convex hulls. The first is the convex hull of the projected control points of $\mathbf{f}_1$, the second is from $\mathbf{f}_2$ and so on.

    ii. Intersect each convex hull with the horizontal axis (that is, $u_{l+1} = 0$). Because the polygon is convex, the intersection may be either a closed interval (which may degenerate to a point) or empty. If it is empty, then no root of the system exists within the given search box.

    iii. Intersect the intervals with one another. Again, if the result is empty, no root exists within the given search box.

c) Construct an l-dimensional box by taking the Cartesian product of each one of these intervals in order. In other words, the u_1 side of the box is the interval resulting from the intersection in the (u_1, u_{l+1})-plane, and so forth.

2. Using the scaling relationship between our current box and the initial box of search, see if the new sub-box represents a sufficiently small box in $\mathbf{R}^l$. If it does not, then go to step 3. If it does, then check the convex hulls of the hypersurface in the new box. If the convex hulls cross each variable axis, conclude that there is a root or at least an approximate root in the new box, and put the new box into a root list. Otherwise the new box is discarded.

3. If any dimension of this sub-box is not much smaller than 1 unit in length (i.e., the box has not decreased much in size along one or more sides), split the box evenly along each dimension which is causing trouble (not reducing in size). Continue on to the next iteration with several independent sub-problems.

4. If none of the boxes is left, then the root-finding process is over. Otherwise, perform an appropriate affine parameter transformation to the functions f_k, so that the box becomes $[0, 1]^l$, and go back to step 1 for each new box. This transformation can be performed with the multivariate de Casteljau subdivision algorithm which is an extension of similar algorithms for 1 and 2 dimensions given in [92]. However, keep track of the scaling relationship between this box and the initial box of search.

If we assume that each equation in (4.14) is of degree m in each variable and the system is n-dimensional, then the total asymptotic time per step is of $O(nlm^{l+1})$. The number of steps depends primarily on the accuracy required [392]. The Projected Polyhedron algorithm achieves quadratic convergence in one dimension, while for higher dimensions, it exhibits at best linear convergence [392]. Once roots have been isolated via the PP algorithm, local quadratically convergent Newton-type algorithms can be used to compute the roots to high precision more efficiently. An extension of the algorithm described above for a set of simultaneous piecewise polynomial nonlinear equations expressed in terms of tensor product B-splines can be found in

[132]. A novel feature of this extension is the normalization of the equations in the range [-1,1] and normalization of the knot vector in each subdomain in range [0,1] at each iteration step of the process to capitalize on the higher density of floating point numbers in this range, thereby improving numerical robustness of the algorithm.

Because the PP algorithm depends only on the convex hull property and ability to perform subdivision and multiplication, in theory one could implement the algorithm for rational B-spline entities without subdividing them into their rational Bézier components. Subdivision algorithms for B-splines are well-known, and Mørken [270] has developed an algorithm for multiplying two piecewise polynomials expressed in the B-spline basis. However, Zhou et al. [461] indicate that this approach sometimes tends to be more time-consuming than subdividing into rational polynomials and applying direct algebraic operations of addition and multiplication of two Bernstein forms (see Sect. 1.3.2). Piegl and Tiller [315] provide detailed description of the procedures that can handle algebraic operators of NURBS curves and surfaces such as dot and cross products, sum/difference and derivative operators. They start with decomposing the B-splines into their Bézier components using knot insertion, and applying the algebraic operators to the Bézier functions and finally recomposing the resulting Bézier functions into B-spline form using knot removal.

4.5 Auxiliary variable method for nonlinear systems with square roots of polynomials

In this section we will focus on how to compute the real roots of systems of irrational equations involving nonlinear polynomials and square roots of nonlinear polynomials within a finite box. Square roots of nonlinear polynomials in the context of shape interrogation arise from normalization of the normal vector and analytical expressions of the principal curvatures of the surface (see (2.24), (3.3), (3.49), (3.50)). They often appear in the form of

$$f(\mathbf{x}) + g(\mathbf{x})\sqrt{h(\mathbf{x})} = 0 , \qquad (4.24)$$

where $\mathbf{x}$ is the unknown vector of l variables, and $f(\mathbf{x})$, $g(\mathbf{x})$ and $h(\mathbf{x})$ are multivariate polynomials over the box $\mathbf{x} \in [0, 1]^l$.

These polynomials can be expressed in the Bernstein basis as

$$f(\mathbf{x}) = \sum_{I}^{M_f} f_I B_{I,M_f}(\mathbf{x}) , \qquad (4.25)$$

$$g(\mathbf{x}) = \sum_{I}^{M_g} g_I B_{I,M_g}(\mathbf{x}) , \qquad (4.26)$$

$$h(\mathbf{x}) = \sum_I^{M_h} h_I B_{I,M_h}(\mathbf{x}) \ . \tag{4.27}$$

Since the square root is involved we cannot use the convex hull property of the Bernstein polynomial directly.

One might consider a *squaring method* to square out the square root, so that the equation becomes

$$f^2(\mathbf{x}) - g^2(\mathbf{x})h(\mathbf{x}) = 0 \ . \tag{4.28}$$

This leads to a higher degree equation, also providing extraneous roots which are not typically necessary. The disadvantages of this squaring method are discussed in [255]. The alternative is the *auxiliary variable method* which will transform the problem into a problem of higher dimensionality. The higher dimensional formulation has been studied by Hoffmann [169] for surface interrogation problems. First we will introduce the auxiliary variable τ such that

$$\tau^2 = h(\mathbf{x}) \ . \tag{4.29}$$

Bounds $a \le \tau \le b$ can be obtained by

$$a = \sqrt{\min_I h_I} \ , \tag{4.30}$$

$$b = \sqrt{\max_I h_I} \ . \tag{4.31}$$

When $\min_I h_I$ is negative, we just set $a=0$. For convenience, we also scale τ such that $\sigma \equiv \frac{\tau-a}{b-a}$, so that $0 \le \sigma \le 1$. Consequently, the system of irrational equations involving nonlinear polynomials and square roots of nonlinear polynomials (4.24), which consists of one equation with l unknowns, has been transformed to a system of nonlinear polynomial equations which consists of two equations with $l+1$ unknowns as follows:

$$f(\mathbf{x}) + g(\mathbf{x}) \left[a + \sigma(b - a) \right] = 0 \ , \tag{4.32}$$

$$\left[a + \sigma(b - a) \right]^2 - h(\mathbf{x}) = 0 \ , \tag{4.33}$$

where $0 \le \sigma \le 1$ and $\mathbf{x} \in [0,1]^l$. Note that even though we transformed the problem into a problem of higher dimensionality, the degree of the new variable σ is only two. System (4.32) (4.33) of two polynomial equations can be solved using the PP algorithm. A similar procedure can be used when (4.24) involves not only one but n scalar equations of the form (4.24). If the $h(\mathbf{x})$ term is different in each of the n equations, then system (4.32) (4.33) will be transformed into $2n$ nonlinear polynomial equations in $l+n$ unknowns.

4.6 Robustness issues

Current state–of–the–art CAD systems used to create and interrogate curved objects are based on geometric solid modeling methods that typically
operate in *floating point arithmetic* (FPA). Arithmetic operations, especially
division, in FPA lead to significant numerical errors. Division operation can
be avoided by four-dimensional homogeneous processing proposed by Yamaguchi [285, 456]. Furthermore CAD systems will frequently fail as a result of
the *limited precision* that is inherent to the internal representation of floating point numbers [167, 168]. One has to keep in mind that any sequence of
operations on a digital computer is essentially equivalent to a finite sequence
of manipulations on a discrete grid of points. For example, a floating point
(FP) number in general form is given by [126]

$$(\pm).b_1 b_2 \cdots b_p \cdot 2^E, \tag{4.34}$$

where $b_1 \cdots b_p$ is the *mantissa* made up of binary digits 0 or 1, $b_i = 0$ or 1,
with $b_1 \neq 0$, and p is the number of *significant digits*, and E is an integer *exponent*. If $p = 2$ and $-2 \leq E \leq 3$ then a list of positive numbers in this system is

$$
\begin{array}{ll}
.10 * 2^{-2} = \frac{1}{8}, & .11 * 2^{-2} = \frac{3}{16}, \\
.10 * 2^{-1} = \frac{1}{4}, & .11 * 2^{-1} = \frac{3}{8}, \\
.10 * 2^{0} = \frac{1}{4}, & .11 * 2^{0} = \frac{3}{4}, \\
.10 * 2^{1} = 1, & .11 * 2^{1} = \frac{3}{2}, \\
.10 * 2^{2} = 2, & .11 * 2^{2} = 3, \\
.10 * 2^{3} = 4, & .11 * 2^{3} = 6,
\end{array}
$$

and are plotted in Fig. 4.7. Obviously, the resulting set of FP numbers is
a finite subset of the rational numbers $\frac{m}{n}$ (where m, n are integers) in the
interval $[\frac{1}{8}, 6]$ and they are distributed non-uniformly in this interval.

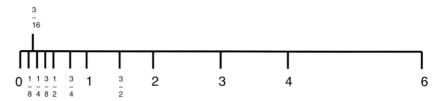

Fig. 4.7. Nonnegative floating point numbers on the interval [0,6] (adapted from
[126])

Nonlinear polynomial solvers operating in *rational arithmetic* (RA), where
the arithmetic is done with rational numbers without approximation [204],
are robust, but are generally memory intensive and time consuming due to
the growth of the number of digits needed to represent rational numbers

that result from arithmetic operations on other rational numbers. On the other hand, nonlinear solvers operating in floating point arithmetic are faster, but generally not robust. Interval methods, which are described in Sect. 4.7, effectively solve these two problems, namely, nonlinear polynomial solvers operating in *interval arithmetic* (IA) are inexpensive compared to rational arithmetic, and they are robust in eliminating regions not containing roots.

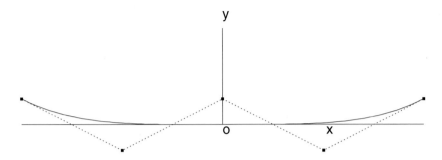

Fig. 4.8. Curves $y = x^4$ and $y = 0$ contact tangentially at the origin (adapted from [179])

Example 4.6.1. Suppose we have a degree four planar Bézier curve whose control points are given by

$$(-0.5, 0.0625), (-0.25, -0.0625), (0, 0.0625), (0.25, -0.0625), (0.5, 0.0625),$$

as shown in Fig. 4.8. This Bézier curve is equivalent to the explicit curve $y = x^4$ ($-0.5 \leq x \leq 0.5$). Apparently the curve intersects with x-axis tangentially at $(x, y) = (0, 0)$. However, if the curve has been translated by $+1$ in the y direction and translated back to the original position by moving by $-\frac{1}{3}$ three times during a geometric processing session, the curve will generally not be the same as the original curve if floating point arithmetic is used for the computation. For illustration, let us assume a decimal computer with a four-digit mantissa, and the computer rounds off intelligently rather than truncating. Then the rational number $-\frac{1}{3}$ will be stored in the decimal computer as -0.3333×10^0 and after the processing the new control points will be

$$(-0.5, 0.0631), (-0.25, -0.0624), (0, 0.0631), (0.25, -0.0624), (0.5, 0.0631).$$

If we evaluate the curve at parameter value $t = 0.5$, we obtain $(0, 0.00035)$ instead of $(0,0)$. Therefore there exists a numerical gap which could later lead to inconsistency between topological structures and geometric equations. For example, if these new control points are used for computing intersections with the x-axis, the algorithm will return no solutions when the tolerance for the function value is smaller than 0.00035.

The above problem illustrates the case when the error is created during the formulation of the governing equations by various algebraic transformations.

Example 4.6.2. This example finds the roots of a cubic polynomial equation $(x - 0.1)(x - 0.6)(x - 0.7) = 0$ by the PP method over an interval $0 \leq x \leq 1$ operating in FPA. Conversion to the Bernstein form was performed in exact arithmetic. This particular example was run at a tolerance of 10^{-4} and the binary subdivision was conducted when the box size did not reduce more than 5% from the previous step. The algorithm output is listed in Table 4.1. At iteration 9, PP algorithm loses the root 0.7 due to floating point rounding.

Table 4.1. $(x - 0.1)(x - 0.6)(x - 0.7) = 0$ solved by PP method operating in FPA (adapted from [255])

Iter	Bounding Box (FPA)	Message
1	[0,1]	
2	[0.0763636363636364, 0.856]	
3	[0.098187732239346, 0.770083868323999]	
4	[0.0999880766853688, 0.72387404781026]	Binary Subdivision
5	[0.402239977003124, 0.704479954527487]	
6	[0.550441290533288, 0.700214508664293]	
7	[0.591018492648952, 0.700000534482207]	
8	[0.599458794784619, 0.700000000003332]	Binary Subdivision
9	[0.649998841568898, 0.699999999999999]	No Root in Box
10	[0.599997683137796, 0.649998841568898]	Root Found in Box
11	[0.099999999478761, 0.402239977003124]	Root Found in Box

This example illustrates another serious problem which arises from the usage of FPA in shape interrogation. To remedy such problems interval arithmetic research in geometric modeling has become quite active as we will see in subsequent Sects. 4.7 to 4.9.

4.7 Interval arithmetic

Interval techniques, primarily interval Newton's methods combined with bisection to ensure convergence, have been the focus of significant attention, see for example Kearfott [192], Neumaier [284]. Interval methods have been applied in geometric modeling and CAD. For example, Mudur and Koparkar [277], Toth [422], Enger [90], Duff [80] and Snyder [400, 399] applied interval algorithms to geometry processing, whereas Sederberg and Farouki [377],

Sederberg and Buehler [375] and Tuohy et al. [425] applied interval methods in approximation problems. In [377] Sederberg and Farouki introduced the concept of interval Bézier curve. Tuohy and Patrikalakis [426] applied interval methods in the representation of functions with uncertainty, such as geophysical property maps. Tuohy et al. [424] and Hager [147] applied interval methods in robotics. Bliek [27] studied interval Newton methods for design automation and inclusion monotonicity properties in interval arithmetic for solving the consistency problem associated with a hierarchical design methodology. Interval methods are also applied in the context of solving systems of nonlinear polynomial equations [178, 179, 254, 255], which we will briefly review in Sect. 4.8. More recently, Hu et al. [180, 181] extended the concept of interval Bézier curves [377] to interval non-uniform rational B-splines (INURBS) curves and surfaces. INURBS differ from classical NURBS in that the real numbers representing control point coordinates are replaced by *interval* numbers. In other words, the control point vectors are replaced by rectangular boxes. This implies that in 3-D space an INURBS curve represents a slender tube and an INURBS surface patch represents a thin shell, if the intervals are chosen sufficiently small. The numerical and geometric properties of interval B-spline curves and surfaces are analyzed in Shen and Patrikalakis [387], while their use in solid modeling is presented in Hu et al. [179, 178, 180, 181], and boundary representation model rectification in Shen [386], Shen et al. [389], Patrikalakis et al. [303] and Sakkalis et al. [360].

An *interval* is a set of real numbers defined below [273]:

$$[a, b] = \{x | a \le x \le b\} \,. \tag{4.35}$$

Two intervals $[a, b]$ and $[c, d]$ are said to be *equal* if

$$a = c \quad and \quad b = d \,. \tag{4.36}$$

The *intersection* of two intervals is *empty* or $[a, b] \cap [c, d] = \emptyset$, if either

$$a > d \quad or \quad c > b \,, \tag{4.37}$$

otherwise,

$$[a, b] \cap [c, d] = [max(a, c), min(b, d)] \,. \tag{4.38}$$

The *union* of the two intersecting intervals is

$$[a, b] \cup [c, d] = [min(a, c), max(b, d)] \,. \tag{4.39}$$

An *order* of intervals is defined by

$$[a, b] < [c, d] \quad if \ and \ only \ if \ b < c \,. \tag{4.40}$$

The width of an interval $[a, b]$ is $b - a$.
The *absolute value* is

$$|[a, b]| = max(|a|, |b|) \,. \tag{4.41}$$

Example 4.7.1.

$$[2,4] \cap [3,5] = [max(2,3), min(4,5)] = [3,4]$$
$$[2,4] \cup [3,5] = [min(2,3), max(4,5)] = [2,5]$$
$$|[-7,-2]| = max(|-7|,|-2|) = 7$$

Interval arithmetic operations are defined by

$$[a,b] \circ [c,d] = \{x \circ y \mid x \in [a,b] \; and \; y \in [c,d]\}, \tag{4.42}$$

where $\circ$ represents an arithmetic operation $\circ \in \{+, -, \cdot, /\}$. Using the end points of the two intervals, we can rewrite equation (4.42) as follows:

$$
\begin{aligned}
&[a,b] + [c,d] = [a+c, b+d], \\
&[a,b] - [c,d] = [a-d, b-c], \\
&[a,b] \cdot [c,d] = [min(ac,ad,bc,bd), max(ac,ad,bc,bd)], \\
&[a,b]/[c,d] = [min(a/c,a/d,b/c,b/d), max(a/c,a/d,b/c,b/d)],
\end{aligned}
\tag{4.43}
$$

provided $0 \notin [c,d]$ in the division operation.

Example 4.7.2.

$$[2,4] + [3,5] = [2+3, 4+5] = [5,9]$$
$$[2,4] - [3,5] = [2-5, 4-3] = [-3,1]$$
$$[2,4] \cdot [3,5] = [min(2\cdot 3, 2\cdot 5, 4\cdot 3, 4\cdot 5), max(2\cdot 3, 2\cdot 5, 4\cdot 3, 4\cdot 5)] = [6,20]$$
$$[2,4]/[3,5] = [min(2/3, 2/5, 4/3, 4/5), max(2/3, 2/5, 4/3, 4/5)] = [2/5, 4/3]$$

Now let us introduce the algebraic properties of interval arithmetic. Interval arithmetic is *commutative*,

$$[a,b] + [c,d] = [c,d] + [a,b], \tag{4.44}$$
$$[a,b] \cdot [c,d] = [c,d] \cdot [a,b], \tag{4.45}$$

and *associative*

$$[a,b] + ([c,d] + [e,f]) = ([a,b] + [c,d]) + [e,f], \tag{4.46}$$
$$[a,b] \cdot ([c,d] \cdot [e,f]) = ([a,b] \cdot [c,d]) \cdot [e,f]. \tag{4.47}$$

But it is not *distributive*; however, it is *subdistributive*

$$[a,b] \cdot ([c,d] + [e,f]) \subseteq [a,b] \cdot [c,d] + [a,b] \cdot [e,f]. \tag{4.48}$$

Example 4.7.3.

$$[1,2] \cdot ([1,2] - [1,2]) = [1,2] \cdot ([-1,1]) = [-2,2] \subseteq [1,2] \cdot [1,2] - [1,2] \cdot [1,2]$$
$$= [1,4] - [1,4] = [-3,3]$$

4.8 Rounded interval arithmetic and its implementation

If floating point arithmetic is used to evaluate these interval arithmetic equations there is no guarantee that the roundings of the bounds are performed conservatively.[1] Rounded interval arithmetic (RIA) [254, 255, 4] ensures that the computed end points always contain the exact interval as follows:

$$
\begin{aligned}
[a, b] + [c, d] &= [(a + c) - \varepsilon_\ell, (b + d) + \varepsilon_u] \,, \\
[a, b] - [c, d] &= [(a - d) - \varepsilon_\ell, (b - c) + \varepsilon_u] \,, \\
[a, b] \cdot [c, d] &= [\min(a{\cdot}c, a{\cdot}d, b{\cdot}c, b \cdot d) - \varepsilon_\ell, \max(a{\cdot}c, a{\cdot}d, b{\cdot}c, b{\cdot}d) + \varepsilon_u] \,, \\
[a, b] \, / \, [c, d] &= [\min(a/c, a/d, b/c, b/d) - \varepsilon_\ell, \max(a/c, a/d, b/c, b/d) + \varepsilon_u] \,,
\end{aligned}
\tag{4.49}
$$

where ε_ℓ and ε_u are the units–in–last–place denoted by ulp_ℓ and ulp_u for each separate floating point number resulting from the floating point operations giving the lower and upper bounds in (4.49). When performing standard operations for interval numbers using RIA, the lower bound is extended to include its previous consecutive FP number, which is smaller than the lower bound by ulp_ℓ. Similarly, the upper bound is extended by ulp_u to include its next consecutive FP number. Thus, the width of the result is enlarged by $ulp_\ell + ulp_u$ and the resulting enlarged interval contains the exact interval. The RIA concept has been applied to topologically reliable approximation of curves and surfaces [57, 58], robust visualization [427], and approximation of uncertain measured data [425].

Before describing the details of the PP algorithm in RIA, let us briefly summarize the IEEE standard binary representation for double precision floating point numbers [4].

4.8.1 Double precision floating point arithmetic

Most commercial processors implement floating point arithmetic using the representation defined by ANSI/IEEE Std 754–1985, *Standard for Binary Floating Point Arithmetic* [10]. This standard defines the binary representation of the floating point number X in terms of a sign bit s, an integer exponent E, for $E_{min} \leq E \leq E_{max}$, and a p–bit significand B, where

$$
X = (-1)^s 2^E B \,.
\tag{4.50}
$$

The significand B is a sequence of p bits $b_0 b_1 \cdots b_{p-1}$, where $b_i = 0$ or 1, with an implied binary point (analogous to a decimal point) between bits b_0 and b_1. Thus, the value of B is calculated as:

---

[1] This statement is true only for the default IEEE-754 rounding mode of *round towards nearest* [10]. The subject of hardware rounding modes will be discussed thoroughly later.

$$B = b_0.b_1 b_2 \cdots b_{p-1} = b_0 2^0 + \sum_{i=1}^{p-1} b_i 2^{-i} . \tag{4.51}$$

For double precision arithmetic, the standard defines $p = 53$, $E_{min} = -1022$, and $E_{max} = 1023$. The number X is represented as a 64–bit quantity with a 1-bit sign s, an 11–bit biased exponent $e = E + 1023$, and a 52–bit fractional mantissa m composed of the bit string $b_1 b_2 \cdots b_{52}$. Since the exponent can always be selected such that $b_0 = 1$ (and thus, $1 \le B < 2$), the value of b_0 is constant and it does not need to be stored in the binary representation.

63	62	$\cdots$	52	51		$\cdots$		0
s		e				m		

The integer value of the 11-bit biased exponent e is calculated as:

$$e = e_0 e_1 \cdots e_{10} = \sum_{i=0}^{10} e_i 2^{10-i} . \tag{4.52}$$

The standard divides the set of representable numbers into the following five categories:

1. If $e = 2047$ and $m \neq 0$, then the value of X is the special flag NaN (not a number).
2. If $e = 2047$ and $m = 0$, then the value of X is $\pm\infty$ depending upon the sign bit: positive if $s = 0$ and negative if $s = 1$.
3. If $0 < e < 2047$, then X is called a *normalized* number, and

$$X = (-1)^s 2^{e-1023} 1.m = (-1)^s 2^{e-1023} \left(1 + \sum_{i=1}^{52} b_i 2^{-i}\right) . \tag{4.53}$$

4. If $e = 0$ and $m \neq 0$, then X is called a *denormalized* number, and

$$X = (-1)^s 2^{-1022} 0.m = (-1)^s 2^{-1022} \left(\sum_{i=1}^{52} b_i 2^{-i}\right) . \tag{4.54}$$

5. If $e = 0$ and $m = 0$, then the value of X is ± 0 depending upon the sign bit. Although they have unique binary representations, arithmetically $-0 \equiv +0$.

Table 4.2 summarizes all of the representable double precision numbers. The binary representation is presented with spaces separating the four 16–bit subsets of the 64–bit value, and the symbol $\cdot$ separating the sign bit, exponent bits, and mantissa bits. The numbers in the first column refer to the aforementioned five categories of representable numbers.

Table 4.2. Representable double–precision numbers and special values (adapted from [4])

1	NaN in binary representation 1·11111111111·1111 1111111111111111 1111111111111111 1111111111111111 $\cdots$ NaN in binary representation 1·11111111111·0000 0000000000000000 0000000000000000 0000000000000001
2	$-\infty$ in binary representation 1·11111111111·0000 0000000000000000 0000000000000000 0000000000000000
3	$-1.7976931348623157 \times 10^{+308}$ in binary representation 1·11111111110·1111 1111111111111111 1111111111111111 1111111111111111 $\cdots$ $-8.9884656743115795 \times 10^{+307}$ in binary representation 1·11111111110·0000 0000000000000000 0000000000000000 0000000000000000 $\cdots$ $-4.4501477170144023 \times 10^{-308}$ in binary representation 1·00000000001·1111 1111111111111111 1111111111111111 1111111111111111 $\cdots$ $-2.2250738585072014 \times 10^{-308}$ in binary representation 1·00000000001·0000 0000000000000000 0000000000000000 0000000000000000
4	$-2.2250738585072009 \times 10^{-308}$ in binary representation 1·00000000000·1111 1111111111111111 1111111111111111 1111111111111111 $\cdots$ $-4.9406564584124654 \times 10^{-324}$ in binary representation 1·00000000000·0000 0000000000000000 0000000000000000 0000000000000001
5	-0.0 in binary representation 1·00000000000·0000 0000000000000000 0000000000000000 0000000000000000 $+0.0$ in binary representation 0·00000000000·0000 0000000000000000 0000000000000000 0000000000000000
4	$+4.9406564584124654 \times 10^{-324}$ in binary representation 0·00000000000·0000 0000000000000000 0000000000000000 0000000000000001 $\cdots$ $+2.2250738585072009 \times 10^{-308}$ in binary representation 0·00000000000·1111 1111111111111111 1111111111111111 1111111111111111
3	$+2.2250738585072014 \times 10^{-308}$ in binary representation 0·00000000001·0000 0000000000000000 0000000000000000 0000000000000000 $\cdots$ $+4.4501477170144023 \times 10^{-308}$ in binary representation 0·00000000001·1111 1111111111111111 1111111111111111 1111111111111111 $\cdots$ $+8.9884656743115795 \times 10^{+307}$ in binary representation 0·11111111110·0000 0000000000000000 0000000000000000 0000000000000000 $\cdots$ $+1.7976931348623157 \times 10^{+308}$ in binary representation 0·11111111110·1111 1111111111111111 1111111111111111 1111111111111111
2	$+\infty$ in binary representation 0·11111111111·0000 0000000000000000 0000000000000000 0000000000000000

It is possible that the result of an operation on two normalized numbers will not itself be representable as a normalized number. Consider the normalized numbers $x = 1.25 \times 10^{-306}$ and $y = 1.23 \times 10^{-306}$. Clearly, $x \neq y$. However, in finite precision normalized floating point arithmetic $x - y = 0$ because $x - y = 0.02 \times 10^{-306} = 2.0 \times 10^{-308}$, which is too small to be represented as a normalized number. It is therefore rounded to the value of 0 [128, pp. 23–24].

The use of denormalized numbers ensures that the relationship

$$x = y \iff x - y = 0, \tag{4.55}$$

always holds true for all normalized numbers. It will also hold true for denormalized numbers where $|x - y| \geq 4.9406564584124654 \times 10^{-324}$, the smallest positive representable denormalized number.

The IEEE standard can represent $2046 \cdot 2^{52} \approx 9.2 \times 10^{18}$ normalized numbers, but only $2^{52} - 1 \approx 4.5 \times 10^{15}$ denormalized numbers. Denormalized numbers are generally not encountered in routine calculations. The ratio of denormalized to normalized numbers is $1/2046 \approx 4.8 \times 10^{-4}$. Furthermore, the denormalized numbers are not uniformly distributed throughout the representable floating point space; rather, they occupy two contiguous groups on either side of 0. Certain operations, however, such as root finding, iteratively generate numbers that are increasingly close to 0. Therefore it is important to allow for the possibility of encountering denormalized numbers when creating robust arithmetic software.

4.8.2 Extracting the exponent from the binary representation

To calculate ulp it is necessary to extract the integer value of the exponent from the binary representation. Recall that the value of the significand B of a double precision number X is:

$$B = 1 + b_1 2^{-1} + b_2 2^{-2} + \cdots + b_{52} 2^{-52}, \tag{4.56}$$

and that the double precision value $X = (-1)^s 2^E B$. The value of the least significant bit b_{52} is 2^{-52}. Thus, the value of ulp is $2^E 2^{-52} = 2^{E-52}$.

The value of ulp can be computed using the standard C mathematical functions $frexp()$ and $ldexp()$ [195, pp. 250-51] as follows [246, 4]:

Algorithm 4.1

```
#include <math.h>              /* standard C math library header */

double ulp(double x)
{
   double ulp;                 /* ulp of x */
   int    exp;                 /* exponent of x, where exp = E+1 */
```

```
frexp(x, &exp);              /* extract exponent of x */
ulp = ldexp(0.5, exp-52);    /* calculate ulp = 0.5^(exp-52) */

return ulp                   /* return ulp */
}
```

(Note that the function *frexp*() assumes that $0.5 \leq B < 1$. Recall that the unbiased exponent E defined by IEEE-754 assumes that $1 \leq B < 2$, thus $exp = E + 1$. In papers [57, 58, 178, 179, 180, 181, 246, 254, 255, 425, 427] the convention of assuming $0.5 \leq B < 1$ is followed.)

Because of the use of standard library functions, this implementation is slow. To avoid using the library functions and to construct the *ulp* directly, recall that the biased exponent e occupies bits 62 through 52. If we could manipulate the binary representation as a 64–bit integer, we could extract e by dividing by 2^{52}, which would right–shift the bit pattern by 52 bits, placing e in bits 10 through 0. The sign bit s, which would then occupy bit 11, could be removed by performing a bitwise logical AND with the 64–bit mask $0\cdots011111111111$ [4].

Most commercially available processors and programming languages, however, do not support 64–bit integers; generally, only 8, 16, and 32–bit integers are available. To overcome this, we can overlay the storage location of the 64–bit double precision number with an array of four 16–bit (short) integers:

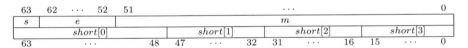

In C or C++ this can be accomplished using the *union* data structure:

```
typedef union {
    double         dp;     /* the 64-bit double precision value */
    unsigned short sh[4];  /* overlay an array of   */
} Double;                  /* 4 16-bit  integers */
```

After the assignment:

```
double x;      /* the double precision value */
Double D;      /* copy of x */

D.dp = x;
```

the exponent of the variable x can be extracted from D.sh[0], whose 16 bits contain the sign bit s (bit 15), the 11–bit biased exponent e (bits 14 through 4), and the 4 most significant bits $b_1 b_2 b_3 b_4$ of the mantissa m (bits 3 through 0):

15	14 ··· 4	3 2 1 0
s	e	$b_1 b_2 b_3 b_4$

The biased exponent e can be extracted from D.sh[0] by performing a bitwise logical AND with the 16–bit mask 0111111111110000 to zero–out the sign bit and the four most significant bits of the mantissa:

15	14 $\cdots$ 4	3 2 1 0
0	e	0 0 0 0

and then right–shifting e by 4 bits:

15 $\cdots$ 11	10 $\cdots$ 0
0 $\cdots$ 0	e

Then the unbiased exponent $E = e - 1023$.

Some processor architectures order the four 16–bit elements of the array sh in the reverse order, in other words, sh[0] is the rightmost (least significant) 16–bit word, not the leftmost (most significant). To avoid the problem, we can define the constant MSW to indicate the proper index of the left-most 16 bit array element:

```
#define MSW 0      /* 0 if the left-most 16-bit short is sh[0] */
                   /* 3 if the left-most 16-bit short is sh[3] */
```

When the ulp is a denormalized number a special case needs to be taken. Recall that $ulp = 2^{E-52}$ and E must be greater than -1023. Thus, if $E \leq -971$ (or equivalently, $e \leq 52$, since $e = E + 1023$) then 2^{E-52} can only be represented as a denormalized number with biased exponent $e_{ulp} = 0$ and mantissa $m_{ulp} = b_i$, for $0 \leq i \leq 51$, where:

$$b_i = \begin{cases} 1 \text{ if } i = e - 1 \\ 0 \text{ otherwise} . \end{cases} \tag{4.57}$$

If $e > 52$ then ulp can be represented by the normalized number with $e_{ulp} = e - 52$ and $m_{ulp} = 0 \cdots 0$.

The following function directly constructs ulp as the appropriate normalized or denormalized number [4]:

Algorithm 4.2

```
static unsigned short mask[16] = { /* bit masks for bits 0 - 15 */
    0x0001, 0x0002, 0x0004, 0x0008, 0x0010, 0x0020, 0x0040, 0x0080,
    0x0100, 0x0200, 0x0400, 0x0800, 0x1000, 0x2000, 0x4000, 0x8000};

double ulp(double x)
{
    Double U,      /* ulp of x */
           X;      /* working copy of x */
    int    bit,    /* position of bit e-1 in 16-bit word */
           e1,     /* biased exponent - 1 */
           word;   /* index of 16-bit word containing bit e-1 */

    X.dp = x;
    X.sh[MSW] &= 0x7ff0;    /* isolate exponent in 16-bit word */

    /* X.sh[0] now holds the exponent in bits 14-4 */
```

```
U.dp = 0.0;        /* initialize exponent and mantissa to 0 */

if (X.sh[MSW] > 0x0340)       /* ulp is normalized number */
   U.sh[MSW] = X.sh[MSW]-0x0340; /* set exponent to e-52 */

/* the value 0x0340 is 52 left-shifted 4 bits,
   i.e. 0x0340 = 832 = 52<<4 */

else {                        /* ulp is denormalized number */
   e1 = (X.sh[MSW]>>4) - 1;         /* biased exponent - 1 */
   word = e1>>4;   /* find 16-bit word containing bit e-1 */
   if (MSW == 0) word = 3 - word;    /* compensate for word
                                          ordering */
   bit  = e1%16;   /* find the bit position in this word */
   U.sh[word] |= mask[bit];          /* set the bit to 1 */
}

   return U.dp;        /* return ulp */
}
```

(Note that the C right-shift operation $n >> m$ is equivalent to integer division, $n/2^m$. Similarly, the left-shift $n << m$ is equivalent to integer multiplication, $n \cdot 2^m$.)

This implementation correctly and efficiently computes the ulp. For example, for the value $X = +2.2250738585072014 \times 10^{-308}$ (the smallest positive normalized double precision number) the $ulp = +4.9406564584124654 \times 10^{-324}$, which has the denormalized binary representation:

0·00000000000·0000 0000000000000000 0000000000000000 0000000000000001

For the value $X = $ -1.7976931348623157 $\times 10^{+308}$ (the largest negative normalized number) the $ulp = +1.9958403095347198 \times 10^{+292}$, which has the normalized binary representation:

0·11111001010·0000 0000000000000000 0000000000000000 0000000000000000

4.8.3 Comparison of two different $unit - in - the - last - place$ implementations

The following table gives a comparison of the running times between Algorithms 4.1 and 4.2 for computing ulp. The timings (in CPU seconds) were taken on a 100 MHz RISC processor (SGI Indy with MIPS R4000 processor). The reported values are the accumulated CPU times to perform 100,000 calculations of the ulp of various representative values of X.

The time required by Algorithm 4.1 increases as the ulp becomes smaller, while the time required by Algorithm 4.2 is constant for normalized ulp's and the time for denormalized ulp's is also constant, but slower by a factor of $1\frac{2}{3}$. In most of the applications in the context of shape interrogation, RIA

Table 4.3. CPU time (in seconds) for the two implementations (adapted from [4])

X	ulp (approx.)	Algorithm 4.1	Algorithm 4.2
-1.25	$+2.22 \times 10^{-16}$	0.10	0.03
-1.25×10^{-100}	$+2.54 \times 10^{-116}$	1.59	0.03
-1.25×10^{-200}	$+1.45 \times 10^{-216}$	3.11	0.03
-1.25×10^{-285}	$+1.87 \times 10^{-301}$	4.40	0.03
-1.25×10^{-295}	$+2.17 \times 10^{-311}$ †	6.68	0.05

† This ulp is a denormalized number.

implementations are an order of magnitude more expensive than non–robust floating point algorithms [4].

4.8.4 Hardware rounding for rounded interval arithmetic

Since floating point numbers are represented in finite precision, many values may need to be rounded to a representable bit pattern [4]. The IEEE-754 standard defines four rounding modes [10]: 1) Round to nearest (the default mode); 2) Round to positive infinity; 3) Round to negative infinity; and 4) Round to zero. We can examine the effects of these rounding modes by calculating intermediate values between two adjacent exactly representable floating point numbers:

$$X_1 = +2.0000000000000009$$
$$= 0 \cdot 10000000000 \cdot 0000\ 0000000000000000\ 0000000000000000\ 0000000000000010$$
$$X_2 = +2.0000000000000013$$
$$= 0 \cdot 10000000000 \cdot 0000\ 0000000000000000\ 0000000000000000\ 0000000000000011$$

which differ only in the last bit of the mantissa.

Since IEEE-754 represents the mantissa with 52 bits, to exactly represent the three uniformly spaced intermediary values, $X_1 + \frac{1}{4}(X_2 - X_1)$, $X_1 + \frac{1}{2}(X_2 - X_1)$, and $X_1 + \frac{3}{4}(X_2 - X_1)$, would require two additional bits in the mantissa, as shown in Table 4.4. To represent the negative values $-X_1$ and $-X_2$ only the sign bit is changed from 0 to 1; the exponent and mantissa bit patterns remain the same. The rounding mode *round to zero* is not depicted in the table since it is not relevant to our application. Round to zero is equivalent to round to negative infinity for positive values, and to round to positive infinity for negative values.

For a given *unlimited precision* floating- point value x, which may not be exactly representable under IEEE-754 (i.e. it may require more than 52 bits to represent the mantissa of x), we want to construct the tightest possible interval $[x_\ell, x_u]$ such that the lower bound x_ℓ is the largest possible representable number not greater than x, and the upper bound x_u is the smallest possible representable number not less than x:

Table 4.4. Comparison of rounding modes (adapted from [4])

Actual			Round To Nearest		Round To $+\infty$	
Value	Mantissa		Rounded	Represented	Rounded	Represented
	52 bits	+2	Mantissa	Value	Mantissa	Value
$+2.0\ldots009$	$00\ldots010$	00	$00\ldots010$	$+2.0\ldots009$	$00\ldots010$	$+2.0\ldots009$
$+2.0\ldots010$	$00\ldots010$	01	$00\ldots010$	$+2.0\ldots009$	$00\ldots011$	$+2.0\ldots013$
$+2.0\ldots011$	$00\ldots010$	10	$00\ldots010$	$+2.0\ldots009$	$00\ldots011$	$+2.0\ldots013$
$+2.0\ldots012$	$00\ldots010$	11	$00\ldots011$	$+2.0\ldots013$	$00\ldots011$	$+2.0\ldots013$
$+2.0\ldots013$	$00\ldots011$	00	$00\ldots011$	$+2.0\ldots013$	$00\ldots011$	$+2.0\ldots013$
$-2.0\ldots009$	$00\ldots010$	00	$00\ldots010$	$-2.0\ldots009$	$00\ldots010$	$-2.0\ldots009$
$-2.0\ldots010$	$00\ldots010$	01	$00\ldots010$	$-2.0\ldots009$	$00\ldots010$	$-2.0\ldots009$
$-2.0\ldots011$	$00\ldots010$	10	$00\ldots010$	$-2.0\ldots009$	$00\ldots010$	$-2.0\ldots009$
$-2.0\ldots012$	$00\ldots010$	11	$00\ldots011$	$-2.0\ldots013$	$00\ldots010$	$-2.0\ldots009$
$-2.0\ldots013$	$00\ldots011$	00	$00\ldots011$	$-2.0\ldots013$	$00\ldots011$	$-2.0\ldots013$

Actual			Round To $-\infty$	
Value	Mantissa		Rounded	Represented
	52 bits	+2	Mantissa	Value
$+2.0\ldots009$	$00\ldots010$	00	$00\ldots010$	$+2.0\ldots009$
$+2.0\ldots010$	$00\ldots010$	01	$00\ldots010$	$+2.0\ldots009$
$+2.0\ldots011$	$00\ldots010$	10	$00\ldots010$	$+2.0\ldots009$
$+2.0\ldots012$	$00\ldots010$	11	$00\ldots010$	$+2.0\ldots009$
$+2.0\ldots013$	$00\ldots011$	00	$00\ldots011$	$+2.0\ldots013$
$-2.0\ldots009$	$00\ldots010$	00	$00\ldots010$	$-2.0\ldots009$
$-2.0\ldots010$	$00\ldots010$	01	$00\ldots011$	$-2.0\ldots013$
$-2.0\ldots011$	$00\ldots010$	10	$00\ldots011$	$-2.0\ldots013$
$-2.0\ldots012$	$00\ldots010$	11	$00\ldots011$	$-2.0\ldots013$
$-2.0\ldots013$	$00\ldots011$	00	$00\ldots011$	$-2.0\ldots013$

$$x_\ell \le x \le x_u . \qquad (4.58)$$

This condition is satisfied by rounding to negative infinity when calculating the lower bound, and rounding to positive infinity when calculating the upper bound. Note that if x is exactly representable, then $x_\ell = x = x_u$.

4.8.5 Implementation of rounded interval arithmetic

For implementational simplicity when switching between *ulp* rounding and hardware rounding, we have developed a C++ class (shown in fragmentary form below) for interval numbers operating in floating point numbers [4]:

```
class Interval {
    private:
        double low;    // lower bound of interval
        double upp;    // upper bound of interval
    public:
        Interval() { low = upp = 0.0; }    // class constructors

        friend Interval add(Interval, Interval, Interval &);
        // utility function
```

```
};
Interval operator + (Interval a, Interval b)
        // overloaded addition operator
{
    Interval c;
    add(a, b, c);            // call appropriate utility function
    return c;                // return sum of a and b
}
```

Software rounding using the ulp is implemented by overloading the arithmetic operators as shown in the following example for addition:

```
Interval add(Interval a, Interval b, Interval &c)
{
  double low = a.low + b.low;  // calculate the lower bound
  double upp = a.upp + b.upp;  // calculate the upper bound

  c.low = low - ulp(low);      // extend the lower bound by ulp
  c.upp = upp + ulp(upp);      // extend the upper bound by ulp
}
```

where $ulp()$ is the function described previously for calculating the ulp.

Hardware rounding is implemented by overloading the arithmetic operators as follows:

```
Interval add(Interval a, Interval b, Interval &c)
{
  swapRM(ROUND_TO_MINUS_INFINITY);  // set round to -infinity mode
  c.low = a.low + b.low;            // calculate the lower bound
  swapRM(ROUND_TO_PLUS_INFINITY);   // set round to +infinity mode
  c.upp = a.upp + b.upp;            // calculate the upper bound
}
```

where $swapRM()$ is the SGI-specific function for setting the IEEE-754 rounding mode. (Although requiring the implementation of the four rounding modes, the standard does not specify the mechanism by which the modes are set.)

The software rounding method is computationally more expensive than hardware rounding, requiring an extra addition and subtraction and the computation of the ulp of two values. Note that the software rounding method extends the upper and lower bounds of the interval during *every* arithmetic operation; the hardware rounding method only extends the bounds when the result of the operation cannot be exactly represented, producing tighter interval bounds. Thus, the relationship between an infinite precision value x and its interval under ulp rounding is $x_\ell < x < x_u$, while for hardware rounding it is $x_\ell \leq x \leq x_u$.

4.9 Interval Projected Polyhedron algorithm

Maekawa [246] and Maekawa and Patrikalakis [254, 255] extended the PP algorithm to operate in rounded interval arithmetic (RIA) in order to solve a nonlinear polynomial system *robustly*, which we refer to as *Interval Projected Polyhedron (IPP) algorithm*. Rounded interval arithmetic operations can be implemented effectively in object-oriented languages such as C++ as we discussed in Sect. 4.8. Other than overloading the arithmetic operations, we need to pay attention in intersecting each convex hull with the horizontal axis (see Sect. 4.4). The computed parametric values result in interval numbers $u_{low} = [u_a, u_b]$ and $u_{up} = [u_c, u_d]$. We simply replace them by $u_{low} = [u_a, u_a]$ and $u_{up} = [u_d, u_d]$ to keep the parameter as real numbers or in other words degenerate interval numbers.

We illustrated the effectiveness of the IPP algorithm using the single polynomial equation $(x - 0.1)(x - 0.6)(x - 0.7) = 0$ that we used in Example 4.6.2. The output of this computation is listed in Table 4.5. If we compare the bounding boxes of Tables 4.1 and 4.5 for each iteration, we can easily recognize that the bounding boxes of the RIA are always conservative with respect to the FPA. Also at iteration 9, FPA loses the root 0.7 due to floating point error, while RIA finds it.

Table 4.5. (x-0.1)(x-0.6)(x-0.7)=0 solved by IPP algorithm (adapted from [255])

Iter	Bounding Box (RIA)	Message
1	[0, 1]	
2	[0.076363636363635, 0.856000000000001]	
3	[0.0981877322393447, 0.770083868324001]	
4	[0.0999880766853675, 0.723874047810262]	Binary Sub.
5	[0.402239977003124, 0.704479954527489]	
6	[0.550441290533286, 0.700214508664294]	
7	[0.591018492648947, 0.700000534482208]	
8	[0.599458794784611, 0.700000000003333]	Binary Sub.
9	[0.649998841568894, 0.7]	Root Found
10	[0.599997683137788, 0.649998841568895]	Root Found
11	[0.0999999994787598, 0.402239977003124]	Root Found

4.9.1 Formulation of the governing polynomial equations

As we have seen in Example 4.6.1, we may introduce numerical errors during the formulation of the governing equations in a shape interrogation problem. Formulation of the governing simultaneous nonlinear polynomial equations in multivariate Bernstein form for shape interrogation usually involves arithmetic operations in Bernstein form (see Sect. 1.3.2) starting from the given input Bézier curve or surface. Therefore to achieve an accurate formulation [255, 254], we suggest:

- Use of *rational arithmetic* (RA) or *rounded interval arithmetic* (RIA) [273] (see also Sect. 4.8), if the control points of the given curve or surface are floating point numbers to maintain a pristine or guaranteed precision statement of the problem, respectively.
- Use of RIA if the control points of the given curve or surface are irrational numbers to avoid any numerical contamination by standard FPA. This happens, for example, when the curve or surface is rotated, since the rotation matrix involves cosines and sines, which are generally irrational.
- Conversion of the coefficients of the nonlinear equations in Bernstein form into intervals with FP number boundaries if rational arithmetic is used in the formulation.

Rational and rounded interval arithmetic operations can be implemented effectively in object-oriented languages such as C++. Computation time comparison for various combinations of arithmetic for the formulation of the governing equations and their solution is presented in [246].

4.9.2 Comparison of software and hardware rounding

We have compared the software and hardware rounding methods in solving the following two examples using the IPP solver. The first example is the degree 20 Wilkinson polynomial, which we introduced in Sect. 1.3.3, whose roots are known for their numerical instability. For a tolerance of $\pm 10^{-8}$ both methods found all 20 roots. However, as reported in Table 4.6 the hardware rounding method was 2.4% faster (24.7 versus 25.3 CPU seconds) and had marginally tighter interval bounds.

For a second example, we used the IPP solver to find the self-intersection points of the offset curve of a planar degree six Bézier curve originally investigated by Maekawa and Patrikalakis [254]. For a tight tolerance of $\pm 10^{-12}$ both methods correctly find two pairs of roots for each of the two self-intersection points (see Fig. 11.11 (a)). However, as shown in Table 4.7 the hardware rounding method was 25.8% faster than the software *ulp* method.

We have compared two methods for performing robust rounded interval arithmetic. The intervals produced by the software *ulp* method are slightly larger, as this method performs the rounding conservatively, extending the upper and lower bounds by *ulp* during every arithmetic operation. The hardware rounding method only extends the bounds when the result of the operation cannot be exactly represented.

Table 4.6. Root finding for degree 20 Wilkinson polynomial for software and hardware rounding. The reported times are the accumulated CPU seconds necessary to solve for the roots 50 times (adapted from [4])

Root	Software rounding
1.0	[0.9999999949999983, 1.000000005]
0.95	[0.9499999949999989, 0.9500000050000008]
0.9	[0.8999999949959929, 0.9000000049959949]
0.85	[0.8499999949999413, 0.8500000049999432]
0.8	[0.7999999900304406, 0.8000000000304426]
0.75	[0.7499999949438847, 0.7500000049438866]
0.7	[0.6999999949920326, 0.7000000049920345]
0.65	[0.6499999949869547, 0.6500000049869566]
0.6	[0.5999999949383569, 0.6000000049383588]
0.55	[0.549999949774965, 0.5500000049774985]
0.5	[0.4999999949594102, 0.5000000049594113]
0.45	[0.4499999950181615, 0.4500000050181623]
0.4	[0.3999999950865321, 0.4000000050865329]
0.35	[0.3499999950804608, 0.3500000050804616]
0.3	[0.2999999950142164, 0.3000000050142173]
0.25	[0.2499999949978854, 0.2500000049978859]
0.2	[0.1999999950002637, 0.2000000050002642]
0.15	[0.1499999950005055, 0.1500000050005059]
0.1	[0.0999999950001587, 0.1000000050001589]
0.05	[0.04999999524378046, 0.05000000524378057]
CPU	25.3

Root	Hardware rounding
1.0	[0.9999999949999999, 1.000000005]
0.95	[0.9499999949999994, 0.9500000049999997]
0.9	[0.899999994995992, 0.9000000049959923]
0.85	[0.8499999949999332, 0.8500000049999337]
0.8	[0.7999999900306577, 0.8000000000306582]
0.75	[0.7499999949787219, 0.7500000049787222]
0.7	[0.6999999949980986, 0.7000000049980991]
0.65	[0.6499999949944346, 0.6500000049944349]
0.6	[0.599999994956996, 0.6000000049569965]
0.55	[0.5499999949777089, 0.5500000049777094]
0.5	[0.4999999949696504, 0.5000000049696507]
0.45	[0.4499999950891842, 0.4500000050891845]
0.4	[0.3999999950232258, 0.400000005023226]
0.35	[0.3499999950964761, 0.3500000050964763]
0.3	[0.2999999950216165, 0.3000000050216168]
0.25	[0.249999950006767, 0.250000005000677]
0.2	[0.199999995000106, 0.2000000050001061]
0.15	[0.1499999950003841, 0.1500000050003842]
0.1	[0.0999999950001635, 0.1000000050001635]
0.05	[0.04999999524378028, 0.05000000524378029]
CPU	24.7

The differences in the running times of the two methods reflect the relative times required to compute the *ulp* versus setting the hardware rounding mode flag. In our experiments performed on an SGI Indy workstation the hardware rounding method is consistently faster than the *ulp* method, with problem-specific performance increases between 2 and 25%. Other researchers have found that hardware rounding is approximately 15% slower than the *ulp* method on a Power Macintosh [311]. Thus, it appears that the computational efficiency of the two methods is dependent on the host system architecture.

Table 4.7. Results of finding self-intersections of offset of degree six Bézier curve (adapted from [4]). Timings are reported in CPU seconds

Method	CPU	Roots
Software	168.0	12
Hardware	124.6	12

5. Intersection Problems

5.1 Overview of intersection problems

Intersections are fundamental in computational geometry, geometric modeling and design, analysis and manufacturing applications [276, 19, 175, 295, 300]. Examples of intersection problems include:

- Contouring of surfaces through intersection with a series of parallel planes or coaxial circular cylinders or cones for visualization (see Fig. 5.1).
- Numerical control machining (milling) involving intersection of offset surfaces with a series of parallel planes, to create machining paths for ball (spherical) cutters (see Fig. 5.2, and Sect. 11.1.2).
- Representation of complex geometries in the *Boundary Representation (B-rep)* scheme; for example, the description of the internal geometry and of structural members of automobiles, airplanes, and ships involves
  - Intersections of free-form parametric surfaces with low order algebraic surfaces (planes, quadrics, torii, cyclides [83]);
  - Intersections of low order algebraic surfaces (see Fig. 5.26);
  in a process called *boundary evaluation*, in which the Boundary Representation is created by evaluating a Constructive Solid Geometry (CSG) model of the object [343, 261, 56, 167, 342, 344]. During this process, intersections of the surfaces of primitives (see Fig. 5.3) must be found during Boolean operations. Boolean operations on point sets A, B include (see Fig. 5.4) union $A \cup B$, intersection $A \cap B$, and difference $A - B$.

All such operations involve intersections of surfaces to surfaces. In order to solve general surface to surface (S/S) intersection problems, the following auxiliary intersection problems need to be considered:

1. point/point (P/P)
2. point/curve (P/C)
3. point/surface (P/S)
4. curve/curve (C/C)
5. curve/surface (C/S)

All above six types of intersection problems are also useful in geometric modeling, robotics, collision avoidance, manufacturing simulation, scientific visualization, etc. When the geometric elements involved in intersections are

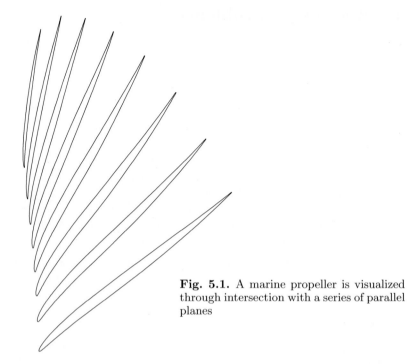

Fig. 5.1. A marine propeller is visualized through intersection with a series of parallel planes

nonlinear (curved), intersection problems typically reduce to solving systems of nonlinear equations, which may be either polynomial or more general functions.

Solution of nonlinear systems is a very complex process in general in numerical analysis and there are specialized textbooks on the topic [293, 69, 274]. However, geometric modeling applications pose severe robustness, accuracy, automation, and efficiency requirements on solvers of nonlinear systems. Therefore, geometric modeling researchers have developed specialized solvers to address these requirements explicitly using geometric formulations as we have seen in Chap. 4.

When studying intersection problems, the type of curves and surfaces that we consider can be classified as follows:

- Rational polynomial parametric (RPP)
- Procedural parametric (PP)
- Implicit algebraic (IA)
- Implicit procedural (IP)

where *procedural* curves and surfaces are defined by means of an evaluation method without explicit use of the specific analytical properties of the defining formula. For example, procedural curves include offsets and evolutes, procedural surfaces include offsets, evolutes, blends and generalized cylinders (e.g. pipe and canal surfaces). However, some of the above procedural curves

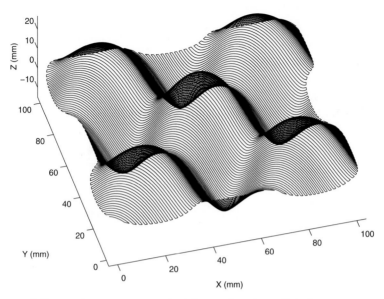

Fig. 5.2. Offset surface is intersected with a series of parallel planes to generate a tool path for 3-D NC machining (adapted from [223])

and surfaces under special conditions can be expressed in the RPP or the IA form, in which case the corresponding methods can be used (see for example [101, 241, 256, 238, 239]).

5.2 Intersection problem classification

The fundamental issue in intersection problems is the efficient discovery and description of *all* features of the solution with high precision commensurate with the tasks required from the underlying geometric modeler [295, 300]. Reliability of intersection algorithms is a basic prerequisite for their effective use in any geometric modeling system and is closely associated with the way features of the solution such as constrictions (near singular or singular situations), small loops and partial surface overlap are handled. The solutions resulting from most present techniques, implemented in practical systems, are further complicated by imprecisions introduced by numerical errors present in finite precision computations.

Intersection problems can be classified according to the dimension of the problem and according to the type of geometric equations involved in defining the various geometric elements (points, curves and surfaces). The solution

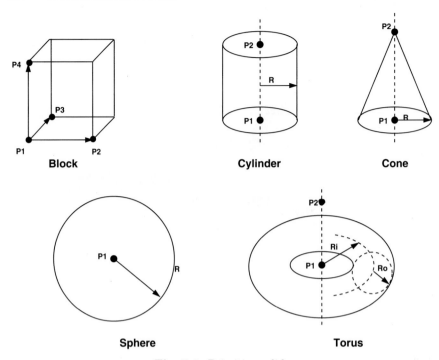

Fig. 5.3. Primitive solids

of intersection problems can also vary according to the number system in which the input is expressed and the solution algorithm is implemented. Such intersection problem classification is addressed in the next three subsections. Only the most important intersection problems are addressed in detail in Sects. 5.3 to 5.8.

5.2.1 Classification by dimension

Using the abbreviation in Sect. 5.1, intersection problems can be classified in three subcategories, where one intersecting entity is a point or curve or surface as below:

1. P/P, P/C, P/S
2. C/C, C/S
3. S/S

5.2.2 Classification by type of geometry

In this subsection, we classify the various types of geometric specification of points, curves and surfaces that we will use in formulating various intersection problems:

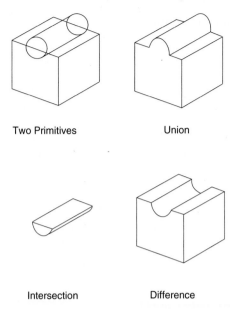

Fig. 5.4. Example of Boolean operations

1. *Points*
   a) Explicit: $\mathbf{r}_0 = (x_0, y_0, z_0)^T$.
   b) Procedural: Intersection of two procedural curves, a procedural curve and a procedural surface, or three procedural surfaces.
   c) Implicit algebraic: Intersection of three implicit surfaces, or equivalently $f(\mathbf{r}) = g(\mathbf{r}) = h(\mathbf{r}) = 0$, where f, g, h are polynomial functions and $\mathbf{r} = (x, y, z)^T$.
2. *Curves*
   a) Parametric: $\mathbf{r} = \mathbf{r}(t)$, $0 \le t \le 1$.
      i. (Rational) (piecewise) polynomial: Bézier, rational Bézier, B-spline, NURBS.
      ii. Procedural: offsets, evolutes, etc.
   b) Implicit algebraic: A 2-D planar curve is given by $z = 0$, $f(x, y) = 0$, while a 3-D space curve is given by intersection of two implicit algebraic surfaces $f(\mathbf{r}) = g(\mathbf{r}) = 0$.
3. *Surfaces*
   a) Parametric: $\mathbf{r} = \mathbf{r}(u, v)$, $0 \le u, v \le 1$.
      i. (Rational) (piecewise) polynomial: Bézier, rational Bézier, B-spline, NURBS.
      ii. Procedural: offsets, blends, generalized cylinders, etc.
   b) Implicit algebraic: $f(\mathbf{r}) = 0$.

5.2.3 Classification by number system

In our discussion of intersection problems, we will refer to various classes of numbers:

1. Rational numbers, m/n, $n \neq 0$, where m, n are integers.
2. Floating point (FP) numbers in a computer (which are a subset of rational numbers, see Sect. 4.8.1 and [4]).
3. Algebraic numbers (roots of polynomials with integer coefficients).
4. Real numbers, e.g. transcendental numbers such as e, π, trigonometric, etc.
5. Interval numbers, $[a, b]$, where a, b are real numbers.
6. Rounded interval numbers, $[c, d]$, where c, d are FP numbers.

Issues relating to floating point and interval numbers affecting the robustness of intersection algorithms were addressed in Chap. 4 in the context of nonlinear solvers as well as in [4, 105, 179, 178, 392].

5.3 Point/point intersection

Point/point intersection problems reduce to checking the Euclidean distance between two points $\mathbf{r}_1$ and $\mathbf{r}_2$, i.e.

$$|\mathbf{r}_1 - \mathbf{r}_2| < \varepsilon , \qquad (5.1)$$

where ε represents the maximum allowable tolerance. Choice of tolerances in a geometric modeler is a difficult open question [309]. For example it may cause *incidence intransitivity*. Figure 5.5 gives an example of three points $\mathbf{r}_1$, $\mathbf{r}_2$ and $\mathbf{r}_3$ where $\mathbf{r}_1 = \mathbf{r}_2$ since $|\mathbf{r}_1 - \mathbf{r}_2| < \varepsilon$, $\mathbf{r}_2 = \mathbf{r}_3$ since $|\mathbf{r}_2 - \mathbf{r}_3| < \varepsilon$, but $\mathbf{r}_1 \neq \mathbf{r}_3$, since $|\mathbf{r}_1 - \mathbf{r}_3| > \varepsilon$. When points are represented procedurally or via implicit algebraic equations, P/P intersection can be typically reduced to comparison of intervals which contain such isolated points. In Hu et al. [180, 181] interval point *equality* is defined in an alternate manner: if the intervals representing the points intersect, assuming these intervals are very small, then the points are considered coincident and a new interval point (the minimal rectangular box with faces parallel to the coordinate planes) is used to replace the two coincident points. With this construction, incidence transitivity is guaranteed in the context of interval solid modeling but at the cost of reduced resolution (accuracy).

5.4 Point/curve intersection

5.4.1 Point/implicit algebraic curve intersection

Intersection between a point and a planar implicit curve is defined as:

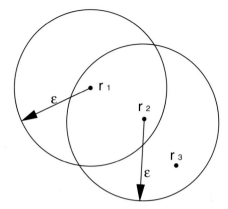

Fig. 5.5. Intersection of points within a tolerance is intransitive

$$\mathbf{r}_0 \cap \{z = 0, \ f(x, y) = 0\} \ , \tag{5.2}$$

where $f(x, y)$ is usually a polynomial and $f(x, y) = 0$ represents an algebraic curve. In an exact arithmetic context, we can substitute $\mathbf{r}_0$ in $\{z, \ f(x, y)\}$ and verify if the results are zero. Similarly, we could handle

$$\mathbf{r}_0 \cap \{f(\mathbf{r}) = g(\mathbf{r}) = 0\} \ , \tag{5.3}$$

where $f(\mathbf{r}) = g(\mathbf{r}) = 0$ represents an implicit 3-D space curve. However, if floating point arithmetic is used in evaluating $f(x, y)$, the result will not be exactly zero due to round off errors.

Now let us examine the distance between a point $\mathbf{r}_0 = (x_0, y_0)^T$ and a planar implicit curve $f(x, y) = 0$. The geometric distance is given by:

$$d = min|\mathbf{r} - \mathbf{r}_0| \ , \tag{5.4}$$

where $\mathbf{r} = (x, y)^T$ must satisfy $f(\mathbf{r}) = 0$. The true geometric distance is difficult and expensive to compute (especially if we need to deal with a large set of points as in inspection problems). As an alternative, we can compute an approximate distance. We Taylor expand $f(x, y) = 0$ about (x_0, y_0) up to the first order as follows:

$$f(x, y) = f(x_0, y_0) + f_x(x_0, y_0)\Delta x + f_y(x_0, y_0)\Delta y = 0 \ , \tag{5.5}$$

where $\Delta x = x - x_0$ and $\Delta y = y - y_0$. From the stationary condition of the distance $|\mathbf{r} - \mathbf{r}_0|$, we can deduce the orthogonality condition

$$f_y(x, y)\Delta x - f_x(x, y)\Delta y = 0 \ . \tag{5.6}$$

Since we do not know the footpoint (x, y) on the implicit curve which gives the minimum distance, we will also Taylor expand $f_x(x, y)$ and $f_y(x, y)$ about (x_0, y_0) up to the first order as follows:

$$f_x(x,y) = f_x(x_0, y_0) + f_{xx}(x_0, y_0)\Delta x + f_{xy}(x_0, y_0)\Delta y \,, \qquad (5.7)$$
$$f_y(x,y) = f_y(x_0, y_0) + f_{yx}(x_0, y_0)\Delta x + f_{yy}(x_0, y_0)\Delta y \,. \qquad (5.8)$$

After substituting (5.7) and (5.8) into (5.6) and neglecting the second order terms we have

$$f_y(x_0, y_0)\Delta x - f_x(x_0, y_0)\Delta y = 0 \,. \qquad (5.9)$$

Equations (5.5) and (5.9) form a linear system in Δx and Δy which can be solved as

$$\Delta x = -\frac{f(x_0, y_0)f_x(x_0, y_0)}{f_x^2(x_0, y_0) + f_y^2(x_0, y_0)}, \qquad \Delta y = -\frac{f(x_0, y_0)f_y(x_0, y_0)}{f_x^2(x_0, y_0) + f_y^2(x_0, y_0)} \,,$$

$$(5.10)$$

provided the denominators are not zero. Therefore the first order approximation to the true geometric distance (5.4) reduces to

$$d = \sqrt{(\Delta x)^2 + (\Delta y)^2} = \frac{|f(x_0, y_0)|}{|\nabla f(x_0, y_0)|} \,, \qquad (5.11)$$

provided that $|\nabla f(x_0, y_0)| \neq 0$. Consequently if the *algebraic distance* $|f(x_0, y_0)|$ $< \varepsilon \ll 1$ where ε is a small positive constant and f is normalized so that $|f(x,y)| \leq 1$ in the domain of interest including $\mathbf{r}_0$, then an approximate minimum distance check can be performed by evaluating the *non-algebraic distance* (5.11).

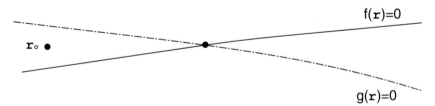

Fig. 5.6. Algebraic curves meet at small angle

Having evaluated the approximate minimum distance (5.11), we can now address a complex point to point intersection problem which further elucidates the notion of geometric distance for use in intersection problems. We discuss an intersection problem, where we need to check if the intersection point of the two planar algebraic curves crossing at a small angle, intersects a point $\mathbf{r}_0$ as illustrated in Fig. 5.6 or more precisely

$$\mathbf{r}_0 \cap \{z = 0, f(x,y) = g(x,y) = 0\} \,, \qquad (5.12)$$

where

$$\left| \frac{\nabla f}{|\nabla f|} \cdot \frac{\nabla g}{|\nabla g|} \right| \simeq 1 \,, \tag{5.13}$$

evaluated at the intersection point $f = g = 0$. Even if $f(x, y)$ and $g(x, y)$ satisfy

$$|f(x_0, y_0)| < \varepsilon \ll 1 \quad \text{and} \quad \delta_1 = \frac{|f(x_0, y_0)|}{|\nabla f(x_0, y_0)|} < \delta \ll 1 \,, \tag{5.14}$$

$$|g(x_0, y_0)| < \varepsilon \ll 1 \quad \text{and} \quad \delta_2 = \frac{|g(x_0, y_0)|}{|\nabla g(x_0, y_0)|} < \delta \ll 1 \,, \tag{5.15}$$

it is not enough to guarantee proximity of $\mathbf{r}_0$ to the intersection of $f(x, y) = 0$, $g(x, y) = 0$ as shown in Fig. 5.6.

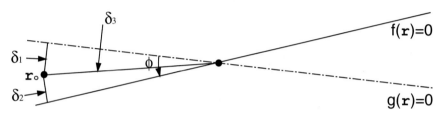

Fig. 5.7. Algebraic curves approximated by straight lines

In such cases, using a linear approximation, and letting

$$\phi \simeq \cos^{-1} \left| \frac{\nabla f(x_0, y_0)}{|\nabla f(x_0, y_0)|} \cdot \frac{\nabla g(x_0, y_0)}{|\nabla g(x_0, y_0)|} \right| \,, \tag{5.16}$$

be the angle of intersection as in Fig. 5.7 near the intersection point, a better criterion for evaluating if $\mathbf{r}_0$ is near the intersection of $f(x, y) = 0$ and $g(x, y) = 0$ is given by

$$\delta_3 = \frac{1}{\phi} \left(\frac{|f(x_0, y_0)|}{|\nabla f(x_0, y_0)|} + \frac{|g(x_0, y_0)|}{|\nabla g(x_0, y_0)|} \right) < \delta \ll 1 \,. \tag{5.17}$$

5.4.2 Point/rational polynomial parametric curve intersection

Mathematically, an intersection between a point and a rational polynomial parametric (RPP) curve is defined as

$$\mathbf{r}_0 \cap \mathbf{r} = \mathbf{r}(t) = \left(\frac{X(t)}{W(t)}, \frac{Y(t)}{W(t)}, \frac{Z(t)}{W(t)} \right)^T \,, \quad 0 \le t \le 1 \,, \tag{5.18}$$

where $X(t)$, $Y(t)$, $Z(t)$ and $W(t)$ are polynomials.

Elementary method. We solve each of the following three nonlinear polynomial equations separately using a numerical scheme such as Newton's method or Laguerre's iteration method [69] and we search for common real roots in $0 \leq t \leq 1$:

$$x(t) - x_0 = 0, \qquad y(t) - y_0 = 0, \qquad z(t) - z_0 = 0 . \qquad (5.19)$$

In principle, this elementary approach is easy, however in practice, this process is complex and inefficient and prone to numerical inaccuracies.

Bounding box and subdivision followed by minimization method. We use a bounding box of $\mathbf{r}(t)$ to eliminate easily resolvable cases, with some level of subdivision to reduce box size. For a rational polynomial curve with control points (x_i, y_i, z_i), the bounding box is given by

$$\min(x_i) \leq x(t) \leq \max(x_i) , \qquad (5.20)$$
$$\min(y_i) \leq y(t) \leq \max(y_i) , \qquad (5.21)$$
$$\min(z_i) \leq z(t) \leq \max(z_i) , \qquad (5.22)$$

as shown in Fig. 5.8 for a planar curve case. A tighter bounding box can be obtained by axis reorientation [208]. To eliminate numerical error in the subdivision process (which can lead to erroneous decisions), rational arithmetic may be employed (if the input coefficients of $\mathbf{r}(t)$ are rational or floating point numbers). This can be easily done in object-oriented languages such as C++ using operator overloading. We continue subdivision until the bounding box is small. Then, we could use a numerical technique [69], for example:

$$F(t) = \min\{|\mathbf{r}_0 - \mathbf{r}(t)|^2\}, \quad t \in I , \qquad (5.23)$$

and use some values of t from the interval I as the initial approximation. Use of the square of the distance function is necessary to avoid possible divergence of the derivative of the distance function, if it approaches zero. If the minimization process converges to t_0 and $\sqrt{F(t_0)} < \varepsilon$, then $t = t_0$ is the desired solution.

Distance function method. A more robust method than the above is to compute the stationary points of the squared distance function $D = D(t)$ between a point and a variable point on a rational polynomial curve. We search for zeros of the derivative $\dot{D}(t)$ and then examine if at those zeros the squared distance function attains a minimum. Detailed formulation is given in Chap. 7, and a robust solution method based on the IPP algorithm is provided in Chap. 4.

Implicitization. As we have discussed in Sect. 5.4.1, point/implicit curve intersection problem is conceptually very simple. Therefore it is natural to consider conversion of the curve equation from a parametric form to an implicit form. Sederberg et al. [371, 372, 374, 378, 380] used *implicitization*, which

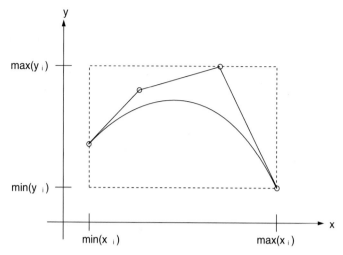

Fig. 5.8. Bounding box of a rational polynomial curve

originates in classical algebraic geometry [437, 217, 2, 413], to compute such intersections.

Let us consider two polynomial equations $x = x(t)$, $y = y(t)$ of degree m and n respectively as follows:

$$x = a_m t^m + a_{m-1} t^{m-1} + \cdots + a_1 t + a_0 , \qquad (5.24)$$
$$y = b_n t^n + b_{n-1} t^{n-1} + \cdots + b_1 t + b_0 , \qquad (5.25)$$

where $a_m \neq 0$ and $b_n \neq 0$. Or equivalently

$$a_m t^m + a_{m-1} t^{m-1} + \cdots + a_1 t + a_0 - x = 0 , \qquad (5.26)$$
$$b_n t^n + b_{n-1} t^{n-1} + \cdots + b_1 t + b_0 - y = 0 . \qquad (5.27)$$

Now we form the following $(m + n) \times (m + n)$ matrix by generating auxiliary equations by multiplying (5.26) and (5.27) by appropriate monomials as follows:

$$
\begin{matrix}
n\ rows \left\{ \vphantom{\begin{matrix}a\\a\\a\end{matrix}} \right. \\
\\
m\ rows \left\{ \vphantom{\begin{matrix}a\\a\\a\end{matrix}} \right.
\end{matrix}
\begin{pmatrix}
a_0 - x & a_1 & \cdots & a_m & & & \\
 & a_0 - x & a_1 & \cdots & a_m & & \\
\cdots & \cdots & \cdots & \cdots & \cdots & \cdots & \\
 & & a_0 - x & a_1 & \cdots & a_m \\
b_0 - y & b_1 & \cdots & b_n & & & \\
 & b_0 - y & b_1 & \cdots & b_n & & \\
\cdots & \cdots & \cdots & \cdots & \cdots & \cdots & \\
 & & b_0 - y & b_1 & \cdots & b_n
\end{pmatrix}
\begin{pmatrix}
1 \\ t \\ t^2 \\ \cdots \\ \cdots \\ \cdots \\ \cdots \\ t^{m+n-1}
\end{pmatrix}
$$

$$
=
\begin{pmatrix}
0 \\
0 \\
0 \\
\cdots \\
\cdots \\
\cdots \\
\cdots \\
0
\end{pmatrix},
\tag{5.28}
$$

where all blanks correspond to zeros. The *resultant* of the two polynomials (5.26), (5.27), which is denoted by $f(x,y)$, is the determinant of the above matrix [217]. The vanishing of the resultant, i.e. $f(x,y) = 0$ is a necessary and sufficient condition for the polynomials (5.26) and (5.27) to have a common root. In other words, all x, y values that satisfy $f(x,y) = 0$ lie on the parametric curve, and hence $f(x,y) = 0$ is the implicit form of the parametric equation $x = x(t)$, $y = y(t)$.

Evidently, $f(x,y)$ has degree n,m in x,y, respectively. Furthermore, McKay and Wang [266] proved that the leading form $f^+(x,y)$, which consists of the terms $r_{ij}x^i y^j$ with $mi + nj = mn$, is equal to $(-1)^m [a_m^{n/d} y^{m/d} - b_n^{m/d} x^{n/d}]^d$, where d is the greatest common divisor of m and n. Therefore, if $m = n$, $f^+(x,y)$ has the form $(-1)^m [a_m y - b_m x]^m$, and thus the highest total degree of the implicit algebraic representation of the curve is m. An alternate geometric way of visualizing this is to consider the maximum number of intersections of a straight line $\alpha x + \beta y + \gamma = 0$, with the planar parametric curve $x = a_m t^m + \cdots + a_1 t + a_0, y = b_n t^n + \cdots + b_1 t + b_0$, which upon substitution into the (linear) equation of the line leads to the conclusion that the highest total degree is $max(m,n)$. Note that the curve $x = t^m$, $y = t^m$ can be represented implicitly using resultants as $(-1)^m (y - x)^m = 0$. The exponent m of $(y - x)$ arises in order to reflect the m complex roots for the parameter t of $x = t^m$ or $y = t^m$ for given x or y.

Once we have the implicit form, we check if $f(x_0, y_0) = 0$ in an exact arithmetic context to verify if (x_0, y_0) is on the initial curve. Resultants can be computed exactly (in rational arithmetic) in symbolic manipulation programs such as MATHEMATICA [446], MAPLE [51], but the evaluation procedure tends to be slow for high degree cases.

Finally we may need to determine the corresponding parameter value t on the parametric curve at the point (x_0, y_0). This process is called *inversion*. We set $x = x_0$ and $y = y_0$ in (5.28) and discard any row of the resultant matrix and solve for any of the ratio of t^p/t^{p-1} $(1 \le p \le m + n - 1)$ which gives the parameter value t [374]. The implicitization and inversion methods are efficient for low degree polynomials but there are no guarantees on accuracy and robustness, if these methods are implemented in floating point arithmetic. Subdivision methods are preferable for higher degrees, and as we have seen in Chap. 4 when coupled with rounded interval arithmetic, they become robust and accurate. Intersection of points (x_0, y_0, z_0) and 3-D polynomial curves

$\mathbf{r} = \mathbf{r}(t)$ via implicitization involves a process of projection on xy-plane and finding t_0 by inversion and verification of $z_0 = z(t_0)$.

5.4.3 Point/procedural parametric curve intersection

Mathematically, an intersection between a point and a procedural parametric (PP) curve is defined as:

$$\mathbf{r}_0 \cap \mathbf{r}(t), \quad 0 \le t \le 1 . \tag{5.29}$$

In general there is no known and easily computable convex box decreasing in size arbitrarily with subdivision for a procedural parametric curve. An approximate solution method may involve minimization of

$$F(t) = |\mathbf{r}(t) - \mathbf{r}_0|^2 , \tag{5.30}$$

where $t \in [0, 1]$. This would also involve the checking of end points, i.e. if $F(0)$ or $F(1)$ are zero. Initial estimate for the possible minima, may be found by using linear approximation of $\mathbf{r}(t)$ to start the process. However, convergence of the above minimization processes is not guaranteed in general and there may exist more than one minima. Furthermore convergence to local and not global minimum (where $F(t) \ne 0$) is possible.

For certain classes of procedural curves such as offsets and evolutes of rational curves involving radicals of polynomials, it is possible to use the auxiliary variable method described in Sect. 4.5 [169, 254, 253] to reduce the problem to a set of nonlinear polynomial equations. Such systems can be solved robustly and efficiently using the IPP algorithm described in Chap. 4. Alternatively, some procedural curves admit a rational parametrization (e.g. offsets [238, 239, 324]) in which case the problem reduces to the formulation of Sect. 5.4.2.

5.5 Point/surface intersection

5.5.1 Point/implicit algebraic surface intersection

The mathematical description for a point and an implicit surface $f(\mathbf{r}) = 0$ intersection is given by

$$\mathbf{r}_0 \cap f(\mathbf{r}) = 0 . \tag{5.31}$$

Similar to the point implicit curve intersection case, the condition for intersection is given by

$$|f(\mathbf{r}_0)| < \varepsilon, \quad \frac{|f(\mathbf{r}_0)|}{|\nabla f(\mathbf{r}_0)|} < \delta , \tag{5.32}$$

where ε, δ are small positive constants, provided $|\nabla f(\mathbf{r}_0)| \ne 0$.

5.5.2 Point/rational polynomial parametric surface intersection

Point/rational polynomial parametric (RPP) surface intersection is defined as:

$$\mathbf{r}_0 \cap \mathbf{r} = \mathbf{r}(u, v) = \left(\frac{X(u, v)}{W(u, v)}, \frac{Y(u, v)}{W(u, v)}, \frac{Z(u, v)}{W(u, v)} \right)^T, \quad 0 \le u, v \le 1 .$$

$$(5.33)$$

Implicitization. Implicitization is possible for RPP surfaces but it is computationally expensive for high degree surfaces and with the necessary use of exact rational arithmetic for robustness. For a rational polynomial surface with maximum degrees in u and v equal to m and n, i.e. $\mathbf{r} = \mathbf{r}(u^m, v^n)$, the implicit equation is of the form $f(\mathbf{r}) = 0$ where the highest total degree of the polynomial $f(\mathbf{r})$ is $q \le 2mn$ [374]. Therefore, for $m = n = 3$, $q \le 18$ and for $m = n = 2$, $q \le 8$. The implicitization method is useful for special surfaces such as cylindrical and conical ruled surfaces, surfaces of revolution, etc:

1. *Implicitization of a surface of revolution*

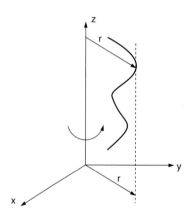

Fig. 5.9. Surface of revolution

Let us consider a planar profile curve to be a rational polynomial of degree n

$$\mathbf{r}(t) = (r(t), z(t))^T ,$$

$$(5.34)$$

as illustrated in Fig. 5.9. By simple implicitization of $\mathbf{r} = \mathbf{r}(t)$, we obtain

$$f(r, z) = a_n r^n + a_{n-1}(z)r^{n-1} + \cdots + a_0(z) = 0 ,$$

$$(5.35)$$

where f is a polynomial in r and z of total degree n. Also,

$$x^2 + y^2 - r^2 = 0 . \tag{5.36}$$

Next we eliminate r from (5.35) and (5.36) by implicitization. The resultant of the two polynomial equations is

$$R = \begin{vmatrix} a_0(z) & a_1(z) & \cdots & a_{n-1}(z) & a_n(z) & 0 \\ 0 & a_0(z) & \cdots & a_{n-2}(z) & a_{n-1}(z) & a_n(z) \\ x^2+y^2 & 0 & -1 & \cdots & & \\ & x^2+y^2 & 0 & -1 & & \\ & & & & \ddots & \\ & & x^2+y^2 & & 0 & -1 \end{vmatrix} = 0 ,$$

$$\tag{5.37}$$

and $R = f(x, y, z) = 0$ is a polynomial in x, y and z of total degree $2n$. For example, a torus results in a degree four algebraic surface.

2. *Implicitization of a cylindrical ruled surface*

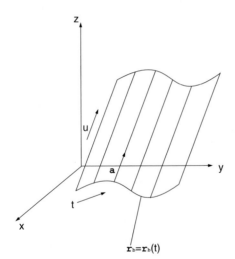

Fig. 5.10. Cylindrical ruled surface

Let

$$x_b = x_b(t), \quad y_b = y_b(t) , \tag{5.38}$$

be a planar base curve or directrix of a ruled surface as shown in Fig. 5.10. The directrix is a rational polynomial curve of degree n in the xy plane. The resulting implicit equation of the curve

$$f(x_b, y_b) = 0 , \tag{5.39}$$

is a polynomial in x_b and y_b of total degree n. Let

$$\mathbf{a} = (a_1, \ a_2, \ a_3)^T \ , \tag{5.40}$$

be a constant direction unit vector which gives the direction of the ruling at each point on the directrix, then the three equations

$$x = x_b + ua_1 \ , \tag{5.41}$$
$$y = y_b + ua_2 \ , \tag{5.42}$$
$$z = ua_3 \ , \tag{5.43}$$

describe a cylindrical ruled surface. If we assume $a_3 \neq 0$, we can eliminate $u = \frac{z}{a_3}$ by substituting into the first two equations. Then solving for x_b and y_b the implicit cylindrical ruled surface equation becomes:

$$f\left(x - \frac{z}{a_3}a_1, y - \frac{z}{a_3}a_2\right) = 0 \ . \tag{5.44}$$

This equation can be expanded to a standard form using a symbolic manipulation program such as MATHEMATICA [446], MAPLE [51] etc.

3. *Implicitization of a conical ruled surface*

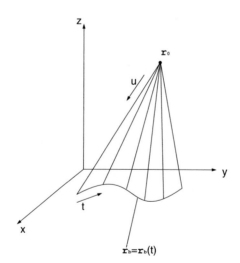

Fig. 5.11. Conical ruled surface

Let a conical ruled surface be defined by its apex

$$\mathbf{r}_0 = (x_0, \ y_0, \ z_0)^T \ , \tag{5.45}$$

and its planar base curve

$$x_b = x_b(t), \quad y_b = y_b(t) \ , \tag{5.46}$$

which is a degree n planar rational polynomial curve on the xy plane. Its implicit equation

$$f(x_b, y_b) = 0 , \qquad (5.47)$$

is a total degree n polynomial. The equation of the resulting conical ruled surface is

$$x = x_0(1 - u) + x_b u , \qquad (5.48)$$
$$y = y_0(1 - u) + y_b u , \qquad (5.49)$$
$$z = z_0(1 - u) . \qquad (5.50)$$

Eliminating u by substituting $u = 1 - \frac{z}{z_0}$ into (5.48) and (5.49), where we assume $z \neq z_0$, and solving for x_b, y_b yields:

$$f\left(\frac{z_0}{z_0 - z} x - \frac{x_0}{z_0 - z} z, \frac{z_0}{z_0 - z} y - \frac{y_0}{z_0 - z} z \right) = 0 . \qquad (5.51)$$

This equation can be expanded to the standard form using a symbolic manipulation program such as MATHEMATICA [446], MAPLE [51] etc.

Newton's method. We start with preprocessing using the bounding box of the RPP surface patch coupled with some level of subdivision. Then we solve a system $x_0 = x(u, v)$, $y_0 = y(u, v)$ using Newton's method and verify the results with the third equation $z_0 = z(u, v)$. Projection of $\mathbf{r}_0$ onto a planar approximation of the surface (faceting) may provide a good initial approximation.

Bounding box and subdivision followed by minimization method. We can easily extend the point/curve intersection case to the point/surface intersection case.

Distance function method. Similarly to the point/curve intersection case, detailed formulation of the squared distance function between a point and a rational polynomial parametric surface is given in Chap. 7, and the solution method is provided in Chap. 4.

5.5.3 Point/procedural parametric surface intersection

Procedural surfaces may include offset surfaces, generalized cylinder surfaces, blending surfaces etc. The typical solution method is minimization [69]. In this case, no convex box assistance is available in general, and we need a dense sampling for an initial approximation which may be expensive, and no rigorous guarantees for the solution's reliability are generally available.

For certain classes of procedural surfaces such as offsets and evolutes of rational surfaces involving radicals of polynomials, it is possible to use the auxiliary variable method, described in Sect. 4.5, to remove radicals from the formulation, followed by solution via the IPP algorithm of Chap. 4. Alternatively, some procedural surfaces admit a rational parametrization (e.g. offsets [240, 324], pipe and canal surfaces [241, 256]) in which case the problem reduces to the formulation of Sect. 5.5.2.

5.6 Curve/curve intersection

Curve to curve intersection cases are identified in Table 5.1. Conceptually, case D3 (RPP/IA curve intersections) is the simplest of the above cases of intersection to describe and use for illustrating various general difficulties of intersection problems (see Sect. 5.6.1). The next case of interest is case D1 (RPP/RPP curve intersections) from Table 5.1 and this is analyzed in Sect. 5.6.2. Next cases D2 and D5 (RPP/PP and PP/PP curve intersections) from Table 5.1 are analyzed in Sect. 5.6.3. These are followed by cases D6 (PP/IA curve intersections) and D8 (IA/IA curve intersections) analyzed in Sects. 5.6.4 and 5.6.5, respectively. Cases D4, D7, D9 and D10 are not addressed here, however the reader should be able to analyze those cases based on the cases treated in this section.

Table 5.1. Classification of curve/curve intersections

Curve type	Curve type			
	RPP	PP	IA	IP
RPP	D1	D2	D3	D4
PP		D5	D6	D7
IA			D8	D9
IP				D10

5.6.1 Rational polynomial parametric/implicit algebraic curve intersection (Case D3)

2-D planar case. We start with an intersection problem of a planar RPP curve and an implicit algebraic curve which is defined as:

$$\mathbf{r} = \mathbf{r}(t) = \left(\frac{X(t)}{W(t)}, \frac{Y(t)}{W(t)} \right)^T \cap \quad f(x,y) = 0, \qquad 0 \le t \le 1. \quad (5.52)$$

Let us denote the degree of planar RPP curve by n and the total degree of the implicit algebraic curve by m. Furthermore we describe the implicit curve by

$$f(x,y) = \sum_{i=0}^{m} \sum_{j=0}^{m-i} c_{ij} x^i y^j = 0. \quad (5.53)$$

Substitution of $x = \frac{X(t)}{W(t)}$ and $y = \frac{Y(t)}{W(t)}$ into the implicit form and multiplication of $W^m(t)$ leads to a polynomial of degree up to mn

$$F(t) = \sum_{i=0}^{m} \sum_{j=0}^{m-i} c_{ij} X^i(t) Y^j(t) W^{m-i-j}(t) = \sum_{i=0}^{mn} a_i t^i = 0 . \quad (5.54)$$

Therefore the problem of intersection is equivalent to finding the real roots of $F(t)$ in $0 \le t \le 1$. The most usual form of $F(t)$ is the power basis. The coefficients can be evaluated symbolically by substitution and collection of terms. This can be readily done in a standard symbolic manipulation program such as MATHEMATICA [446], MAPLE [51] etc. This can be followed by numerical evaluation of the coefficients ideally in exact rational arithmetic. Symbolic manipulation programs are oriented to processing rational numbers exactly.

Example 5.6.1. Let the algebraic curve be an ellipse $\frac{x^2}{4} + y^2 - 1 = 0$, and the parametric curve be a cubic Bézier curve with control points $(0,1)$, $(1, -4)$, $(2,1)$ and $(2,0)$ as illustrated in Fig. 5.12. The ellipse and Bézier curve are chosen to be tangent at $(2,0)$.

Using a symbolic manipulation program and simplifying, we get in exact arithmetic mode:

$$F(t) = 1025t^6 - 3840t^5 + 5514t^4 - 3728t^3 + 1149t^2 - 120t = 0 .$$

Next we find the real roots of $F(t)$ in $t \in [0,1]$ using factoring over the integers, which leads to

$$F(t) \equiv t(t-1)^2 G(t) ,$$

where

$$G(t) = 1025t^3 - 1790t^2 + 909t - 120 .$$

Using a standard numerical solver for polynomials in floating point such as [134, 135], we obtain the following numbers as solutions of $G(t) = 0$

$$t = 0.9228 \cdots, \; 0.61843 \cdots, \; 0.2051 \cdots .$$

Alternately solving $F(t) = 0$ using the same routine leads to the following six complex and real roots

t_R	1	1	0.9228	0.6183	0.2051	0
t_I	-0.22×10^{-6}	0.22×10^{-6}	0	0	0	0

where $t = t_R + i t_I$, and i is the imaginary unit.

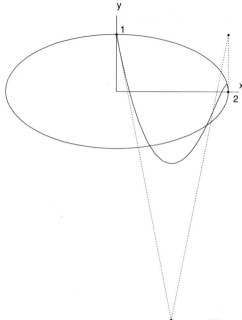

Fig. 5.12. Intersection of ellipse and cubic Bézier curve

Notice the sensitivity to errors for the $6th$ degree polynomial, especially for the double root $t = 1$. In floating point arithmetic, such roots may split into conjugate complex numbers. Obviously, complex roots are not usable as we require only the real intersection points. The consequence is lost roots, which implies an erroneous solution of the intersection problem.

An alternate basis for the representation of $F(t) = 0$ is the Bernstein basis, which leads to better stability for its real roots under perturbations of its coefficients than the power form [105] as we discussed in Sect. 1.3.3.

Setting $F(t) = \sum_{i=0}^{6} c_i B_{i,6}(t)$ in the above example of the ellipse and cubic Bézier curve, we have $c_0 = 0$, $c_1 = -20$, $c_2 = \frac{183}{50}$, $c_3 = -\frac{83}{50}$, $c_4 = \frac{8}{5}$, $c_5 = 0$, and $c_6 = 0$. Here the conversion is done exactly using rational arithmetic (given that the conversion itself is not in general numerically well conditioned [106]). By the use of linear precision property

$$t = \sum_{i=0}^{mn} \frac{i}{mn} B_{i,mn}(t) , \tag{5.55}$$

we can construct a graph

$$\mathbf{f}(t) = (t, F(t))^T = \sum_{i=0}^{mn} \begin{pmatrix} \frac{i}{mn} \\ c_i \end{pmatrix} B_{i,mn}(t) , \tag{5.56}$$

which is now a degree mn Bézier curve as shown in Fig. 5.13. Now we can
apply the IPP method introduced in Chap. 4 which converts the problem of
finding roots of polynomials into a problem of finding the intersection of the
Bézier curve with the parameter axis.

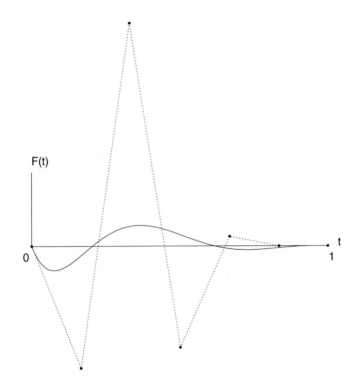

Fig. 5.13. Intersection of a Bézier curve/straight line

Notice that in our example $c_0 = 0$ which implies that $t = 0$ is a root. Also
$c_5 = c_6 = 0$ implies that $t = 1$ is a double root.

3-D space curve. The intersection problem of a 3-D rational polynomial
parametric curve and a 3-D implicit curve is defined as:

$$\mathbf{r} = \mathbf{r}(t) = \left(\frac{X(t)}{W(t)}, \frac{Y(t)}{W(t)}, \frac{Z(t)}{W(t)} \right)^T \cap \ f(\mathbf{r}) = g(\mathbf{r}) = 0, \qquad 0 \le t \le 1 \,.$$

$$(5.57)$$

If we denote the total degree of implicit algebraic surfaces $f(x, y, z) = 0$
and $g(x, y, z) = 0$ as m, and substituting $x = \frac{X(t)}{W(t)}$, $y = \frac{Y(t)}{W(t)}$ and $z = \frac{Z(t)}{W(t)}$

into the implicit forms $f(x, y, z) = 0$ and $g(x, y, z) = 0$ and multiplying by $W^m(t)$, we obtain two univariate nonlinear polynomial equations $F_1(t) = 0$ and $F_2(t) = 0$. One way to solve this problem is to compute the resultant of $F_1(t)$, $F_2(t)$, where all the coefficients of the two polynomials are known. If the resultant $R(F_1(t), F_2(t)) \equiv 0$, then there is a common root between the two polynomials and hence we can use the inversion algorithm to find t.

A robust way to solve this overconstrained problem $F_1(t) = F_2(t) = 0$ (2 equations with 1 unknown) is to use the IPP algorithm (see Chap. 4). In such cases, the substitution must be conducted in exact arithmetic, to maintain a pristine or guaranteed precision statement of the problem.

Alternatively one could directly solve the overconstrained five-equation system in four variables (x, y, z, t)

$$X(t) - x\, W(t) = 0 , \tag{5.58}$$
$$Y(t) - y\, W(t) = 0 , \tag{5.59}$$
$$Z(t) - z\, W(t) = 0 , \tag{5.60}$$
$$f(x, y, z) = 0 , \tag{5.61}$$
$$g(x, y, z) = 0 , \tag{5.62}$$

using the IPP algorithm as in Chap. 4.

5.6.2 Rational polynomial parametric/rational polynomial parametric curve intersection (Case D1)

We can define the intersection problem as:

$$\mathbf{r} = \mathbf{r}_1(t) = \left(\frac{X_1(t)}{W_1(t)}, \frac{Y_1(t)}{W_1(t)}, \frac{Z_1(t)}{W_1(t)} \right)^T , \quad 0 \le t, \le 1 , \tag{5.63}$$

$$\cap \quad \mathbf{r} = \mathbf{r}_2(\sigma) = \left(\frac{X_2(\sigma)}{W_2(\sigma)}, \frac{Y_2(\sigma)}{W_2(\sigma)}, \frac{Z_2(\sigma)}{W_2(\sigma)} \right)^T , \quad 0 \le \sigma \le 1 .$$

Setting $\mathbf{r}_1(t) = \mathbf{r}_2(\sigma)$ leads to 3 nonlinear polynomial equations with 2 unknowns t, σ (overconstrained system). There are several approaches to solve this intersection problem.

1. The bounding box and subdivision method followed by minimization that we described in Sect. 5.4.2 can be applied to this problem. If two bounding boxes intersect, and they are of finite size, we can find roots using linear approximation. However, in the presence of tangential intersection the following cases may happen. The boxes intersect but the linear approximations do not, and the curves intersect as illustrated in Fig. 5.14 (a) for planar curves. Similar behavior is observed in (b) where polygon is used as the curve approximation. *Hodographs* indicate the range of tangent directions of a Bézier curve. Sederberg and Meyers [379] construct a

bounding wedge, which is a bounding angular sector of a hodograph, containing all tangent vectors of the given Bézier curve (see shaded triangle in Fig. 5.15). The bounding wedges are useful in predicting the number of intersection points of two curves. We first translate the bounding wedges of the two Bézier curves so that their vertices are coincident. If the two wedges do not overlap, the curves cannot intersect more than once. This is a sufficient condition but not a necessary condition. Figure 5.15 (a) shows non-overlapping wedges, while (b) shows overlapping wedges. In both figures, hodographs together with their control polygons for each Bézier curve are superposed on the corresponding wedges.

If we combine the bounding box subdivision technique with the bounding wedge technique, we are able to locate the intersection points more effectively. For a precisely tangential intersection, this method would lead to infinite subdivision steps.

2. Another possible approach is to choose 2 equations to solve for t, σ, and then substitute the results into the third equation for verification. We can implicitize $x = \frac{X_2(\sigma)}{W_2(\sigma)}$ and $y = \frac{Y_2(\sigma)}{W_2(\sigma)}$ to obtain $f(x,y) = 0$ followed by the substitution of $x = \frac{X_1(t)}{W_1(t)}$ and $y = \frac{Y_1(t)}{W_1(t)}$ into $f(x,y)$ yielding $F(t) = 0$. Then solve it for real roots in $[0,1]$ and use the inversion algorithm to find σ. Finally we verify the solution by checking if $\frac{Z_1(t)}{W_1(t)}$ becomes equal to $\frac{Z_2(\sigma)}{W_2(\sigma)}$. This method needs to be implemented in rational arithmetic for robustness.

3. Hawat and Piegl [156] studied the application of a *genetic algorithm* to the curve/curve intersection problems. First a number of points are selected on each curve in a random manner, then a pair of points from each curve are coupled. Finally the genetic algorithm is applied to create generations of couples that minimize the distance between them. The probabilistic nature of genetic algorithms is discussed in [267].

4. Solve directly the overconstrained 3-equations with 2-unknowns system using the IPP algorithm described in Chap. 4. If the parametric curves are in integral/rational B-spline form, the first step is to subdivide the integral/rational B-spline curve into a number of integral/rational Bézier curves. This method guarantees a robust solution (no missing roots).

5.6.3 Rational polynomial parametric/procedural parametric and procedural parametric/procedural parametric curve intersections (Cases D2 and D5)

These intersection problems are defined as:

$$\mathbf{r} = \mathbf{r}_1(t) \cap \mathbf{r} = \mathbf{r}_2(\sigma), \quad 0 \le t, \sigma \le 1, \tag{5.64}$$

where we have 3 equations with 2 unknowns t and σ. Unlike the RPP/RPP curve intersection, there is no known and easily computable convex box decreasing arbitrarily with subdivision for PP curves.

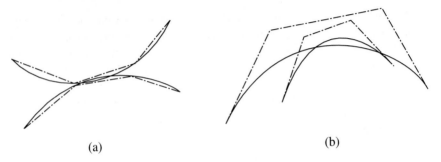

Fig. 5.14. Linear approximation of curves in finding intersections: **(a)** approximation by linear segments, **(b)** approximation by polygon

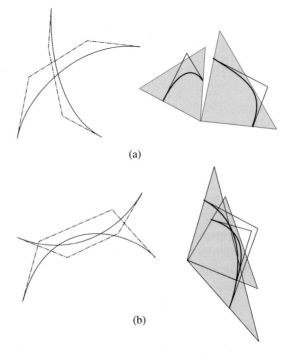

(a)

(b)

Fig. 5.15. Bounding wedges

A possible approach is to minimize the squared distance function D between RPP curve and PP curve or between two PP curves

$$D = F(t, \sigma) = |\mathbf{r}_1(t) - \mathbf{r}_2(\sigma)|^2, \quad 0 \leq t, \sigma \leq 1 , \qquad (5.65)$$

using numerical techniques [69]. Initial approximation may be obtained by using linear approximations of $\mathbf{r}_1(t)$ and $\mathbf{r}_2(\sigma)$. In general there is no guarantee to find all the stationary points.

For offsets and evolutes of rational polynomial curves, we are able to avoid the square roots of polynomials by using the auxiliary variable method, described in Sect. 4.5, so that we can apply the IPP algorithm to enhance robustness.

5.6.4 Procedural parametric/implicit algebraic curve intersection (Case D6)

The intersection problem is defined as:

$$\mathbf{r} = \mathbf{r}(t) \cap f(\mathbf{r}) = g(\mathbf{r}) = 0, \qquad 0 \le t \le 1 . \qquad (5.66)$$

This can be reduced to PP curve/IA surface intersection (see Sect. 5.7.4), i.e. $\mathbf{r} = \mathbf{r}(t) \cap f(\mathbf{r}) = 0$ and $\mathbf{r} = \mathbf{r}(t) \cap g(\mathbf{r}) = 0$, and comparison of solutions.

5.6.5 Implicit algebraic/implicit algebraic curve intersection (Case D8)

The planar case is of interest in processing trimmed patches and the definition of this intersection problem is given as

$$f(u, v) = 0 \cap g(u, v) = 0, \qquad 0 \le u, v \le 1 . \qquad (5.67)$$

Implicitization. We can eliminate v to form the resultant $F(u)$, then solve $F(u) = 0$ for u and use the inversion algorithm to obtain v.

Example 5.6.2. Let us consider an ellipse and a circle

$$f = \frac{x^2}{4} + y^2 - 1 = 0 ,$$

$$g = (x - 1)^2 + y^2 - 1 = 0 ,$$

as in Fig. 5.16.

First we eliminate y from these two equations. This leads to

$$3x^2 - 8x + 4 = 0 ,$$

which has two real roots $x = 2$ and $x = \frac{2}{3}$. These lead to $y^2 = 0$ and $y^2 = \frac{8}{9}$, respectively.

However there are possible numerical problems at the tangential intersection point $x = 2$, $y = 0$. Let us assume that due to error $\epsilon > 0$, we have

$$x = 2 + \epsilon ,$$

hence

$$y^2 = -\epsilon(1 + \frac{\epsilon}{4}) < 0 .$$

This implies that y is imaginary and that no real roots exist. This would have as a consequence missing an intersection solution, leading to a robustness problem.

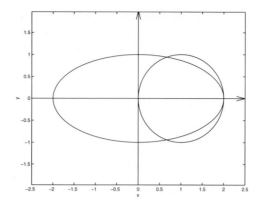

Fig. 5.16. Ellipse and circle intersection

Newton's method. After tracing of $f(u, v) = 0$ and $g(u, v) = 0$, based on the technique which will be discussed in Sect. 5.8.1, linear approximation of each algebraic curve is available. By finding intersections of linear approximations and minimum distance points between them, we can initiate a Newton's method on the system $f = g = 0$ or a minimization of $F = f^2 + g^2$. However, no general robustness guarantees exist with such method.

Interval Projected Polyhedron solver. A robust and efficient method is based on use of the IPP algorithm, described in Chap. 4 for two equations with two unknowns.

5.7 Curve/surface intersection

Curve to surface intersections are classified in Table 5.2. Such intersection problems are useful in solving the more general surface to surface intersection problems. When the curve is a straight line, the curve/surface intersection is useful for

- ray tracing in computer graphics and visualization;
- point classification in solid modeling;
- procedural surface interrogation.

In Sects. 5.7.1 to 5.7.6 several of the most frequent curve to surface intersection problem cases E3, E1, E2/E6, E7, E11, E9 are analyzed in some detail. The remaining cases are not analyzed in detail, but the reader should be able to analyze them via the cases addressed in this section. We will start with case $E3$ (RPP curve to IA surface intersection), which is quite representative of the complexities of this type of problem.

Table 5.2. Classification of curve/surface intersections

	Surface type			
Curve type	RPP	PP	IA	IP
RPP	E1	E2	E3	E4
PP	E5	E6	E7	E8
IA	E9	E10	E11	E12
IP	E13	E14	E15	E16

5.7.1 Rational polynomial parametric curve/implicit algebraic surface intersection (Case E3)

The intersection problem is defined as:

$$\mathbf{r} = \mathbf{r}(t) = \left(\frac{X(t)}{W(t)}, \frac{Y(t)}{W(t)}, \frac{Z(t)}{W(t)} \right)^T \cap f(\mathbf{r}) = 0, \quad 0 \le t \le 1 . \quad (5.68)$$

Let us consider an implicit algebraic surface of total degree m

$$f(x, y, z) = \sum_{i=0}^{m} \sum_{j=0}^{m-i} \sum_{k=0}^{m-i-j} c_{ijk} x^i y^j z^k = 0 . \quad (5.69)$$

We substitute $x = \frac{X(t)}{W(t)}$, $y = \frac{Y(t)}{W(t)}$ and $z = \frac{Z(t)}{W(t)}$ of degree n into the implicit equation and multiply by $W^m(t)$ leading to

$$F(t) = \sum_{i=0}^{m} \sum_{j=0}^{m-i} \sum_{k=0}^{m-i-j} c_{ijk} X^i(t) Y^j(t) Z^k(t) W^{m-i-j-k}(t) = 0 , \quad (5.70)$$

of degree $\le mn$ in t. We then find its real roots in $[0, 1]$, as described in Sect. 5.6.1.

Alternatively, the problem can be formulated as a nonlinear polynomial system of four equations in four unknowns (x, y, z, t) and solved using the IPP algorithm.

5.7.2 Rational polynomial parametric curve/rational polynomial parametric surface intersection (Case E1)

The intersection problem between a rational polynomial parametric curve and a rational polynomial parametric surface is defined as:

$$\mathbf{r} = \mathbf{r}_1(t) = \left(\frac{X_1(t)}{W_1(t)}, \frac{Y_1(t)}{W_1(t)}, \frac{Z_1(t)}{W_1(t)} \right)^T , \quad 0 \le t \le 1 , \quad (5.71)$$

$$\cap \quad \mathbf{r} = \mathbf{r}_2(u, v) = \left(\frac{X_2(u, v)}{W_2(u, v)}, \frac{Y_2(u, v)}{W_2(u, v)}, \frac{Z_2(u, v)}{W_2(u, v)} \right)^T , \quad 0 \le u, v \le 1 .$$

The equation consists of three nonlinear equations $\mathbf{r}_1(t) = \mathbf{r}_2(u, v)$ in three unknowns t, u, v. A preprocessing step of checking bounding boxes for absence of intersection is helpful.

Implicitization. Implicitization of $\mathbf{r}_2(u, v)$ (which is recommended for low degree surfaces) reduces this problem to Case E3 described in Sect. 5.7.1.

Bounding box and subdivision followed by minimization method. Use of bounding boxes coupled with recursive subdivision will lead us to small bounding boxes which may contain intersection points. Then we use a linear approximations for $\mathbf{r}_1$ and $\mathbf{r}_2$ to obtain approximate initial solutions, which can be used to initiate a Newton's method on $\mathbf{r}_1(t) - \mathbf{r}_2(u, v) = 0$ or a minimization method on $F(t, u, v) = |\mathbf{r}_1(t) - \mathbf{r}_2(u, v)|^2$. However, no general robustness guarantees exist with such method.

Interval Projected Polyhedron solver. A robust and efficient way is to solve the three equations with three unknowns using the IPP algorithm discussed in Chap. 4.

5.7.3 Rational polynomial parametric/procedural parametric and procedural parametric/procedural parametric curve/surface intersections (Cases E2/E6)

The intersection problem between a rational polynomial parametric curve or a procedural parametric curve and a procedural parametric surface is defined as:

$$\mathbf{r} = \mathbf{r}_1(t) \cap \mathbf{r} = \mathbf{r}_2(u, v), \quad 0 \le t, u, v \le 1 . \tag{5.72}$$

Similar to Case E1, this problem reduces to three nonlinear equations in three unknowns involving non-polynomial functions, for which bounds are not generally available.

A possible approach is to use the minimization technique

$$F(t, u, v) = |\mathbf{r}_1(t) - \mathbf{r}_2(u, v)|^2 , \tag{5.73}$$

in a cube $0 \le t, u, v \le 1$. Comments under the point/PP curve intersection case (see Sect. 5.4.3) also apply to this problem.

5.7.4 Procedural parametric curve/implicit algebraic surface intersection (Case E7)

The intersection problem between a procedural parametric curve and an implicit algebraic surface is defined as:

$$\mathbf{r} = \mathbf{r}(t) \cap f(\mathbf{r}) = 0, \quad 0 \le t \le 1 . \tag{5.74}$$

This leads to four nonlinear equations in four unknowns t, $\mathbf{r}$. We could use Newton's method initiated by a linear approximation of $\mathbf{r} = \mathbf{r}(t)$, which can be intersected more easily with $f(\mathbf{r}) = 0$ using the method of Case E3 (see Sect. 5.7.1). However, no robustness guarantees exist in general.

5.7.5 Implicit algebraic curve/implicit algebraic surface intersection (Case E11)

Implicit algebraic curve and implicit algebraic surface intersection problem is defined as:

$$\underbrace{f(\mathbf{r}) = g(\mathbf{r})}_{curve} = \underbrace{h(\mathbf{r})}_{surface} = 0 . \tag{5.75}$$

The formulation comprises three nonlinear equations in three unknowns $\mathbf{r}$. Possible solution approaches include elimination methods, Newton's method, minimization methods with objective function $F(\mathbf{r}) = f^2 + g^2 + h^2$, approximating $f(\mathbf{r}) = g(\mathbf{r}) = 0$ curve with a linear spline reducing to Case E3 and refinement using minimization, and the IPP algorithm.

5.7.6 Implicit algebraic curve/rational polynomial parametric surface intersection (Case E9)

The implicit algebraic curve and rational polynomial parametric surface intersection is defined as:

$$f(\mathbf{r}) = g(\mathbf{r}) = 0 \cap \mathbf{r} = \mathbf{r}(u, v) = \left(\frac{X(u, v)}{W(u, v)}, \frac{Y(u, v)}{W(u, v)}, \frac{Z(u, v)}{W(u, v)} \right)^T ,$$
$$0 \le u, v \le 1 . \tag{5.76}$$

By substituting $\mathbf{r} = \mathbf{r}(u, v)$ into $f(\mathbf{r}) = 0$ and $g(\mathbf{r}) = 0$ we obtain two algebraic curves $F(u, v) = 0$ and $G(u, v) = 0$. This formulation reduces to IA/IA curve intersection, Case D8 in Sect. 5.6.5 (see Fig. 5.17). A more detailed discussion of algebraic curves is given in Sect. 5.8.1.

5.8 Surface/surface intersections

Surface to surface intersection cases are identified in Table 5.3. The solution of a surface/surface intersection problem may be empty, or include a curve (possibly made of several branches), a surface patch, or a point. In Sects. 5.8.1 to 5.8.3 several of the most frequent surface to surface intersection problem cases F3, F1, and F8 are studied in detail. The remaining cases are not analyzed, but could be addressed based on cases F3, F1 and F8, although without general robustness guarantees. Conceptually, RPP/IA surface intersection (Case F3) is the simplest of the above cases and may serve as illustrating general difficulties of surface to surface intersection problems.

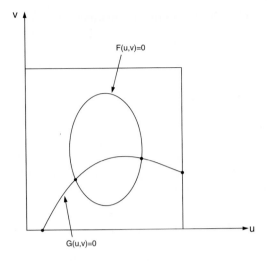

Fig. 5.17. Intersection of two algebraic curves

Table 5.3. Classification of surface/surface intersections

	Surface type			
Surface type	RPP	PP	IA	IP
RPP	F1	F2	F3	F4
PP		F5	F6	F7
IA			F8	F9
IP				F10

5.8.1 Rational polynomial parametric/implicit algebraic surface intersection (Case F3)

We start with a rational polynomial parametric surface to implicit algebraic surface intersection problem defined as:

$$\mathbf{r} = \mathbf{r}(u, v) = \left(\frac{X(u, v)}{W(u, v)}, \frac{Y(u, v)}{W(u, v)}, \frac{Z(u, v)}{W(u, v)} \right)^T \cap \ f(\mathbf{r}) = 0, \quad 0 \le u, v \le 1 \ .$$

(5.77)

This leads to four algebraic equations in five unknowns $\mathbf{r}, u, v$ (underconstrained system). For the usual low degree surfaces $f(\mathbf{r})$ and low degree patches $\mathbf{r}(u, v)$, we can substitute $\mathbf{r}(u, v)$ into $f(\mathbf{r}) = 0$ to obtain an implicit algebraic curve in u, v [124, 333, 211, 302, 212]. Examples of low order implicit algebraic surfaces in practical use are planes (degree 1), the natural quadrics (cylinder, sphere, cone) (degree 2), and torii (degree 4). In fact in a survey of mechanical parts (mechanical elements), over 90% of all surfaces involved are of these types [149]. It is also well known that low order implicit algebraic surfaces have a low degree rational polynomial parametric representation (which can be easily obtained [314]), so that when two such low order implicit algebraic surfaces are intersected, the methods of this section may be also used.

Formulation. Now let us denote the implicit algebraic surface $f(x, y, z) = 0$ of total degree m by

$$f(x, y, z) = \sum_{i=0}^{m} \sum_{j=0}^{m-i} \sum_{k=0}^{m-i-j} c_{ijk} x^i y^j z^k \ . \tag{5.78}$$

By substituting $x = \frac{X(u,v)}{W(u,v)}$, $y = \frac{Y(u,v)}{W(u,v)}$, $z = \frac{Z(u,v)}{W(u,v)}$, where $X(u, v)$, $Y(u, v)$, $Z(u, v)$ and $W(u, v)$ are all of maximum degree p in u and q in v into (5.78) and multiplying by $W^m(u, v)$ leads to an algebraic curve

$$F(u, v) = \sum_{i=0}^{m} \sum_{j=0}^{m-i} \sum_{k=0}^{m-i-j} c_{ijk} X^i(u, v) Y^j(u, v) Z^k(u, v) W^{m-i-j-k}(u, v) = 0 \ ,$$

$$\tag{5.79}$$

of maximum degree $M = mp$ and $N = mq$ in u, v, respectively. Consequently, the problem of intersection reduces to the problem of tracing $F(u, v) = 0$ without omitting any special features of the curve, e.g. small loops, singularities, and accurately computing all its branches. This is a fundamental problem in *algebraic geometry* [437] and much work has been done to understand its solution. In the context of algebraic geometry the coefficients of $F(u, v) = 0$ are integers. In the context of CAD and computer implementation, the coefficients of $F = 0$, and $\mathbf{r} = \mathbf{r}(u, v)$ are floating point numbers. Therefore, if the above substitution is performed in floating point arithmetic the coefficients of $F(u, v) = 0$ involve error, which may considerably modify the problem being solved. To avoid such error, rational arithmetic may be used for robustness. These issues are discussed in the Chap. 4.

The algebraic curve

$$F(u, v) = \sum_{i=0}^{M} \sum_{j=0}^{N} c_{ij}^M u^i v^j = 0 \ , \tag{5.80}$$

can be reformulated in terms of Bernstein polynomials using (4.18) as follows:

$$F(u, v) = \sum_{i=0}^{M} \sum_{j=0}^{N} c_{ij}^B B_{i,M}(u) B_{j,N}(v) = 0 \ , \tag{5.81}$$

where $(u, v) \in [0, 1]^2$.

As an example, consider a plane in an implicit form

$$ax + by + cz + d = 0 \ , \tag{5.82}$$

and a rational Bézier patch of degree m in u, n in v

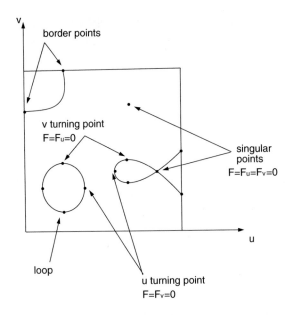

border points

v turning point
F=F_u=0

singular
points
F=F_u=F_v=0

u

loop

u turning point
F=F_v=0

Fig. 5.18. Parameter space
of $\mathbf{r}(u, v)$ and resulting
algebraic curve $F(u, v) = 0$

$$\mathbf{r}(u, v) = \frac{\sum_{i=0}^{m} \sum_{j=0}^{n} w_{ij} \mathbf{b}_{ij} B_{i,m}(u) B_{j,n}(v)}{\sum_{i=0}^{m} \sum_{j=0}^{n} w_{ij} B_{i,m}(u) B_{j,n}(v)} \,, \qquad (5.83)$$

where $\mathbf{b}_{ij} = (x_{ij}, y_{ij}, z_{ij})^T$ and weights $w_{ij} \geq 0$.

The resulting algebraic curve is of the form of equation (5.81) with

$$c_{ij}^B = (ax_{ij} + by_{ij} + cz_{ij} + d)w_{ij} \,. \qquad (5.84)$$

In fact the power basis form of $F(u, v) = 0$ need not be computed at all, if polynomial arithmetic for Bernstein polynomials, described in Sect. 1.3.2, is used (see also [106]).

The advantage of the Bernstein form is the higher numerical stability of the roots in comparison to the power basis and the convex hull property. If $c_{ij}^B > 0$ or $c_{ij}^B < 0$ for all i, j, there is no solution and the two surfaces do not intersect. More precisely, the entire algebraic surface $f(\mathbf{r}) = 0$ does not intersect the surface patch $\mathbf{r} = \mathbf{r}(u, v)$ for $(u, v) \in [0, 1]^2$. Obviously, when all $c_{ij}^B = 0$ the two surfaces coincide in their entirety. A somewhat complex algebraic curve $F(u, v) = 0$ is shown in Fig. 5.18 involving various branches (from border to border), internal loops, and singularities.

Tracing method. Given a point on every branch of an algebraic curve, we are able to trace the curve using differential curve properties. The idea is to find increments δu and δv such that $F(u + \delta u, v + \delta v) = 0$, when we have $F(u, v) = 0$.

Let us Taylor expand $F(u, v)$

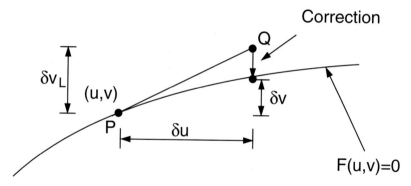

Fig. 5.19. A zoomed view of an algebraic curve near a point (u, v)

$$F(u + \delta u, v + \delta v) = F(u, v) + F_u \delta u + F_v \delta v \tag{5.85}$$
$$+ \frac{1}{2}(F_{uu}\delta u^2 + 2F_{uv}\delta u \delta v + F_{vv}\delta v^2) + \cdots .$$

When F_u and F_v are not both zero or $F_u^2 + F_v^2 > 0$, in order to have $F(u, v) = 0$ and $F(u + \delta u, v + \delta v) = 0$ to the first order approximation, we must have

$$F_u \delta u + F_v \delta v = 0 , \tag{5.86}$$

or

$$\delta v_L = -\frac{F_u}{F_v}\delta u , \tag{5.87}$$

assuming $F_v \neq 0$. However, as illustrated in Fig. 5.19 δv_L leads to a point Q which may be far from the curve $F(u, v) = 0$. Newton's method on $F(u + \delta u, v) = 0$ with initial approximation $v_I = v + \delta v_L$ may be used to compute δv with high accuracy and in an efficient manner. For vertical branches, i.e. when $|F_v|$ is very small, we may use $\delta u_L = -\frac{F_v}{F_u}\delta v$.

To avoid these special stepping procedures, (5.86) may be rewritten as

$$F_u \dot{u} + F_v \dot{v} = 0 , \tag{5.88}$$

where u, v are considered as functions of a parameter t. The solution to the differential equation is given by

$$\dot{u} = \xi F_v(u, v) , \tag{5.89}$$
$$\dot{v} = -\xi F_u(u, v) , \tag{5.90}$$

where ξ is an arbitrary nonzero factor. For example, ξ can be chosen to provide arc length parametrization using the first fundamental form (3.13) of the surface as a normalization condition

$$\xi = \pm \frac{1}{\sqrt{EF_v^2 - 2FF_uF_v + GF_u^2}} , \tag{5.91}$$

where E, F and G are first fundamental form coefficients of the parametric surface evaluated at u, v. Equations (5.89) and (5.90) form a system of two first order nonlinear differential equations which can be solved by the Runge-Kutta or other methods with adaptive step size [69, 126]. For the tracing method to work properly, we must provide all the starting points of all branches in advance. Step size selection is complex and too large a step size may lead to straying or looping [124], as in Fig. 5.20, in the presence of constrictions where $F_u^2 + F_v^2$ is very small. Tracing through singularities ($F_u^2 + F_v^2 = 0$) is also problematic. When $\xi \sqrt{F_u^2 + F_v^2}$ is small then the right hand sides of (5.89) and (5.90) are small and step size needs to be reduced for topologically reliable tracing of the curve.

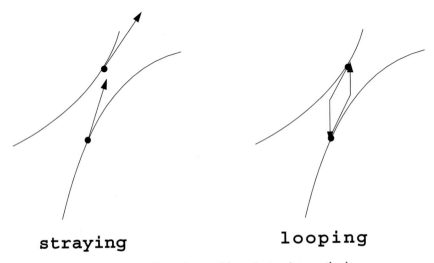

straying   looping

Fig. 5.20. Step size problems in tracing method

Characteristic points. Starting points for tracing algebraic curves are identified by looking for characteristic points defined below:

1. *Border points*: The intersections of $F(u,v) = 0$ with all four boundary edges of the parameter space $[0,1]^2$, e.g. $F(0,v) = 0$, $0 \le v \le 1$.
2. *Turning points*: The u-turning points are the points where the tangent of $F(u,v) = 0$ is parallel to the $u = 0$ axis, which satisfies the simultaneous equations $F = F_v = 0$ (with $F_u \ne 0$). On the other hand the v-turning points are the points where the tangent of $F(u,v) = 0$ is parallel to the $v = 0$ axis, which satisfies the simultaneous equations $F = F_u = 0$ (with $F_v \ne 0$). Both types of turning points are shown in Fig. 5.18. If F has a degree of (M, N) in (u, v), then the degrees of F_u and F_v will be $(M - 1, N)$ and $(M, N - 1)$, respectively. It can be shown that the total number of roots of two simultaneous polynomial equations in two

variables whose degrees are (m, n) and (p, q), respectively, is $mq+np$ [95]. Therefore the number of u-turning points and v-turning points can be at most $2MN - M$ and $2MN - N$, respectively over the entire complex plane. However within the square parameter space $[0, 1]^2$, the number of turning points is typically much reduced in practice and therefore methods that use the $[0, 1]^2$ square as the search space of the roots such as the IPP algorithm in Sect. 4.9 or interval Newton methods [159] would typically outperform other methods.

3. *Singular points*: The points on the curve which satisfy the following three simultaneous equations $F = F_u = F_v = 0$ are called singular points. Noting that $f(x, y, z) = 0$, and $F(u, v) = W^m(u, v)f(x, y, z)$, we deduce

$$F_u = mW^{m-1}W_u f + W^m \left(\frac{\partial f}{\partial x}\frac{\partial x}{\partial u} + \frac{\partial f}{\partial y}\frac{\partial y}{\partial u} + \frac{\partial f}{\partial z}\frac{\partial z}{\partial u} \right) = W^m \nabla f \cdot \mathbf{r}_u .$$

(5.92)

Similarly we obtain $F_v = W^m \nabla f \cdot \mathbf{r}_v$, and hence at singular points $\nabla f \cdot \mathbf{r}_u = \nabla f \cdot \mathbf{r}_v = 0$. This means that $\nabla f \parallel \mathbf{r}_u \times \mathbf{r}_v$ or that the normals of two surfaces are parallel and since $F(u, v) = 0$ at these points the two surfaces intersect tangentially. If F has a degree of (M, N) in (u, v), the degrees of F_u and F_v will be $(M - 1, N)$ and $(M, N - 1)$, respectively, thus the number of singular points can be at most $2MN - M - N + 1$ [95] over the entire complex plane and typically much less in number in $[0, 1]^2$. If the systems $F = F_u = 0$ and $F = F_v = 0$ are already solved, a small extra evaluation can identify their common roots which are the singular points. Alternatively the IPP algorithm for the overconstrained system $F = F_u = F_v = 0$ can be used to find the singular points.

From the above discussions we can get upper bounds for the maximum number of u-turning, v-turning and singular points as listed in Table 5.4. These bounds refer to the maximum possible number of solutions (u, v) in the entire complex plane. Biquadratic and bicubic surfaces in the first column of Table 5.4 are degree 8 and 18 implicit algebraic surfaces. It turns out that the number of such points in the real square $[0, 1]^2$ is much smaller, but can still be quite large. Consequently, methods which focus only on the real solutions in $[0, 1]^2$ are advantageous, such as IPP algorithm described in Chap. 4 or interval Newton's method [159].

Analysis of singular points. Let us construct a parametric equation of a straight line L through a point (u_0, v_0) on the algebraic curve $F(u, v) = 0$

$$u = u_0 + \alpha t, \quad v = v_0 + \beta t ,$$

(5.93)

where α and β are constants and t is a parameter [437, 95, 107]. We find the intersections between L and the algebraic curve $F(u, v) = 0$ by determining the roots of $F(u_0 + \alpha t, v_0 + \beta t) = 0$. Since $F(u_0, v_0) = 0$, Taylor expansion of the left hand side gives

Table 5.4. Maximum number of turning and singular points in various cases

S_1	S_2	algebraic curve $F(u,v)$ degree M, N	max number u-turning pts $2MN - M$	max number v-turning pts $2MN - N$	max number singular points $2MN - M$ $-N + 1$
plane	biquadratic	2, 2	6	6	5
plane	bicubic	3, 3	15	15	13
quadric	biquadratic	4, 4	28	28	25
quadric	bicubic	6, 6	66	66	61
torus	biquadratic	8, 8	120	120	113
torus	bicubic	12, 12	276	276	265
biquadratic	biquadratic	16, 16	496	496	481
bicubic	biquadratic	36, 36	2556	2556	2521
bicubic	bicubic	54, 54	5778	5778	5725

$$(\alpha F_u + \beta F_v)t + \frac{1}{2}(\alpha^2 F_{uu} + 2\alpha\beta F_{uv} + \beta^2 F_{vv})t^2 + \cdots = 0 , \qquad (5.94)$$

where partial derivatives of F are evaluated at (u_0, v_0).

When F_u and F_v are not both zero $(F_u^2 + F_v^2 > 0)$ at (u_0, v_0), (5.94) has a simple root $t = 0$ and every line through (u_0, v_0) has a single intersection with the algebraic curve at (u_0, v_0) except for one case where $\alpha F_u + \beta F_v = 0$ for certain values of α and β. In such cases (5.94) has a double root $t = 0$, provided at least one of the second order partial derivatives is not zero $(F_{uu}^2 + F_{uv}^2 + F_{vv}^2 > 0)$, and L is tangent to the curve at (u_0, v_0).

When (u_0, v_0) is a singular point $(F_u(u_0, v_0) = F_v(u_0, v_0) = F(u_0, v_0) = 0)$, and at least one of F_{uu}, F_{uv}, F_{vv} is not zero $(F_{uu}^2 + F_{uv}^2 + F_{vv}^2 > 0)$, then $t = 0$ is a double root and has at least two intersections at (u_0, v_0) except for the values of α and β which satisfy

$$\alpha^2 F_{uu} + 2\alpha\beta F_{uv} + \beta^2 F_{vv} = 0 . \qquad (5.95)$$

In such cases, $t = 0$ is a triple root, provided at least one of the third order partial derivatives is not zero $(F_{uuu}^2 + F_{uuv}^2 + F_{uvv}^2 + F_{vvv}^2 > 0)$. We can solve the quadratic equation (5.95) for $\frac{\alpha}{\beta}$ or $\frac{\beta}{\alpha}$ which leads to the following three possibilities:

(1) Two real distinct roots: These values correspond to two distinct tangent directions at the singular point, which implies the algebraic curve has a self-intersection. The Folium of Descartes, which is shown in Fig. 1.1, has such singularity at the origin.

(2) One real double root: This value corresponds to one tangent direction at the singular point, which implies a cusp. An illustrative example, which is a semi-cubical parabola, is given in Fig. 2.3.

(3) Two complex roots: No real tangents at the singular point imply an isolated point. An example of an isolated point is given in Example 5.8.1 (see Fig. 5.21).

Example 5.8.1. Let the algebraic curve be $F(u, v) = u^3 + u^2 + v^2 = 0$ [437], then

$$F_u = u(3u + 2), \ F_v = 2v, \ F_{uu} = 6u + 2, \ F_{uv} = 0, \ F_{vv} = 2 ,$$
$$F_{uuu} = 6 \ , F_{uuv} = F_{uvv} = F_{uuv} = F_{vvv} = 0 \ .$$

The u-turning points can be found by finding the roots of $F = F_v = 0$ and $F_u \neq 0$. We immediately deduce $v = 0$. Upon substitution to $F = 0$ we obtain $u = 0, \ -1$. Since $F_u(0, 0) = 0$, $(0,0)$ is not a u-turning point. Therefore $(-1,0)$ is the only u-turning point. On the other hand v-turning points, which satisfy $F = F_u = 0$ and $F_v \neq 0$, have no real solutions. It is apparent from the above discussion that $u = v = 0$ is the only singular point. Tangents at $u = v = 0$ can be obtained from $\alpha^2 F_{uu} + 2\alpha\beta F_{uv} + \beta^2 F_{vv} = 2\alpha^2 + 2\beta^2 = 0$, which gives $(\frac{\alpha}{\beta})^2 + 1 = 0$, and hence no real solution. Therefore, $u = v = 0$ is an isolated point. If the domain of interest is $[-2, 1] \times [-1, 1]$, border points are $(-1.465, \pm 1)$. The above algebraic curve is depicted in Fig. 5.21.

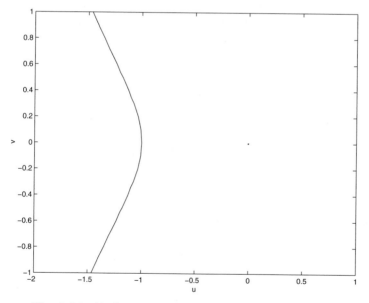

Fig. 5.21. Algebraic curve with an isolated point at $(0,0)$

Example 5.8.2. We have studied the semi-cubical parabola $F(u, v) = u^3 - v^2 = 0$ in Example 2.1.1. The curve has a singular point at $u = v = 0$. Since

$$F_u = 3u^2, \ F_v = -2v, \ F_{uu} = 6u, \ F_{uv} = 0, \ F_{vv} = -2 \ ,$$
$$F_{uuu} = 6 \ , F_{uuv} = F_{uvv} = F_{uuv} = F_{vvv} = 0 \ ,$$

we have $\alpha^2 F_{uu} + 2\alpha\beta F_{uv} + \beta^2 F_{vv} = 6u\alpha^2 - 2\beta^2 = 0$. At (0,0) we have a double root $\beta = 0$. Thus, at the singular point (0,0) we have a cusp whose tangent direction is along the $v = 0$ axis as shown in Fig. 2.3.

Example 5.8.3. Let us consider the equation

$$F(u,v) = (u+1)u(u-1)(v+1)v(v-1) + \frac{1}{20} = 0 \ ,$$

within the domain $[-2,2]^2$, taken from Geisow [124]. This is a degree 6 algebraic curve illustrated in Fig. 5.22. On every border line segment, there are three border points. The curve has no singular points, but involves two (internal) loops and six border-to-border branches. The algebraic curve $F(u,v) = 0$ in this example has degrees $M = 3, N = 3$ in u and v. Consequently, using the previous formulae the number of u turning points, v turning points and singular points (in the entire complex plane) is bounded by $2MN - M = 15$, $2MN - N = 15$, and $2MN - M - N + 1 = 13$. However, as we can see in Fig. 5.22, these numbers overestimate the actual number of such points in the real square $[-2,2]^2$.

Computing starting points for all branches. Starting points for tracing algebraic curves could be border points, turning points and singular points. Border points involve solution of a univariate polynomial equation, e.g. for border along $u = 0$, using (5.81)

$$F(0,v) = \sum_{j=0}^{N} c_{0j}^B B_{j,N}(v) = 0 \ . \tag{5.96}$$

Turning and singular point computation involve the first order partial derivatives:

$$F_u(u,v) = M \sum_{i=0}^{M-1} \sum_{j=0}^{N} (c_{i+1,j}^B - c_{ij}^B) B_{i,M-1}(u) B_{j,N}(v) \ , \tag{5.97}$$

$$F_v(u,v) = N \sum_{i=0}^{M} \sum_{j=0}^{N-1} (c_{i,j+1}^B - c_{ij}^B) B_{i,M}(u) B_{j,N-1}(v) \ . \tag{5.98}$$

Consequently, computation of turning points ($F = F_u = 0$ and $F = F_v = 0$) is equivalent to solving a system of two nonlinear polynomial equations in two variables, and computation of singularities $F = F_u = F_v = 0$ is equivalent to solving an overconstrained system of three nonlinear polynomial equations in two variables. Robust and efficient solution of these systems of nonlinear polynomial equations is addressed in Chap. 4.

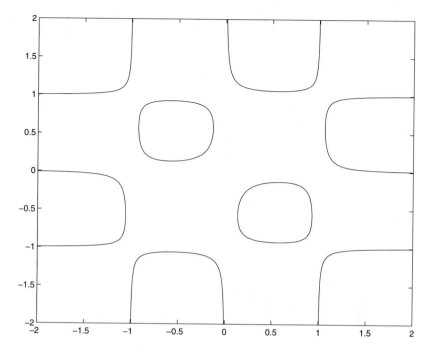

Fig. 5.22. A degree six algebraic curve (adapted from [124])

5.8.2 Rational polynomial parametric/rational polynomial parametric surface intersection (Case F1)

Rational polynomial parametric surface to rational polynomial parametric surface intersection is defined as:

$$\mathbf{r} = \mathbf{r}_1(\sigma, t) = \left(\frac{X_1(\sigma, t)}{W_1(\sigma, t)}, \frac{Y_1(\sigma, t)}{W_1(\sigma, t)}, \frac{Z_1(\sigma, t)}{W_1(\sigma, t)} \right)^T , \quad 0 \le \sigma, t \le 1 , \quad (5.99)$$

$$\cap \ \mathbf{r} = \mathbf{r}_2(u, v) = \left(\frac{X_2(u, v)}{W_2(u, v)}, \frac{Y_2(u, v)}{W_2(u, v)}, \frac{Z_2(u, v)}{W_2(u, v)} \right)^T , \quad 0 \le u, v \le 1 .$$

Formulation can be provided by setting $\mathbf{r}_1(\sigma, t) = \mathbf{r}_2(u, v)$ which leads to three nonlinear polynomial equations for four unknowns σ, t, u, v. It is an underconstrained system with 3 equations and 4 unknowns. This system can be solved by the IPP algorithm of Chap. 4. However, as the solutions are typically not isolated points but curves, such approach is very slow when small tolerances are used. One could also implicitize $\mathbf{r}_1(\sigma, t)$ to the form $f(x, y, z) = 0$ and substitute $x = \frac{X_2(u,v)}{W_2(u,v)}$, $y = \frac{Y_2(u,v)}{W_2(u,v)}$ and $z = \frac{Z_2(u,v)}{W_2(u,v)}$ into f to reduce the problem to Case F3 for low degree surfaces [212]. Heo et al. [160] studied the intersection of two ruled surfaces which is simpler than the general parametric surface to surface intersection problem.

There are three major techniques for solving RPR/RPP surface intersections. Detailed reviews can be found in [295, 300].

Lattice methods. *Lattice method* reduces the dimensionality of surface intersections by computing intersections of a number of iso-parametric curves of one surface with the other surface followed by connection of the resulting discrete intersection points to form different solution branches [352]. For intersections of parametric patches, the method reduces to the solution of a large number of independent systems of nonlinear equations. The reduction of problem dimensionality in lattice methods involves an initial choice of grid resolution, which, in turn, may lead the method to miss important features of the solution, such as small loops and isolated points which reflect near tangency or tangency of intersecting surfaces, and to provide incorrect connectivity. Appropriate methods for the solution of the resulting nonlinear equations in the present context are identified in Chap. 4.

Subdivision methods. *Subdivision methods* in their most basic form, involve recursive decomposition of the problem into simpler similar problems until a level of simplicity is reached, which allows simple direct solution, (e.g. plane/plane intersection [177, 295]). This is followed by a connection phase of the individual solutions to form the complete solution. Dokken [77] transforms surface/surface intersection problems to finding zeroes of functions of four variables using recursive subdivision techniques [221]. Initially conceived in the context of intersections of polynomial parametric surfaces [220], they can be extended to the computation of RPP/IA and IA/IA surface intersections [302]. A simple subdivision algorithm employs uniform subdivision which leads to a uniform quadtree data structure shown in Fig. 5.23. Subdivision techniques do not require starting points as marching methods, an important advantage. General non-uniform subdivision allows selective refinement of the solution providing the basis for an adaptive intersection technique. A disadvantage of subdivision techniques used in the evaluation of the entire intersection set is that, in actual implementations with finite subdivision steps, correct connectivity of solution branches in the vicinity of singular or near-singular points is difficult to guarantee, small loops may be missed (in methods with polyhedral surface approximations) or extraneous loops may be present in the approximation of the solution. Furthermore, if subdivision methods are used for high precision evaluation of the entire intersection set, they lead to data proliferation and are consequently slow, and, therefore, unattractive. There are many applications in CAD/CAM, that require high accuracy, for which pure subdivision methods are impractical. However, adaptive subdivision methods coupled with efficient local techniques to get high accuracy, offer the best known practical approach for the computation of characteristic points. These points can then be used in initiating efficient marching methods for tracing intersection curves.

Marching methods. *Marching methods* involve generation of sequences of points of an intersection curve branch by stepping from a given point on the

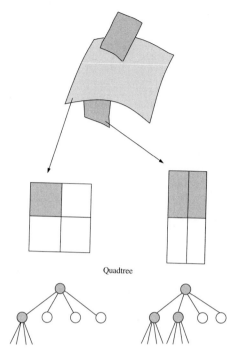

Quadtree

Fig. 5.23. Subdivision method

required curve in a direction prescribed by the local differential geometry [17, 19, 20, 210, 454], as we have studied in tracing the planar algebraic curve $F(u,v) = 0$ in Sect. 5.8.1. However, such methods are by themselves *incomplete* in that they require *starting points* for every branch of the solution. In order to identify all connected components of the intersection curve, a set of important points on the intersection curve (*characteristic points*) can be defined. As seen in Sect. 5.8.1, such a set may include *border, turning and singular points* of the intersection and provides at least one point on any connected intersection segment and identifies all singularities. For a 3-D RPP/RPP intersection case a more convenient set of such points sufficient to discover all connected components of the intersection, includes *border and collinear normal points* between the two surfaces. Collinear normal points provide points inside all intersection loops and all singular points.

Border points are points of the intersection at which at least one of the parametric variables σ, t, u, v takes a value equal to the border of the σ-t or u-v parametric domain. To compute border points, a piecewise rational polynomial curve to piecewise rational polynomial surface intersection capability is required, e.g., $\mathbf{r}_1(0,t) = \mathbf{r}_2(u,v)$, which we discussed in Sect. 5.7.2. Müllenheim [278] addressed a local method to find starting points for two parametric surfaces. Abdel-Malek and Yeh [1] introduced two local methods, iterative optimization and the Moore-Penrose pseudo-inverse method to determine starting points on the intersection curve between two parametric surfaces.

Sederberg et al. [376] first recognized the importance of collinear normal points in detecting the existence of closed intersection loops in intersection problems of two distinct parametric surface patches. These are points on the two parametric surfaces at which the normal vectors are collinear. Collinear normal points are a subset of parallel normal points first used by Sinha et al. [396] in surface intersection loop detection methods.

To simplify the notation, we replace $\mathbf{r}_1(\sigma, t)$ by $\mathbf{p}(\sigma, t)$ and $\mathbf{r}_2(u, v)$ by $\mathbf{q}(u, v)$. Then the collinear normal points satisfy the following equations [376]

$$(\mathbf{p}_\sigma \times \mathbf{p}_t) \cdot \mathbf{q}_u = 0, \quad (\mathbf{p}_\sigma \times \mathbf{p}_t) \cdot \mathbf{q}_v = 0 ,$$
$$(\mathbf{p} - \mathbf{q}) \cdot \mathbf{p}_\sigma = 0, \quad (\mathbf{p} - \mathbf{q}) \cdot \mathbf{p}_t = 0 . \tag{5.100}$$

Equations (5.100) form a system of four nonlinear polynomial equations that can be solved using the robust methods of Chap. 4. Now we split the patches in (at least) one parametric direction at these collinear normal points. Consequently, starting points are only border points on the boundaries of all subdomains created. Grandine and Klein [133] follow a systematic approach for topology resolution of B-spline surface intersections. In this process, they determine the structure of the intersection curves including closed loops prior to numerical tracing (following a marching method based on numerical integration of a differential algebraic system of equations). Topology resolution in this context relies on an extension of the PP algorithm (see Sect. 4.4) to the B-spline case implemented in floating point (with normalization of the equations in the range [-1,1] and normalization of the knot vector in each subdomain in the range [0,1] at each iteration step of the process to capitalize on the higher density of floating point numbers in this range, thereby improving numerical robustness of the algorithm). An alternate way to detect closed intersection loops is to use topological methods [236, 53, 264, 210, 245, 244, 436, 435]. Also bounding pyramids [209, 382] can be used effectively to assure the nonexistence of closed surface to surface intersection loops. These earlier methods need to be implemented in exact or RIA for robustness.

In order to trace the intersection curve, starting points must be located prior to tracing. An intersection curve branch can be traced if its pre-image starts from the parametric domain boundary in either parameter domain [19]. The marching direction coincides with the tangential direction of the intersection curve $\mathbf{c}(s)$ which is perpendicular to the normal vectors of both surfaces. Therefore, the marching direction can be obtained as follows:

$$\mathbf{c}'(s) = \frac{\mathbf{P}(\sigma, t) \times \mathbf{Q}(u, v)}{|\mathbf{P}(\sigma, t) \times \mathbf{Q}(u, v)|} , \tag{5.101}$$

where the normalization forces $\mathbf{c}(s)$ to be arc length parametrized and

$$\mathbf{P}(\sigma, t) = \mathbf{p}_\sigma \times \mathbf{p}_t, \quad \mathbf{Q}(u, v) = \mathbf{q}_u \times \mathbf{q}_v , \tag{5.102}$$

are the normal vectors of $\mathbf{p}$ and $\mathbf{q}$, respectively. When the two surfaces intersect tangentially, we cannot use (5.101) since the denominator vanishes. In such cases we must find the marching direction in an alternate way which will be discussed in Sect. 6.4.

The intersection curve can also be viewed as a curve on the two intersecting surfaces. A curve $\sigma = \sigma(s)$, $t = t(s)$ in the σt-plane defines a curve $\mathbf{r} = \mathbf{c}(s) = \mathbf{p}(\sigma(s), t(s))$ on a parametric surface $\mathbf{p}(\sigma, t)$, as well as a curve $u = u(s)$ $v = v(s)$ in the uv-plane defines a curve $\mathbf{r} = \mathbf{c}(s) = \mathbf{q}(u(s), v(s))$ on a parametric surface $\mathbf{q}(u, v)$. We can derive the first derivative of the intersection curve as a curve on the parametric surface using the chain rule:

$$\mathbf{c}'(s) = \mathbf{p}_\sigma \sigma' + \mathbf{p}_t t', \quad \mathbf{c}'(s) = \mathbf{q}_u u' + \mathbf{q}_v v' . \tag{5.103}$$

Since we know the unit tangent vector of the intersection curve from (5.101), we can find σ' and t' as well as u' and v' by taking the dot product on both hand sides of the first equation of (5.103) with $\mathbf{p}_\sigma$ and $\mathbf{p}_t$ and the second equation with $\mathbf{q}_u$ and $\mathbf{q}_v$, which leads to two linear systems [178]. The solutions are immediately obtained as

$$\sigma' = \frac{det(\mathbf{c}', \mathbf{p}_t, \mathbf{P}(\sigma, t))}{\mathbf{P}(\sigma, t) \cdot \mathbf{P}(\sigma, t)} \ , \quad t' = \frac{det(\mathbf{p}_\sigma, \mathbf{c}', \mathbf{P}(\sigma, t))}{\mathbf{P}(\sigma, t) \cdot \mathbf{P}(\sigma, t)} \ , \tag{5.104}$$

$$u' = \frac{det(\mathbf{c}', \mathbf{q}_v, \mathbf{Q}(u, v))}{\mathbf{Q}(u, v) \cdot \mathbf{Q}(u, v)}, \quad v' = \frac{det(\mathbf{q}_u, \mathbf{c}', \mathbf{Q}(u, v))}{\mathbf{Q}(u, v) \cdot \mathbf{Q}(u, v)} \ , \tag{5.105}$$

where det denotes the determinant.

The points of the intersection curves are computed successively by integrating the initial value problem for a system of nonlinear ordinary differential equations (5.104) and (5.105) using numerical techniques such as the Runge-Kutta or adaptive stepping methods [69, 126]. Figure 5.24 presents the intersection of a rational quadratic-linear B-spline patch (representing a cylinder) with a biquadratic Bézier patch representing an elliptic paraboloid [208]. Figure 5.25 presents the intersection between two biquartic Bézier patches (nearly coincident surfaces) [210].

5.8.3 Implicit algebraic/implicit algebraic surface intersection (Case F8)

Implicit algebraic surface to implicit algebraic surface intersection is defined as follows:

$$f(\mathbf{r}) = 0 \cap g(\mathbf{r}) = 0 , \tag{5.106}$$

where f, g are polynomial functions. Here we have two equations in three unknowns $\mathbf{r}$. Bajaj et al. [17] developed a marching method for IA/IA surface intersection as well as for parametric surfaces.

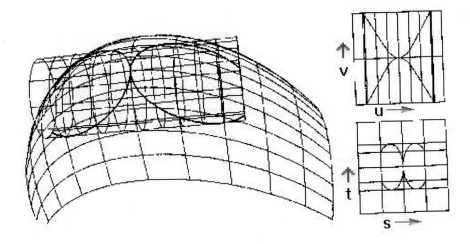

Fig. 5.24. Cylinder - elliptic paraboloid intersection (adapted from [208])

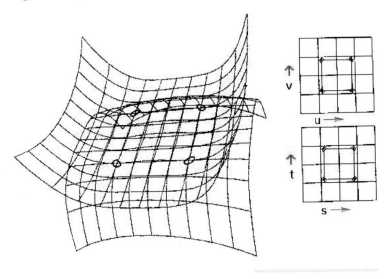

Fig. 5.25. Intersection of two biquartic Bézier patches forming four small loops (adapted from [210])

A method for low order f, g is to eliminate one variable (e.g. z) to find projection of intersection curves on the plane of other two variables (e.g. x, y), then trace the algebraic curve and use the inversion algorithm to find z. Intersections of low degree implicit algebraic surfaces are of special interest in the boundary evaluation of the Constructive Solid Geometry models. A more complete analysis of the special intersections of two quadric surfaces (used frequently in CAD/CAM of mechanical parts) can be found in [233, 234, 367, 104, 443, 390, 268].

Example 5.8.4. Consider the intersection of a sphere and a circular cylinder given by

$$f = x^2 + y^2 + z^2 - 1 = 0 \,,$$
$$g = x^2 + (y - \frac{1}{2})^2 - \frac{1}{4} = 0 \,,$$

as shown in Fig. 5.26. The projection of the intersection curves on the three coordinate planes is illustrated in Fig. 5.27.

Hartmann [155] proposed the idea of *numerical implicitization* which allows treatment of intersection problems of not only parametric surfaces but also non-standard surfaces such as an offset of an implicit surface, a Voronoi surface, an envelope of a one parametric family of spheres etc. The key idea is that in tracing the intersection curve of two implicit surfaces, we are only required to calculate the implicit function values and the gradients of the implicit functions at the intersection points as in Bajaj et al. [17]. In other words, we do not need to know the functions explicitly. Therefore if we can implicitize any two surfaces numerically we are able to trace the intersection curve using the IA/IA surface intersection algorithm of Bajaj et al. [17].

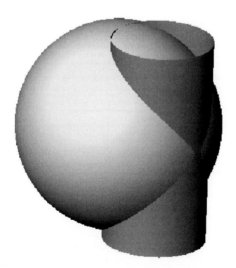

Fig. 5.26. Intersection of two implicit quadrics (sphere and cylinder)

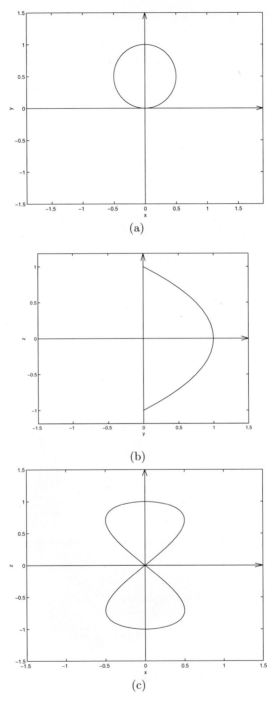

Fig. 5.27. (a) Projection of intersection curve on $z = 0$ (xy) plane, i.e. $x^2 + (y - \frac{1}{2})^2 - \frac{1}{4} = 0$, (b) projection of intersection curve on $x = 0$ (yz) plane, i.e. $y = 1 - z^2$, (c) projection of intersection curve on $y = 0$ (xz) plane, i.e. $x^2 + z^4 - z^2 = 0$

5.9 Overlapping of curves and surfaces

So far we have focused mainly on transversal intersection problems of regular curves and surfaces. However, in real engineering problems we may encounter curve/curve (see Fig. 5.28), curve/surface or surface/surface overlapping of curves and surfaces.

We will illustrate the curve/curve overlapping case using two planar cubic Bézier curves $\mathbf{r}_1(t)$ (AB) and $\mathbf{r}_2(\sigma)$ (CD) whose control points are given by (0,0), (0.8,0.8), (1.6,0.32), (2.4,0.608) and (0.6,0.392), (1.4,0.68), (2.2,0.2), (3,1) (see Fig. 5.28). If we use the IPP algorithm to compute the intersections of the two curves, the rate of convergence of the solver drops significantly due to an extensive amount of binary subdivision [179]. In such cases we may run the IPP solver with a fairly coarse level of accuracy, for example $\epsilon = 10^{-2}$ or 10^{-3}. If we observe a number of boxes overlap to one another, as shown in Fig. 5.29, it is very likely that overlap exists. Figure 5.29 shows the boxes of roots of the intersection points of two curves computed with $\epsilon = 10^{-2}$. We can observe that the curve AB overlaps with curve CD from $t = 0.25$ to $t = 1$ and the curve CD overlaps with curve AB from $\sigma = 0$ to $\sigma = 0.75$.

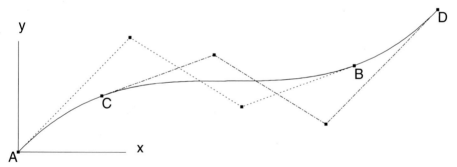

Fig. 5.28. Two cubic Bézier curves AB and CD overlapping each other at CB (adapted from [179])

Hu et al. [179, 178] discuss the treatment of curve/curve, curve/surface as well as surface/surface overlapping problems based on interval polynomial curves and surfaces with the IPP solver. They also introduced the following theorem which can be used to find the starting and end points of the overlapping segment.

Theorem 5.9.1. *If two C^∞ curve segments $\mathbf{r}_1(t)$ and $\mathbf{r}_2(\sigma)$ overlap along a finite part of their length, they must overlap everywhere. Otherwise, they end at boundary points.*

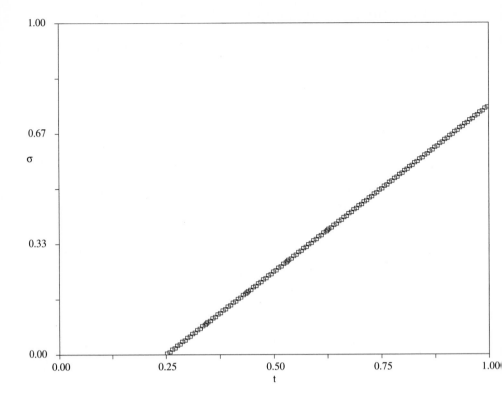

Fig. 5.29. The coarse boxes that contain roots of intersection of two overlapping curves (adapted from [179])

This theorem means it is impossible that two C^∞ curves, such as two parametric polynomial curves, overlap along a finite part of their length and separate from each other at one point, as illustrated in Fig. 5.30. This theorem can be proven contrapositively.

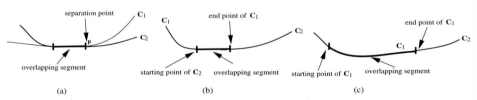

Fig. 5.30. Overlap of two C^∞ curves along a finite segment. (**a**) is an *impossible* configuration, and (**b**) and (**c**) are the two possible configurations (adapted from [178])

Proof: Assume that there exist two C^∞ curves $\mathbf{r}_1(t)$ and $\mathbf{r}_2(\sigma)$ which overlap partially. This means that there is an interior point $\mathbf{p}$ (referred to as separation point) that ends the overlapping segment of $\mathbf{r}_1(t)$ and $\mathbf{r}_2(\sigma)$, as illustrated in Fig. 5.30. Let t_p and σ_p be the parameters of $\mathbf{r}_1(t)$ and $\mathbf{r}_2(\sigma)$ at $\mathbf{p}$, respectively. Suppose further that the two curves are arc length parametrized, and have the same orientation. Since the two curves are C^∞, their overlapping segment should be also C^∞, i.e.,

$$\mathbf{r}_1^{(i)}(t_p) = \mathbf{r}_2^{(i)}(\sigma_p) , \tag{5.107}$$

where superscript (i) denotes derivative of order i valid for all nonnegative integers i. Therefore, from the Taylor expansion theorem, we have

$$\mathbf{r}_1(t_p + \delta) = \mathbf{p} + \sum_{i=1}^{\infty} \frac{\mathbf{r}_1^{(i)}(t_p)}{i!} \delta^i , \tag{5.108}$$

$$\mathbf{r}_2(\sigma_p + \delta) = \mathbf{p} + \sum_{i=1}^{\infty} \frac{\mathbf{r}_2^{(i)}(\sigma_p)}{i!} \delta^i . \tag{5.109}$$

From (5.107), (5.108) and (5.109), we have $\mathbf{r}_1(t_p + \delta) = \mathbf{r}_2(\sigma_p + \delta)$, which means that $\mathbf{p}$ cannot be an interior point. Hence, Theorem 5.9.1 is proven. ∎

Further discussions on curve/surface and surface/surface overlapping problems can be found in [178].

5.10 Self-intersection of curves and surfaces

So far we have focused mainly on intersection problems of regular curves and surfaces without self-intersections. In this section we will show how to compute self-intersections of curves and surfaces.

Self-intersection of a planar rational polynomial parametric curve can be formulated as finding pairs of distinct parameter values $\sigma \neq t$ such that

$$\mathbf{r}(\sigma) = \mathbf{r}(t) , \tag{5.110}$$

or in terms of components as

$$\frac{X(\sigma)}{W(\sigma)} - \frac{X(t)}{W(t)} = 0 , \quad \frac{Y(\sigma)}{W(\sigma)} - \frac{Y(t)}{W(t)} = 0 , \tag{5.111}$$

where

$$X(t) = \sum_{i=0}^{n} w_i x_i B_{i,n}(t), \quad Y(t) = \sum_{i=0}^{n} w_i y_i B_{i,n}(t), \quad W(t) = \sum_{i=0}^{n} w_i B_{i,n}(t) .$$

$$\tag{5.112}$$

Lasser [225] presents an algorithm to find all the self-intersection points of a Bézier curve by subdividing the Bézier polygon instead of the curve itself. Finally the self-intersection points are approximated by straight line intersections of the refined Bézier polygon.

Here we introduce a method to find all the self-intersection points of a planar rational polynomial parametric curve based on the IPP algorithm introduced in Chap. 4. Multiplying by the denominators of (5.111), we obtain

$$X(\sigma)W(t) - X(t)W(\sigma) = 0 , \quad Y(\sigma)W(t) - Y(t)W(\sigma) = 0 . \quad (5.113)$$

These equations can be rewritten as

$$\sum_{i=0}^{n}\sum_{j=0}^{n} w_i w_j x_i [B_{i,n}(\sigma)B_{j,n}(t) - B_{j,n}(\sigma)B_{i,n}(t)] = 0 , \qquad (5.114)$$

$$\sum_{i=0}^{n}\sum_{j=0}^{n} w_i w_j y_i [B_{i,n}(\sigma)B_{j,n}(t) - B_{j,n}(\sigma)B_{i,n}(t)] = 0 . \qquad (5.115)$$

which form a system of two nonlinear polynomial equations in σ and t. Since

$$\frac{B_{i,n}(\sigma)B_{j,n}(t) - B_{j,n}(\sigma)B_{i,n}(t)}{\sigma - t} \qquad (5.116)$$

$$= B_{j,n}(t)\frac{B_{i,n}(\sigma) - B_{i,n}(t)}{\sigma - t} - B_{i,n}(t)\frac{B_{j,n}(\sigma) - B_{j,n}(t)}{\sigma - t} ,$$

we can easily factor out $(\sigma - t)$ from (5.114) and (5.115) to exclude the trivial solutions $\sigma = t$.

Self-intersections of a rational polynomial parametric surface are defined by finding pairs of distinct parameter values $(\sigma, t) \neq (u, v)$ such that

$$\mathbf{r}(\sigma, t) = \mathbf{r}(u, v) . \qquad (5.117)$$

Barnhill et al. [19] compute surface self-intersections by their procedural surface/surface intersection algorithm. Also Lasser [224] introduces a method to compute all the self-intersection curves of a Bézier surface patch by subdividing the Bézier control net instead of the surface patch itself. Finally the self-intersection curves are approximated by the polygons resulting from the plane/plane intersections of the refined Bézier control net. Andersson et al. [6] provide necessary and sufficient conditions to preclude self-intersections of composite Bézier curves and patches.

Unlike the curve self-intersection case, it is inefficient to solve surface self-intersection problems with the IPP solver (see Chap. 4). The key difficulty arises in the removal of the trivial solutions. We cannot divide out factors $\sigma - u$ and $t - v$ from the system directly, since terms $x(\sigma, t) - x(u, v)$, $y(\sigma, t) - y(u, v)$ and $z(\sigma, t) - z(u, v)$ do not necessarily exactly involve the factors $\sigma - u$ and $t - v$. A technique to remove such trivial solutions is given in [253] and in Sect. 11.3.5.

Self-intersections of offsets of curves and surfaces, which are more difficult to compute, are discussed fully in Chap. 11.

5.11 Summary

Some important outstanding issues in the area of intersection problems are summarized below [300]. While solving nonlinear polynomial systems, as a preliminary step in computing characteristic points of surface intersections, it is frequently necessary to deal with solution sets that are not zero-dimensional (e.g. the solution sets are one-dimensional, two-dimensional etc.). Most of the methods experience serious numerical and efficiency difficulties in those cases. Methods to deal effectively with these problems need to be developed, including methods to identify and, if possible, parameterize these higher-dimensional solution sets.

Extension of current intersection methods applied on rational B-spline surfaces, to more general and complex surfaces requires further study. Such surfaces include offset, generalized cylinder (pipe or canal surfaces in particular), blending, and medial surfaces and surfaces arising from the solution of partial differential equations or via recursion techniques (subdivision surfaces [354]). Intersections of such surfaces with the basic low order algebraic and rational B-spline surfaces, commonly used in CAD need to be explored. However, a basic element of a solution of many of these problems is the auxiliary variable method described in [169, 253, 300], where the problem is reduced to a higher dimension nonlinear polynomial system. In some cases, recent research has indicated that some special instances of these general surfaces can be exactly expressed as rational polynomial surfaces [324, 241, 256, 250] of higher degree and therefore these problems are reducible at least in principle to the problems addressed in this chapter. Further research is needed to implement this idea in a practical setting and examine the relative efficiency of competing approaches.

Investigating the effects of floating point arithmetic on the implementation of intersection algorithms has been an important area for basic research during the last decade. Ways to enhance the precision of intersection computation, to monitor numerical error contamination and alternate means of performing arithmetic, not relying on imprecise floating point computation alone, have been explored in some detail. Researchers in surface intersection problems during the last decade have already obtained a good understanding of robustness problems when employing floating point arithmetic and of methods to mitigate these problems based on normalization of the system [133] and rounded interval arithmetic [178]. However, these methods are not a panacea since they cannot resolve effectively non-zero-dimensional solution sets of nonlinear systems or achieve very high precision in reasonable computation times. A related active problem area has been the *rectification* of solid models expressed in the Boundary Representation form, which attempts to resolve intersection inconsistencies in such models and create topologically and geometrically consistent models [303, 360, 388, 389].

As a result of these deficiencies, recent research tends to focus on exact methods involving rational arithmetic [196, 356, 358]. Much research remains to be done in bringing such methods to the CAD practice, generalizing the arithmetic to go beyond rational and algebraic numbers (eg. involving transcendental numbers of trigonometric form), and to explore more efficient alternatives that are generally applicable in low and high degree problems alike. Finally, a general and comprehensive comparison and mapping of the efficiency properties of all available methods for solving nonlinear systems robustly would be valuable as a guide for future research.

6. Differential Geometry of Intersection Curves

6.1 Introduction

In Chap. 5 we have studied the classification, detection, and solution of intersection problems. In this chapter we focus on the differential geometry properties of intersection curves of two surfaces. To compute the intersection curves more accurately and efficiently, higher order approximation of intersection curves may be needed. This requires the computation of not only the tangents of the intersection curves, but also curvature vectors and higher order derivative vectors, i.e. higher order differential properties of the curves.

The two types of surfaces commonly used in geometric modeling systems are parametric and implicit surfaces that lead to three types of surface-surface intersection problems: *parametric-parametric, implicit-implicit* and *parametric-implicit*. While differential geometry of a parametric curve can be found in textbooks such as in [412, 444, 76], there is little literature on differential geometry of intersection curves. Faux and Pratt [116] give a formula for the curvature of an intersection curve of two parametric surfaces. Willmore [444] describes how to obtain the unit tangent $\mathbf{t}$, the unit principal normal $\mathbf{n}$, and the unit binormal $\mathbf{b}$, as well as the curvature κ and the torsion τ of the intersection curve of two implicit surfaces. Hartmann [154] provides formulae for computing the curvature κ of intersection curves for all three types of intersection problems. They all assume *transversal intersections* where the tangential direction at an intersection point can be computed simply by the cross product of the normal vectors of the both surfaces.

However, when the two normals are parallel to each other, the tangent direction cannot be determined by this method. We call such intersection points *tangential intersection points*. Kruppa describes in his book [213] that the tangential direction of the intersection curve at a tangential intersection point corresponds to the direction from the intersection point towards the intersection of the Dupin's indicatrices of the two surfaces. Cheng [53], Markot and Magedson [264, 263] give solutions for parametric surfaces at isolated tangential intersection points, based on the analysis of the plane vector field function defined by the gradient of an oriented distance function of one surface from the other. The plane field function will vanish at the tangential intersection point, and higher-order expansion of the function is required at such points to determine the marching direction for the intersection curve. Kriezis [208] and Kriezis et al. [210] determine the marching direction for tangential intersection curves based on the fact that the determinant of the Hessian matrix of the oriented distance function is zero. Luo et al. [242] present a method to trace such tangential intersection curves for parametric-parametric surfaces employing the marching method. The marching direction is obtained by solving an underdetermined system based on the equality of the differentiation of the two normal vectors and the projection of the Taylor expansion of the two surfaces onto the normal vector at the intersection point. Ye and Maekawa [458] developed algorithms to compute unit tangent vectors, curvature vectors, binormal vectors, curvatures, torsions, and algorithms to evaluate the higher order derivatives for transversal as well as tangential intersections for all three types of intersection problems.

6.2 More differential geometry of curves

Let $x = x(s)$, $y = y(s)$, $z = z(s)$ or in vector form $\mathbf{r} = \mathbf{c}(s)$ be the intersection curve with arc length parametrization. Then from (2.5) and (2.20), we have

$$\mathbf{c}'(s) = \mathbf{t} \, , \tag{6.1}$$
$$\mathbf{c}''(s) = \mathbf{k} = \kappa\mathbf{n} \, , \tag{6.2}$$

where $\mathbf{t}$ is the unit tangent vector and $\mathbf{k}$ is the curvature vector, which is the rate of change of the tangent vector. From (6.2) it follows that

$$\kappa^2 = \mathbf{k} \cdot \mathbf{k} = \mathbf{c}'' \cdot \mathbf{c}'' \, . \tag{6.3}$$

Now let us evaluate the third derivative $\mathbf{c}'''(s)$ by differentiating (6.2)

$$\mathbf{c}'''(s) = \kappa'\mathbf{n} + \kappa\mathbf{n}' \, , \tag{6.4}$$

where we can replace $\mathbf{n}'$ by the second equation of the Frenet-Serret formulae (2.56) yielding

$$\mathbf{c}'''(s) = -\kappa^2\mathbf{t} + \kappa'\mathbf{n} + \kappa\tau\mathbf{b} \, . \tag{6.5}$$

Since the vectors $\mathbf{t}$, $\mathbf{n}$, $\mathbf{b}$ are a right-handed orthonormal triplet, the torsion can be obtained from (6.5) as

$$\tau = \frac{\mathbf{b} \cdot \mathbf{c}'''}{\kappa} \, , \tag{6.6}$$

provided that the curvature does not vanish.

Classical differential geometry textbooks [412, 206, 444, 76] do not cover the case $\kappa = 0$, which is addressed below following Ye and Maekawa [458]. When $\kappa = 0$ (6.2) does not define the unit principal normal vector. To obtain the principal normal vector at points where $\kappa = 0$, higher order derivatives of the curve are involved. If $\kappa \equiv 0$, then the curve is a straight line, and the Frenet frame of the curve is not defined. We assume here that $\kappa = 0$ occurs only at isolated points. In such case, (2.56) is valid. If $\kappa = 0$ and $\kappa' \neq 0$, the third order derivative (6.4) reduces to

$$\mathbf{c}'''(s) = \kappa' \mathbf{n} , \tag{6.7}$$

which defines the unit principal normal vector where κ' is obtained from $(\kappa')^2 = \mathbf{c}''' \cdot \mathbf{c}'''$. If $\kappa = \kappa' = 0$ and $\kappa'' \neq 0$, we need to evaluate the fourth order derivative by differentiating (6.4) yielding

$$\mathbf{c}^{(4)}(s) = \kappa'' \mathbf{n} , \tag{6.8}$$

where $(\kappa'')^2 = \mathbf{c}^{(4)} \cdot \mathbf{c}^{(4)}$. In general, if $\kappa = \kappa' = \cdots = \kappa^{(j-1)} = 0$ and $\kappa^{(j)} \neq 0$, then

$$\mathbf{c}^{(j+2)}(s) = \kappa^{(j)} \mathbf{n} , \tag{6.9}$$

where $(\kappa^{(j)})^2 = \mathbf{c}^{(j+2)} \cdot \mathbf{c}^{(j+2)}$.

The evaluation of torsion when the curvature vanishes can be performed as follows. If $\kappa = 0$ and $\kappa' \neq 0$, we need to evaluate the fourth order derivative of $\mathbf{c}(s)$, i.e. $\mathbf{c}^{(4)}(s)$. This can be obtained by differentiating (6.5) and replacing $\mathbf{t}'$, $\mathbf{n}'$, $\mathbf{b}'$ using the Frenet-Serret formulae which results in:

$$\mathbf{c}^{(4)}(s) = -3\kappa\kappa' \mathbf{t} + (\kappa'' - \kappa\tau^2 - \kappa^3)\mathbf{n} + (2\kappa'\tau + \kappa\tau')\mathbf{b} . \tag{6.10}$$

In this case (6.10) further reduces to

$$\mathbf{c}^{(4)}(s) = \kappa'' \mathbf{n} + 2\kappa'\tau \mathbf{b} , \tag{6.11}$$

thus

$$\tau = \frac{\mathbf{b} \cdot \mathbf{c}^{(4)}}{2\kappa'} . \tag{6.12}$$

Similarly we have

$$\mathbf{c}^{(5)}(s) = (-4\kappa\kappa'' - 3(\kappa')^2 + \kappa^4 + \kappa^2\tau^2)\mathbf{t} + (\kappa''' - 6\kappa^2\kappa' - 3\kappa'\tau^2 - 3\kappa\tau\tau')\mathbf{n}$$
$$+ (3\kappa''\tau + 3\kappa'\tau' - \kappa^3\tau - \kappa\tau^3 + \kappa\tau'')\mathbf{b} . \tag{6.13}$$

and hence, if $\kappa = \kappa' = 0$ and $\kappa'' \neq 0$, then τ becomes

$$\tau = \frac{\mathbf{b} \cdot \mathbf{c}^{(5)}}{3\kappa''} . \tag{6.14}$$

In general, if $\kappa = \kappa' = \cdots = \kappa^{(j-1)} = 0$ and $\kappa^{(j)} \neq 0$, then [458]

$$\tau = \frac{\mathbf{b} \cdot \mathbf{c}^{(j+3)}}{(j+1)\kappa^{(j)}} . \tag{6.15}$$

Let $x = x^A(u_A, v_A)$, $y = y^A(u_A, v_A)$, $z = z^A(u_A, v_A)$ and $x = x^B(u_B, v_B)$, $y = y^B(u_B, v_B)$, $z = z^B(u_B, v_B)$ or in vector form, $\mathbf{r} = \mathbf{r}^A(u_A, v_A)$ and $\mathbf{r} = \mathbf{r}^B(u_B, v_B)$, be the two parametric surfaces. Also, let us denote the two implicit surfaces as $f^A(x, y, z) = 0$ and $f^B(x, y, z) = 0$. We assume that these surfaces are all *regular*. In other words

$$\mathbf{r}^A_{u_A} \times \mathbf{r}^A_{v_A} \neq \mathbf{0}, \qquad \mathbf{r}^B_{u_B} \times \mathbf{r}^B_{v_B} \neq \mathbf{0}, \qquad \nabla f^A \neq \mathbf{0}, \qquad \nabla f^B \neq \mathbf{0} . \quad (6.16)$$

The unit normal vector of a parametric surface and an implicit surface are given by (3.3) and (3.9).

So far, we have studied the intersection curve independent of the two intersecting surfaces. However, the intersecting curve can also be viewed as a curve on the two intersecting surfaces. A curve $u = u(s)$, $v = v(s)$ in the uv-plane defines a curve $\mathbf{r} = \mathbf{c}(s) = \mathbf{r}(u(s), v(s))$ on a parametric surface $\mathbf{r}(u, v)$, while a curve $x = x(s)$, $y = y(s)$, $z = z(s)$ with constraint $f(x(s), y(s), z(s)) = 0$ defines a curve on an implicit surface $f(x, y, z) = 0$.

We can easily derive the first three derivatives of the intersection curve $\mathbf{c}'(s)$, $\mathbf{c}''(s)$, $\mathbf{c}'''(s)$ as a curve on a parametric surface using the chain rule:

$$\mathbf{c}'(s) = \mathbf{r}_u u' + \mathbf{r}_v v' , \tag{6.17}$$

$$\mathbf{c}''(s) = \mathbf{r}_{uu}(u')^2 + 2\mathbf{r}_{uv}u'v' + \mathbf{r}_{vv}(v')^2 + \mathbf{r}_u u'' + \mathbf{r}_v v'' , \tag{6.18}$$

$$\mathbf{c}'''(s) = \mathbf{r}_{uuu}(u')^3 + 3\mathbf{r}_{uuv}(u')^2 v' + 3\mathbf{r}_{uvv}u'(v')^2 + \mathbf{r}_{vvv}(v')^3 \tag{6.19}$$
$$+ 3(\mathbf{r}_{uu}u'u'' + \mathbf{r}_{uv}(u''v' + u'v'') + \mathbf{r}_{vv}v'v'') + \mathbf{r}_u u''' + \mathbf{r}_v v''' .$$

Similarly we can evaluate $\frac{df}{ds}$, $\frac{d^2f}{ds^2}$ and $\frac{d^3f}{ds^3}$ as follows:

$$\frac{df}{ds} = f_x x' + f_y y' + f_z z' = 0 , \tag{6.20}$$

$$\frac{d^2 f}{ds^2} = f_{xx}(x')^2 + f_{yy}(y')^2 + f_{zz}(z')^2 + 2(f_{xy}x'y' + f_{yz}y'z' + f_{xz}x'z') \tag{6.21}$$
$$+ f_x x'' + f_y y'' + f_z z'' = 0 ,$$

$$\frac{d^3 f}{ds^3} = f_{xxx}(x')^3 + f_{yyy}(y')^3 + f_{zzz}(z')^3 + 3(f_{xxy}(x')^2 y' + f_{xxz}(x')^2 z' \tag{6.22}$$
$$+ f_{xyy}x'(y')^2 + f_{yyz}(y')^2 z' + f_{xzz}x'(z')^2 + f_{yzz}y'(z')^2 + 2f_{xyz}x'y'z')$$
$$+ 3(f_{xx}x'x'' + f_{yy}y'y'' + f_{zz}z'z'' + f_{xy}(x''y' + x'y''))$$
$$+ f_{yz}(y''z' + y'z'') + f_{xz}(x''z' + x'z'')) + f_x x''' + f_y y''' + f_z z''' = 0 .$$

6.3 Transversal intersection curve

6.3.1 Tangential direction

The tangent vector of the transversal intersection curve $\mathbf{c}(s)$ lies on the tangent planes of both surfaces. Therefore it can be obtained as the cross product of the unit surface normal vectors of the two surfaces at P as illustrated in Fig. 6.1:

$$\mathbf{t} = \frac{\mathbf{N}^A \times \mathbf{N}^B}{|\mathbf{N}^A \times \mathbf{N}^B|} , \qquad (6.23)$$

where $\mathbf{N}^A$ and $\mathbf{N}^B$ are the unit surface normal vectors of the two surfaces which are given either by (3.3) or by (3.9) according to the type of two intersecting surfaces. When the two normals are parallel to each other, the tangent direction cannot be determined by (6.23). This happens when the two surfaces intersect tangentially and the tangent direction must be treated in a different way. We will investigate the tangential intersection case in Sect. 6.4.

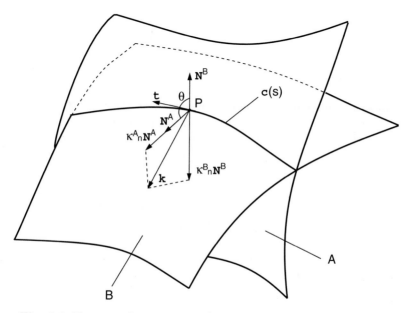

Fig. 6.1. Transversal intersection of two surfaces (adapted from [458])

6.3.2 Curvature and curvature vector

The curvature vector of the intersection curve at P, being perpendicular to $\mathbf{t}$, must lie in the normal plane spanned by $\mathbf{N}^A$ and $\mathbf{N}^B$. Thus we can express it as

$$\mathbf{k} = \alpha \mathbf{N}^A + \beta \mathbf{N}^B , \qquad (6.24)$$

where α and β are the coefficients that we need to determine. The normal curvature at P in direction $\mathbf{t}$ is the projection of the curvature vector $\mathbf{k}$ onto the unit surface normal vector $\mathbf{N}$ at P given by

$$\kappa_n = \mathbf{k} \cdot \mathbf{N} . \tag{6.25}$$

By projecting (6.24) onto the normals of both surfaces (see Fig. 6.1) we have

$$\kappa_n^A = \alpha + \beta \cos \theta ,$$
$$\kappa_n^B = \alpha \cos \theta + \beta , \tag{6.26}$$

where θ is the angle between $\mathbf{N}^A$ and $\mathbf{N}^B$ and is evaluated by

$$\cos \theta = \mathbf{N}^A \cdot \mathbf{N}^B . \tag{6.27}$$

Solving the coefficients α and β from linear system (6.26), and substituting into (6.24) yields

$$\mathbf{k} = \frac{\kappa_n^A - \kappa_n^B \cos\theta}{\sin^2\theta} \mathbf{N}^A + \frac{\kappa_n^B - \kappa_n^A \cos\theta}{\sin^2\theta} \mathbf{N}^B . \tag{6.28}$$

It follows that if we can evaluate the two normal curvatures κ_n^A and κ_n^B at P, we are able to obtain the curvature vector of the intersection curve at P from (6.28). Note that (6.28) does not depend on the type of surfaces. Let us first derive the normal curvature for a parametric surface. Recall that the curvature vector of the intersection curve is also given by (6.18) considered as a curve on the parametric surface. The normal curvature is obtained by projecting (6.18) onto the unit surface normal

$$\kappa_n = L(u')^2 + 2Mu'v' + N(v')^2 , \tag{6.29}$$

where L, M, N are the second fundamental form coefficients (3.28).

We still need to evaluate u', v' to compute (6.29). Since we know the unit tangent vector of the intersection curve from (6.23), we can find u' and v' by taking the dot product on both hand sides of (6.17) with $\mathbf{r}_u$ and $\mathbf{r}_v$, which leads to a linear system

$$Eu' + Fv' = \mathbf{r}_u \cdot \mathbf{t} , \tag{6.30}$$
$$Fu' + Gv' = \mathbf{r}_v \cdot \mathbf{t} , \tag{6.31}$$

where E, F, G are the first fundamental form coefficients given in (3.12). Thus,

$$u' = \frac{(\mathbf{r}_u \cdot \mathbf{t})G - (\mathbf{r}_v \cdot \mathbf{t})F}{EG - F^2} , \qquad v' = \frac{(\mathbf{r}_v \cdot \mathbf{t})E - (\mathbf{r}_u \cdot \mathbf{t})F}{EG - F^2} , \tag{6.32}$$

where $EG - F^2 \neq 0$, since we are assuming regular surfaces (see (6.16)). Similarly we can compute the normal curvature of the implicit surface by using (6.21). The projection of curvature vector $\mathbf{c}'' = (x'', y'', z'')$ onto the unit normal vector $\frac{\nabla f}{|\nabla f|}$ of the surface, from (6.21), is given by

$$\kappa_n = \frac{f_x x'' + f_y y'' + f_z z''}{\sqrt{f_x^2 + f_y^2 + f_z^2}} \tag{6.33}$$

$$= -\frac{f_{xx}(x')^2 + f_{yy}(y')^2 + f_{zz}(z')^2 + 2(f_{xy}x'y' + f_{yz}y'z' + f_{xz}x'z')}{\sqrt{f_x^2 + f_y^2 + f_z^2}} ,$$

where x', y', z' are the three components of $\mathbf{t}$ given by (6.23).

Consequently, the curvature of the intersection curve $\mathbf{c}$ at P can be calculated using (6.3), (6.27) and (6.28) as follows:

$$\kappa = \sqrt{\mathbf{k} \cdot \mathbf{k}} = \frac{1}{|\sin\theta|}\sqrt{(\kappa_n^A)^2 + (\kappa_n^B)^2 - 2\kappa_n^A \kappa_n^B \cos\theta} . \tag{6.34}$$

6.3.3 Torsion and third order derivative vector

Since $\mathbf{N}^A$ and $\mathbf{N}^B$ lie on the normal plane, the terms $\kappa'\mathbf{n} + \kappa\tau\mathbf{b}$ in (6.5) can be replaced by $\gamma\mathbf{N}^A + \delta\mathbf{N}^B$. Thus

$$\mathbf{c}'''(s) = -\kappa^2\mathbf{t} + \gamma\mathbf{N}^A + \delta\mathbf{N}^B . \tag{6.35}$$

Now, if we project $\mathbf{c}'''(s)$ onto the unit surface normal vector $\mathbf{N}$ at P and denote by λ_n, we have

$$\lambda_n^A = \gamma + \delta\cos\theta ,$$
$$\lambda_n^B = \gamma\cos\theta + \delta . \tag{6.36}$$

Solving the linear system for γ and δ, and substituting them into (6.35) yields

$$\mathbf{c}''' = -\kappa^2\mathbf{t} + \frac{\lambda_n^A - \lambda_n^B\cos\theta}{\sin^2\theta}\mathbf{N}^A + \frac{\lambda_n^B - \lambda_n^A\cos\theta}{\sin^2\theta}\mathbf{N}^B . \tag{6.37}$$

Similar to the curvature vector case in Sect. 6.3.2, we need to provide λ_n^A and λ_n^B to evaluate $\mathbf{c}'''$. For a parametric surface, λ_n can be obtained by projecting $\mathbf{c}'''$, which is the third order derivative of the intersection curve as a curve on a parametric surface, i.e. (6.19), onto the unit surface normal vector $\mathbf{N}$, resulting in

$$\lambda_n = \mathbf{c}''' \cdot \mathbf{N} = 3[Lu'u'' + M(u''v' + u'v'') + Nv'v''] + III , \tag{6.38}$$

where

$$III = \mathbf{r}_{uuu} \cdot \mathbf{N}(u')^3 + 3\mathbf{r}_{uuv} \cdot \mathbf{N}(u')^2v' + 3\mathbf{r}_{uvv} \cdot \mathbf{N}u'(v')^2 + \mathbf{r}_{vvv} \cdot \mathbf{N}(v')^3 , \tag{6.39}$$

and u'' and v'' in (6.38) are evaluated by taking the dot product on the both sides of (6.18) with $\mathbf{r}_u$ and $\mathbf{r}_v$. Noting that $\mathbf{c}'' = \mathbf{k}$ leads to a linear system

$$Eu'' + Fv'' = \mathbf{k} \cdot \mathbf{r}_u - \frac{E_u}{2}(u')^2 - E_v u' v' - \left(F_v - \frac{G_u}{2}\right)(v')^2 , \quad (6.40)$$

$$Fu'' + Gv'' = \mathbf{k} \cdot \mathbf{r}_v - \left(F_u - \frac{E_v}{2}\right)(u')^2 - G_u u' v' - \frac{G_v}{2}(v')^2 , \quad (6.41)$$

which can be solved for u'' and v''.

For an implicit surface, the projection of $\mathbf{c}''' = (x''', y''', z''')$ onto the unit normal vector of the surface $\frac{\nabla f}{|\nabla f|}$ can be obtained from (6.22) as

$$\lambda_n = \frac{f_x x''' + f_y y''' + f_z z'''}{\sqrt{f_x^2 + f_y^2 + f_z^2}} = -\frac{F_1 + F_2 + F_3}{\sqrt{f_x^2 + f_y^2 + f_z^2}} , \quad (6.42)$$

where

$$F_1 = f_{xxx}(x')^3 + f_{yyy}(y')^3 + f_{zzz}(z')^3 , \quad (6.43)$$

$$F_2 = 3(f_{xxy}(x')^2 y' + f_{xxz}(x')^2 z' + f_{xyy}x'(y')^2 + f_{yyz}(y')^2 z' + f_{xzz}x'(z')^2$$
$$+ f_{yzz}y'(z')^2 + 2f_{xyz}x'y'z') , \quad (6.44)$$

$$F_3 = 3(f_{xx}x'x'' + f_{yy}y'y'' + f_{zz}z'z'' + f_{xy}(x''y' + x'y'') + f_{yz}(y''z' + y'z''))$$
$$+ f_{xz}(x''z' + x'z'')) , \quad (6.45)$$

and (x'', y'', z'') are given by (6.28).

Finally, the torsion can be obtained from (6.6) and (6.37) as follows

$$\tau = \frac{1}{\kappa \sin^2 \theta} \{ [\lambda_n^A - \lambda_n^B \cos \theta](\mathbf{b} \cdot \mathbf{N}^A) + [\lambda_n^B - \lambda_n^A \cos \theta](\mathbf{b} \cdot \mathbf{N}^B) \} , \quad (6.46)$$

where the binormal vector and curvature are evaluated by (2.40) and (6.34).

6.3.4 Higher order derivative vector

The algorithm introduced in Sect. 6.3.3 to compute the third order derivative vector of the intersection curve can be generalized to compute the higher order derivative vectors $\mathbf{c}^{(m)}$ $(m \geq 4)$, under the assumption that we have evaluated $\mathbf{c}^{(j)}$ for $1 \leq j \leq m - 1$, $u^{(j-1)}$ and $v^{(j-1)}$ for $2 \leq j \leq m - 1$. The algorithm is as follows:

1. Evaluate the m-th $(m \geq 4)$ order derivative vector $\mathbf{c}^{(m)}$ by successively differentiating (6.5). At each differentiation step replace $\mathbf{t}'$, $\mathbf{n}'$, and $\mathbf{b}'$ by $\kappa \mathbf{n}$, $-\kappa \mathbf{t} + \tau \mathbf{b}$, and $-\tau \mathbf{n}$ using the Frenet-Serret formulae (2.56), which leads to the equation

$$\mathbf{c}^{(m)} = c_t \mathbf{t} + c_n \mathbf{n} + c_b \mathbf{b} , \quad (6.47)$$

where c_t, c_b and c_n are the coefficients that depend exclusively on κ and τ and their derivatives (see (6.10), (6.13) for reference). As we will see in step 2, it is not necessary to evaluate c_n and c_b. The coefficient c_t consists of κ, τ and their derivatives of order up to $m - 3$ ($m \geq 4$) and $m - 5$ ($m \geq 6$), respectively, which have already been evaluated in the earlier stages of the computation. For example κ', τ', κ'', τ'' can be obtained by taking the dot product between the curve derivative vectors with $\mathbf{n}$ or $\mathbf{b}$, thus from (6.5), (6.10) and (6.13):

$$\kappa' = \mathbf{c}''' \cdot \mathbf{n} \,, \tag{6.48}$$

$$\tau' = (\mathbf{c}^{(4)} \cdot \mathbf{b} - 2\kappa'\tau)/\kappa \,, \tag{6.49}$$

$$\kappa'' = \mathbf{c}^{(4)} \cdot \mathbf{n} + \kappa^3 + \kappa\tau^2 \,, \tag{6.50}$$

$$\tau'' = (\mathbf{c}^{(5)} \cdot \mathbf{b} + \kappa^3\tau + \kappa\tau^3 - 3\kappa''\tau - 3\kappa'\tau')/\kappa \,. \tag{6.51}$$

2. Replace the terms $c_n\mathbf{n} + c_b\mathbf{b}$ in (6.47) by $\gamma\mathbf{N}^A + \delta\mathbf{N}^B$ since they both lie on the normal plane:

$$\mathbf{c}^{(m)} = c_t\mathbf{t} + \gamma\mathbf{N}^A + \delta\mathbf{N}^B \,. \tag{6.52}$$

3. Evaluate λ_n by projecting $\mathbf{c}^{(m)}$, which is the m-th derivative of the intersection curve evaluated as a curve on a surface, onto the unit surface normal.
   - *Parametric surface*: Differentiate (6.19) with respect to the arc length using the chain rule to evaluate $\mathbf{c}^{(m)}$ as a curve on the parametric surface. To compute $\mathbf{c}^{(m)}$ we also need to obtain $u^{(m-1)}$ and $v^{(m-1)}$, which can be done by taking the dot product on the both sides of $\mathbf{c}^{(m-1)}$ with $\mathbf{r}_u$ and $\mathbf{r}_v$ and solving the linear system. Project $\mathbf{c}^{(m)}$ onto two surface normal to obtain λ_n.
   - *Implicit surface*: Differentiate (6.22) with respect to the arc length successively. The resulting expression always involves the terms of the form $f_x x^{(m)} + f_y y^{(m)} + f_z z^{(m)}$, which is the projection of $\mathbf{c}^{(m)}$ onto the surface normal ∇f. Thus, by moving the rest of the terms to the right hand side we have

$$\lambda_n = \frac{f_x x^{(m)} + f_y y^{(m)} + f_z z^{(m)}}{\sqrt{f_x^2 + f_y^2 + f_z^2}} \tag{6.53}$$

$$= -\frac{\frac{\partial^m f}{\partial x^m}(x')^m + \frac{\partial^m f}{\partial y^m}(y')^m + \frac{\partial^m f}{\partial z^m}(z')^m + \cdots}{\sqrt{f_x^2 + f_y^2 + f_z^2}} \,.$$

The numerator involves terms $x^{(j)}$, $y^{(j)}$, $z^{(j)}$, for $1 \leq j \leq m - 1$ (not explicitly expressed here), are obtained by the components of $\mathbf{c}^{(j)}$.

4. Project (6.52) onto both the unit surface normal vectors yielding

$$\lambda_n^A = \gamma + \delta\cos\theta, \qquad \lambda_n^B = \gamma\cos\theta + \delta \,, \tag{6.54}$$

where $\cos\theta = \mathbf{N}^A \cdot \mathbf{N}^B$.

5. Substitute λ_n^A and λ_n^B, obtained from Step 3, into (6.54), and solve the linear system for γ and δ and substitute into (6.52), resulting in

$$\mathbf{c}^{(m)} = c_t \mathbf{t} + \frac{\lambda_n^A - \lambda_n^B \cos\theta}{\sin^2\theta} \mathbf{N}^A + \frac{\lambda_n^B - \lambda_n^A \cos\theta}{\sin^2\theta} \mathbf{N}^B . \qquad (6.55)$$

6.4 Intersection curve at tangential intersection points

Now, let us assume that the two surfaces A and B intersect tangentially at a point P on the intersection curve $\mathbf{c}(s)$, i.e. $\mathbf{N}^A \parallel \mathbf{N}^B$ at P. By orienting the surfaces appropriately we can assume that $\mathbf{N}^A = \mathbf{N}^B \equiv \mathbf{N}$ (see Fig. 6.2). In this case, (6.23) is invalid. Therefore, we have to find new methods to compute the differential geometry properties of $\mathbf{c}(s)$. In the following Sect. 6.4.1 we also classify these tangential contact points P in several categories.

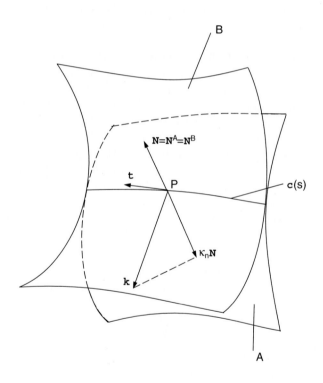

Fig. 6.2. Tangential intersection of two surfaces (adapted from [458])

6.4.1 Tangential direction

The unit tangential vector $\mathbf{t}$ of $\mathbf{c}(s)$ at P must lie on the common tangent plane of A and B. Therefore, $\mathbf{t}$ can be represented as a linear combination of $\mathbf{r}_{u_A}^A$ and $\mathbf{r}_{v_A}^A$, as well as $\mathbf{r}_{u_B}^B$ and $\mathbf{r}_{v_B}^B$, as in (6.17), i.e.

$$\mathbf{t} = \mathbf{r}_{u_A}^A u_A' + \mathbf{r}_{v_A}^A v_A' = \mathbf{r}_{u_B}^B u_B' + \mathbf{r}_{v_B}^B v_B' \ . \tag{6.56}$$

Equation (6.56) consists of two linear equations with four unknowns (u_A', v_A', u_B', v_B'), since the tangent vector is constrained in the tangent plane and does not have a normal component. Since $\mathbf{N}^A = \mathbf{N}^B = \mathbf{N}$ at P, we find that $\kappa_n^A = \kappa_n^B$ from (6.25). Thus, from (6.29) we have

$$L^A(u_A')^2 + 2M^A u_A' v_A' + N^A(v_A')^2 = L^B(u_B')^2 + 2M^B u_B' v_B' + N^B(v_B')^2 \ . \tag{6.57}$$

This equation is a quadratic equation in (u_A', v_A', u_B', v_B'). Thus together with the unit length constraint of the tangent vector, (6.56) and (6.57) form a system of four nonlinear equations in four unknowns. This nonlinear system can be solved by representing u_B' and v_B' in terms of linear combinations of u_A' and v_A' from (6.56), and then substituting the results into (6.57). By taking the cross product of both sides of (6.56) with $\mathbf{r}_{u_B}^B$ and $\mathbf{r}_{v_B}^B$, and projecting the resulting equations onto the common surface normal vector $\mathbf{N}$ at P, u_B' and v_B' can be represented as the following linear combinations of u_A' and v_A'

$$u_B' = a_{11} u_A' + a_{12} v_A' \ , \tag{6.58}$$
$$v_B' = a_{21} u_A' + a_{22} v_A' \ , \tag{6.59}$$

where

$$a_{11} = \frac{(\mathbf{r}_{u_A}^A \times \mathbf{r}_{v_B}^B) \cdot \mathbf{N}}{(\mathbf{r}_{u_B}^B \times \mathbf{r}_{v_B}^B) \cdot \mathbf{N}} = \frac{det(\mathbf{r}_{u_A}^A, \mathbf{r}_{v_B}^B, \mathbf{N})}{\sqrt{E^B G^B - (F^B)^2}} \ , \tag{6.60}$$

$$a_{12} = \frac{(\mathbf{r}_{v_A}^A \times \mathbf{r}_{v_B}^B) \cdot \mathbf{N}}{(\mathbf{r}_{u_B}^B \times \mathbf{r}_{v_B}^B) \cdot \mathbf{N}} = \frac{det(\mathbf{r}_{v_A}^A, \mathbf{r}_{v_B}^B, \mathbf{N})}{\sqrt{E^B G^B - (F^B)^2}} \ , \tag{6.61}$$

$$a_{21} = \frac{(\mathbf{r}_{u_B}^B \times \mathbf{r}_{u_A}^A) \cdot \mathbf{N}}{(\mathbf{r}_{u_B}^B \times \mathbf{r}_{v_B}^B) \cdot \mathbf{N}} = \frac{det(\mathbf{r}_{u_B}^B, \mathbf{r}_{u_A}^A, \mathbf{N})}{\sqrt{E^B G^B - (F^B)^2}} \ , \tag{6.62}$$

$$a_{22} = \frac{(\mathbf{r}_{u_B}^B \times \mathbf{r}_{v_A}^A) \cdot \mathbf{N}}{(\mathbf{r}_{u_B}^B \times \mathbf{r}_{v_B}^B) \cdot \mathbf{N}} = \frac{det(\mathbf{r}_{u_B}^B, \mathbf{r}_{v_A}^A, \mathbf{N})}{\sqrt{E^B G^B - (F^B)^2}} \ . \tag{6.63}$$

Substituting (6.58) and (6.59) into (6.57), we have

$$b_{11}(u_A')^2 + 2b_{12}(u_A')(v_A') + b_{22}(v_A')^2 = 0 \ , \tag{6.64}$$

where

$$b_{11} = a_{11}^2 L^B + 2a_{11} a_{21} M^B + a_{21}^2 N^B - L^A \ , \tag{6.65}$$
$$b_{12} = a_{11} a_{12} L^B + (a_{11} a_{22} + a_{21} a_{12}) M^B + a_{21} a_{22} N^B - M^A \ ,$$
$$b_{22} = a_{12}^2 L^B + 2a_{12} a_{22} M^B + a_{22}^2 N^B - N^A \ .$$

If we denote $\omega = \frac{u'_A}{v'_A}$ when $b_{11} \neq 0$ or $\mu = \frac{v'_A}{u'_A}$ when $b_{11} = 0$ and $b_{22} \neq 0$, and solve (6.64) for ω or μ, then $\mathbf{t}$ can be computed as

$$\mathbf{t} = \frac{\omega \mathbf{r}^A_{u_A} + \mathbf{r}^A_{v_A}}{|\omega \mathbf{r}^A_{u_A} + \mathbf{r}^A_{v_A}|}, \tag{6.66}$$

or

$$\mathbf{t} = \frac{\mathbf{r}^A_{u_A} + \mu \mathbf{r}^A_{v_A}}{|\mathbf{r}^A_{u_A} + \mu \mathbf{r}^A_{v_A}|}. \tag{6.67}$$

There are four distinct cases to the solution of (6.64) depending upon the discriminant $b_{12}^2 - b_{11} b_{22}$:

1. *Isolated tangential contact point:* If $b_{12}^2 - b_{11} b_{22} < 0$ then (6.64) does not have any real solution. Thus, P is an isolated contact point of A and B.
2. *Tangential intersection curve:* If $b_{12}^2 - b_{11} b_{22} = 0$ and $b_{11}^2 + b_{12}^2 + b_{22}^2 \neq 0$ then (6.64) has a double root and $\mathbf{t}$ is unique. Thus, A and B intersect at P and at its neighborhood.
3. *Branch Point:* If $b_{12}^2 - b_{11} b_{22} > 0$ then (6.64) has distinct roots. Thus, P is a branch point of the intersection curve $\mathbf{c}(s)$, i.e. there is another intersection branch crossing $\mathbf{c}(s)$ at P.
4. *Higher order contact point:* If $b_{11} = b_{12} = b_{22} = 0$ then (6.64) vanishes for any values of u'_A and v'_A. Thus, A and B has a contact of at least second order (i.e., curvature continuous) at P. In related work by Pegna and Wolter [305], they developed mathematical criteria for curvature continuity between two surfaces. Those criteria were later generalized to arbitrary higher order continuity (contact) in [161].

When P is a flat point of one of the surfaces, say $\mathbf{r}^B$, then L^B, M^B, N^B all vanish, however we can still evaluate (6.64). When P is a flat point of both surfaces, then the two surfaces have a contact of order at least 2 at P which is addressed under case 4.

There is a geometric interpretation to the tangent direction $\mathbf{t}$ at P. Recall that the Dupin's indicatrix of a surface at point P is a conic section (see Sect. 3.6). Since A and B intersect tangentially at P, they have the same tangent-plane at P. Equation (6.57) indicates that along $\mathbf{t}$, the Dupin's indicatrices of A and B at P intersect. Conversely, $\mathbf{t}$ is the vector(s) on the common tangent-plane at P along which the Dupin's indicatrices of A and B intersect. The two Dupin's indicatrices may not intersect at all (isolated tangential contact point), or intersect at two points tangentially, or intersect transversally at four points (branch point), or overlap (higher order contact point). In the case of overlap, they must be the same at P, and A and B are at least curvature continuous at P. Figure 6.3 shows the possible combinations of Dupin's indicatrices of two surfaces for four distinct cases. Although the coordinate system of the two indicatrices are chosen to be the same for simplicity, in general they may have different orientations. At hyperbolic points the Dupin's indicatrix is a set of conjugate hyperbolas depending on which side of the tangent plane the normal section is locally lying. However, for simplicity we have only illustrated the cases for one of the conjugate hyperbolas. The Dupin's indicatrices of case 2 upper right in Fig. 6.3 are parallel to each other and do not intersect. This is the case when two surfaces A and B intersect tangentially at a parabolic point P where they have the same principal directions. We assume without loss of generality that $u-$ and $v-$ parameter curves are in the directions of the principal directions, where $u = constant$ being the principal direction with zero curvature. With these assumptions, we have $M^A = M^B = N^A = N^B = 0$ and $a_{12} = a_{21} = 0$ thus (6.64) reduces to

$$(a_{11}^2 L^B - L^A)(u_A')^2 = 0 , \tag{6.68}$$

Therefore it has a double root with unique direction ($u_A = constant$) for $\mathbf{t}$, provided that $a_{11}^2 L^B - L^A \neq 0$. When $a_{11}^2 L^B - L^A = 0$, the two surfaces are at least curvature continuous at P, and their Dupin's indicatrices overlap.

Implicit-implicit and parametric-implicit intersection cases can be handled in a similar way. For the implicit-implicit intersection case we first equate the normal curvatures of the two implicit surfaces A and B using (6.34), where the unknowns are the unit tangent vector (x', y', z'). We can eliminate one of the component, say z', using (6.20) yielding the quadratic equation in x' and y' similar to (6.64).

For the parametric-implicit intersection case we equate the normal curvatures (6.29) and (6.34) of the parametric and implicit surfaces where the unknowns are x', y', z', u' and v'. We can replace x', y', z' in terms of u' and v' using (6.17), which leads us to a quadratic equation similar to (6.64). Upon solving the quadratic equation and applying the unit length constraint, we obtain the unit tangent vector.

There are also four distinct cases for implicit-implicit and parametric-implicit intersections, depending on the discriminant of the quadratic equation.

6.4.2 Curvature and curvature vector

The curvature vector $\mathbf{k}$ (see (6.2)) of the intersection curve $\mathbf{c}(s)$ at P can be expressed as in (6.18) as follows:

$$\begin{aligned}
\mathbf{c}''(s) &= \mathbf{r}_{u_A u_A}^A (u_A')^2 + 2\mathbf{r}_{u_A v_A}^A u_A' v_A' + \mathbf{r}_{v_A v_A}^A (v_A')^2 + \mathbf{r}_{u_A}^A u_A'' + \mathbf{r}_{v_A}^A v_A'' \\
&= \mathbf{r}_{u_B u_B}^B (u_B')^2 + 2\mathbf{r}_{u_B v_B}^B u_B' v_B' + \mathbf{r}_{v_B v_B}^B (v_B')^2 + \mathbf{r}_{u_B}^B u_B'' + \mathbf{r}_{v_B}^B v_B'' .
\end{aligned} \tag{6.69}$$

To obtain the curvature vector $\mathbf{k}=\mathbf{c}''(s)$, we need to determine the coefficients (u_A'', v_A'') and (u_B'', v_B''). Equation (6.69) introduces two constraints on the four unknowns, since the normal components of both sides of (6.69) are the same (see (6.57)). This can be seen clearly if we rewrite (6.69) as follows

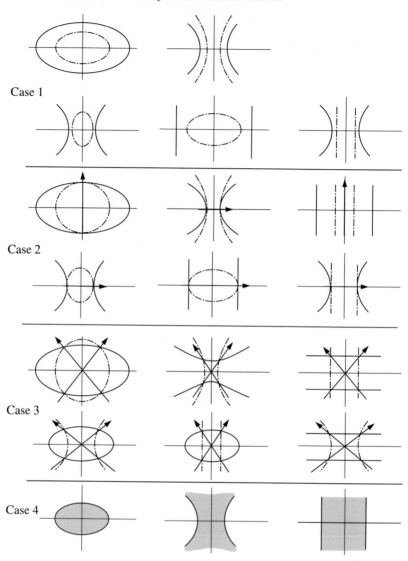

Fig. 6.3. Dupin's indicatrices of two tangentially intersecting surfaces (adapted from [458])

$$\mathbf{r}^A_{u_A} u''_A + \mathbf{r}^A_{v_A} v''_A = \mathbf{r}^B_{u_B} u''_B + \mathbf{r}^B_{v_B} v''_B + \boldsymbol{\Lambda} \,, \tag{6.70}$$

where

$$\boldsymbol{\Lambda} = \mathbf{r}^B_{u_B u_B}(u'_B)^2 + 2\mathbf{r}^B_{u_B v_B} u'_B v'_B + \mathbf{r}^B_{v_B v_B}(v'_B)^2 - \mathbf{r}^A_{u_A u_A}(u'_A)^2 \tag{6.71}$$

$$-2\mathbf{r}^A_{u_A v_A} u'_A v'_A - \mathbf{r}^A_{v_A v_A}(v'_A)^2 \ .$$

From (6.70), (u''_B, v''_B) can be expressed by (u''_A, v''_A) as follows:

$$u''_B = a_{11}u''_A + a_{12}v''_A + a_{13} \ , \tag{6.72}$$
$$v''_B = a_{21}u''_A + a_{22}v''_A + a_{23} \ , \tag{6.73}$$

where a_{11}, a_{12}, a_{21} and a_{22} are coefficients defined in (6.60) through (6.63), a_{13} and a_{23} are coefficients defined as follows:

$$a_{13} = \frac{(\boldsymbol{\Lambda} \times \mathbf{r}^B_{v_B}) \cdot \mathbf{N}}{(\mathbf{r}^B_{u_B} \times \mathbf{r}^B_{v_B}) \cdot \mathbf{N}} = \frac{det(\boldsymbol{\Lambda}, \mathbf{r}^B_{v_B}, \mathbf{N})}{\sqrt{E^B G^B - (F^B)^2}} \ , \tag{6.74}$$

$$a_{23} = \frac{(\mathbf{r}^B_{u_B} \times \boldsymbol{\Lambda}) \cdot \mathbf{N}}{(\mathbf{r}^B_{u_B} \times \mathbf{r}^B_{v_B}) \cdot \mathbf{N}} = \frac{det(\mathbf{r}^B_{u_B}, \boldsymbol{\Lambda}, \mathbf{N})}{\sqrt{E^B G^B - (F^B)^2}} \ . \tag{6.75}$$

We still need two more equations to solve for (u''_A, v''_A). One additional equation can be obtained by differentiating $\mathbf{c}(s)$ from $\mathbf{c}(s) = \mathbf{r}^A(u_A(s), v_A(s)) = \mathbf{r}^B(u_B(s), v_B(s))$ at P three times (see (6.19)) and projecting the resulting vector equation onto the normal vector $\mathbf{N}$, i.e. $\lambda^A_n = \lambda^B_n$ (see (6.38), (6.39)):

$$3[L^A u'_A u''_A + M^A(u''_A v'_A + u'_A v''_A) + N^A v'_A v''_A] + III^A \tag{6.76}$$
$$= 3[L^B u'_B u''_B + M^B(u''_B v'_B + u'_B v''_B) + N^B v'_B v''_B] + III^B \ .$$

Another additional equation can be obtained from the fact that the curvature vector $\mathbf{k}$ is perpendicular to the tangent vector $\mathbf{t}$, i.e.

$$\mathbf{c}'' \cdot \mathbf{t} = (\mathbf{r}_u \cdot \mathbf{t})u''_A + (\mathbf{r}_v \cdot \mathbf{t})v''_A + (\mathbf{r}_{uu} \cdot \mathbf{t})(u'_A)^2 + 2(\mathbf{r}_{uv} \cdot \mathbf{t})u'_A v'_A + (\mathbf{r}_{vv} \cdot \mathbf{t})(v'_A)^2 = 0 \ . \tag{6.77}$$

Upon substituting (6.72) and (6.73) into (6.76) we can solve the linear system (6.76) and (6.77) for u''_A and v''_A, and hence, the curvature vector $\mathbf{k}$ can be computed from (6.69), and the curvature κ follows immediately from (6.34).

The curvature vector of the implicit-implicit intersection case, as well as the parametric-implicit case, can be obtained by a similar procedure. We need to evaluate (x'', y'', z'') for the implicit-implicit case by solving a linear system of three equations. The first linear equation in (x'', y'', z'') is derived using (6.21). The second equation is given by equating the projection of the third derivative onto the unit surface normal vector, i.e. (6.42). Finally, the third equation is obtained from the fact that the curvature vector is perpendicular to the tangent vector, i.e. $x'x'' + y'y'' + z'z'' = 0$.

The parametric-implicit case can be obtained by solving a linear system of two equations in u'' and v''. The first linear equation in (u'', v'') is derived by equating the projection of the third derivative vector of $\mathbf{c}$ onto the unit normal vector, i.e. (6.38) and (6.42). The first and second derivatives (x', y', z') and (x'', y'', z'') appear in (6.43) - (6.45) are replaced in terms of u', v', u'' and v'' using (6.17) and (6.18). The second equation is given by (6.77).

6.4.3 Third and higher order derivative vector

The third and higher order derivative vector $m \geq 3$ can be obtained in a manner similar to the curvature vector case. We assume that $\mathbf{c}^{(j)}$ for $1 \leq j \leq m-1$, and $u^{(j-1)}$ and $v^{(j-1)}$ for $2 \leq j \leq m-1$ are already evaluated. The algorithm is given as follows:

Parametric-parametric.

1. Differentiate $\mathbf{c}(s) = \mathbf{r}^A(u_A(s), v_A(s)) = \mathbf{r}^B(u_B(s), v_B(s))$ m times, from which we can express $u_B^{(m)}$ and $v_B^{(m)}$ as linear combinations of $u_A^{(m)}$ and $v_A^{(m)}$ (see (6.72), (6.73) for $m = 2$).
2. Differentiate $\mathbf{c}(s) = \mathbf{r}^A(u_A(s), v_A(s)) = \mathbf{r}^B(u_B(s), v_B(s))$ $m+1$ times and project the resulting vectors onto the normal vector $\mathbf{N}$, from which we obtain a linear equation in $u_A^{(m)}, v_A^{(m)}, u_B^{(m)}, v_B^{(m)}$ (see (6.76) for $m = 2$). Substitute $u_B^{(m)}$ and $v_B^{(m)}$, which are obtained from Step 1, into the resulting equation.
3. Another additional linear equation is obtained from $\mathbf{c}^{(m)} \cdot \mathbf{t} = c_t$, where $\mathbf{c}^{(m)}$ is the m-th order derivative of $S^A(u_A(s), v_A(s))$ and c_t is defined in (6.47) and depends exclusively on κ and τ and their derivatives (see (6.76) for $m = 2$).
4. Solve the linear system for $(u_A^{(m)}, v_A^{(m)})$ and substitute them into the expression of $\mathbf{c}^{(m)}(s)$ in Step 1.

Implicit-implicit.

1. Total differentiate $f(x, y, z) = 0$ m times with respect to s, which will provide a linear equation in $(x^{(m)}, y^{(m)}, z^{(m)})$.
2. Equate the projections of the $(m+1)$-th order derivative $(x^{(m+1)}, y^{(m+1)}, z^{(m+1)})$ of the two implicit surfaces onto the unit normal vector (see (6.53)) to obtain a linear equation in $(x^{(m)}, y^{(m)}, z^{(m)})$.
3. The third linear equation in $(x^{(m)}, y^{(m)}, z^{(m)})$ can be obtained from $x'x^{(m)} + y'y^{(m)} + z'z^{(m)} = c_t$, where c_t, defined in (6.47), depends exclusively on κ and τ and their derivatives.
4. Solve the system of three linear equations for $(x^{(m)}, y^{(m)}, z^{(m)})$.

Parametric-implicit.

1. Equate the projection of the $(m+1)$-th order derivative vector of $\mathbf{c}$ with respect to the arc length of the parametric and implicit surfaces onto the unit normal vector, i.e. $\mathbf{r}^{(m+1)}(u(s), v(s)) \cdot \mathbf{N}$, and (6.53), to form a linear equation in $u^{(m)}$ and $v^{(m)}$, where $(x^{(j)}, y^{(j)}, z^{(j)})$, $1 \leq j \leq m$, are replaced by the components of $\mathbf{r}^{(j)}(u(s), v(s))$.
2. The second linear equation in $u^{(m)}$ and $v^{(m)}$ is obtained from $\mathbf{c}^{(m)} \cdot \mathbf{t} = c_t$, where $\mathbf{c}^{(m)} = \mathbf{r}^{(m)}(u(s), v(s))$ and c_t, which depends exlusively on κ and τ and their derivatives, is defined in (6.47).
3. Solve the linear system for $(u^{(m)}, v^{(m)})$ and substitute them into $\mathbf{r}^{(m)}(u(s), v(s))$.

6.5 Examples

For illustrative purposes we present two examples, one from transversal intersection of parametric-implicit surfaces and the other from tangential intersection of implicit-implicit surfaces.

6.5.1 Transversal intersection of parametric-implicit surfaces

In this example, the parametric surface A is a hyperbolic paraboloid given by

$$\mathbf{r} = \mathbf{r}(u, v) = (u, v, uv)^T, \quad 0.5 \leq u \leq 2 \ and \ 0 \leq v \leq 2, \quad (6.78)$$

and the implicit surface B is a cone given by

$$f(x, y, z) = xz - y^2 = 0.$$

Figure 6.4 shows the two intersecting surfaces with two intersection curves, one of which coincides with the x-axis. From (6.78) we have $\mathbf{r}_u = (1, 0, v)^T$,

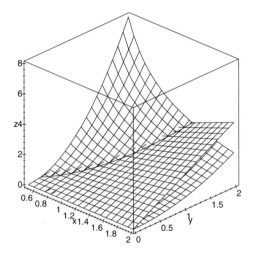

Fig. 6.4. Transversal intersection of parametric-implicit surfaces (adapted from [458])

$\mathbf{r}_v = (0, 1, u)^T$, $\mathbf{r}_{uu} = \mathbf{r}_{vv} = \mathbf{0}$, $\mathbf{r}_{uv} = (0, 0, 1)^T$ and the partial derivatives of order higher than two are all zero. The first and second fundamental form coefficients are readily given by $E = 1 + v^2$, $F = uv$, $G = 1 + u^2$, $L = N = 0$, $M = \frac{1}{\sqrt{u^2 + v^2 + 1}}$.

Similarly we have $f_x = z$, $f_y = -2y$, $f_z = x$, $f_{xx} = f_{xy} = f_{yz} = f_{zz} = 0$, $f_{yy} = -2$, $f_{xz} = 1$ and the partial derivatives of order higher than two are all zero.

The unit normal vectors of surfaces A and B and their dot product are given by

$$\mathbf{N}^A = \frac{(-v, -u, 1)^T}{\sqrt{u^2 + v^2 + 1}}, \quad \mathbf{N}^B = \frac{\nabla f}{|\nabla f|} = \frac{(z, -2y, x)^T}{\sqrt{x^2 + 4y^2 + z^2}},$$

$$\cos\theta = \frac{x + 2yu - zv}{\sqrt{u^2 + v^2 + 1}\sqrt{x^2 + 4y^2 + z^2}}.$$

Hence, the unit tangent vector of the intersection curve becomes

$$\mathbf{t} = (x', y', z')^T = \frac{(-xu + 2y, xv + z, 2yv + zu)^T}{\sqrt{(-xu + 2y)^2 + (xv + z)^2 + (2yv + zu)^2}}.$$

To evaluate the normal curvature of parametric surface A in the direction of $\mathbf{t}$, we start by computing (u', v') using (6.32) yielding

$$(u', v') = \frac{(-xu + 2y, xv + z)}{\sqrt{(-xu + 2y)^2 + (xv + z)^2 + (2yv + zu)^2}},$$

and hence

$$\kappa_n^A = \frac{2(-xu + 2y)(xv + z)}{\sqrt{u^2 + v^2 + 1}[(-xu + 2y)^2 + (xv + z)^2 + (2yv + zu)^2]}.$$

The normal curvature of the implicit surface in the direction $\mathbf{t}$ can be obtained from (6.34)

$$\kappa_n^B = \frac{2(y')^2 z - 2x'z'}{\sqrt{x^2 + 4y^2 + z^2}}.$$

By substituting $\mathbf{N}^A$, $\mathbf{N}^B$, $\cos\theta$, κ_n^A and κ_n^B into (6.28) we obtain the curvature vector, and hence the curvature κ.

The projection of the third order derivatives $\mathbf{c}'''(s)$ onto the unit surface normal vector can be computed using (6.38) and (6.42) for the parametric and implicit surfaces, respectively, yielding

$$\lambda_n^A = \frac{3(u'v'' + u''v')}{\sqrt{u^2 + v^2 + 1}}, \quad \lambda_n^B = \frac{-3(x'z'' + x''z' - 2y'y'')}{\sqrt{x^2 + 4y^2 + z^2}},$$

where (u'', v'') are obtained by solving the linear system (6.40) and (6.41)

$$u'' = \frac{(\mathbf{k} \cdot \mathbf{r}_u - 2vu'v')(1 + u^2) - (\mathbf{k} \cdot \mathbf{r}_v - 2uu'v')uv}{u^2 + v^2 + 1},$$

$$v'' = \frac{(\mathbf{k} \cdot \mathbf{r}_v - 2uu'v')(1 + v^2) - (\mathbf{k} \cdot \mathbf{r}_u - 2vu'v')uv}{u^2 + v^2 + 1}.$$

Knowing κ, $\mathbf{t}$, $\mathbf{N}^A$, $\mathbf{N}^B$, $\cos\theta$, λ_n^A, and λ_n^B, the third order derivative is obtained from (6.37) and the torsion from (6.46).

6.5.2 Tangential intersection of implicit-implicit surfaces

In this example, the two implicit surfaces are ellipsoids given by

$$f^A(x,y,z) = \frac{x^2}{0.6^2} + \frac{y^2}{0.8^2} + \frac{z^2}{1^2} - 1 = 0,$$

$$f^B(x,y,z) = \frac{x^2}{0.45^2} + \frac{y^2}{0.8^2} + \frac{z^2}{1.25^2} - 1 = 0.$$

The partial derivatives of the two implicit functions are readily computed as $f_x^A(x,y,z) = \frac{x}{0.18}$, $f_y^A = 3.125y$, $f_z^A = 2z$, $f_{xx}^A = \frac{1}{0.18}$, $f_{yy}^A = 3.125$, $f_{zz}^A = 2$, $f_{xy}^A = f_{yz}^A = f_{xz}^A = 0$, $f_x^B(x,y,z) = \frac{x}{0.10125}$, $f_y^B = 3.125y$, $f_z^B = 1.28z$, $f_{xx}^B = \frac{1}{0.10125}$, $f_{yy}^B = 3.125$, $f_{zz}^B = 1.28$, $f_{xy}^B = f_{yz}^B = f_{xz}^B = 0$. The partial derivatives higher than order two are all zero. It is clear that at the intersection point P $(0, 0.8, 0)$, ∇f^A and ∇f^B become parallel, thus P is a tangential intersection point (see Fig. 11.17). Now let us compute the tangential direction and curvature vector of the intersection curves at the tangential intersection point P. Since $f_y^A \neq 0$ at P, we can express y' in terms of x' and z'

$$y' = -\frac{f_x^A x' + f_z^A z'}{f_y^A}.$$

Furthermore y' reduces to zero, because $f_x^A = f_z^A = 0$ at P. At P the normal curvatures of both implicit surfaces in direction $\mathbf{t}$ are the same, thus from (6.34) we have

$$\frac{f_{xx}^A (x')^2 + f_{zz}^A (z')^2}{f_y^A} = \frac{f_{xx}^B (x')^2 + f_{zz}^B (z')^2}{f_y^B}.$$

This is a quadratic equation in x' and z' that can be solved for z' in terms of x'. By substituting $x = z = 0$, $y = 0.8$, we have $z' = \pm\frac{25}{27}\sqrt{7}x'$ and hence P is a branch point. Normalization gives the unit tangent vector $(x', y', z') = \left(\frac{27}{4\sqrt{319}}, 0, \pm\frac{25\sqrt{7}}{4\sqrt{319}}\right)$.

Next, we evaluate the curvature vector as described in Sect. 6.4.2. The first linear equation in (x'', y'', z'') is obtained from (6.21)

$$f_x^A x'' + f_y^A y'' + f_z^A z'' = -f_{xx}^A (x')^2 - f_{yy}^A (y')^2 - f_{zz}^A (z')^2. \tag{6.79}$$

The second equation is given by equating the projections of the third derivatives of the two implicit surfaces onto the normal vector of the surface

$$\frac{f_{xx}^A x' x'' + f_{yy}^A y' y'' + f_{zz}^A z' z''}{\sqrt{(f_x^A)^2 + (f_y^A)^2 + (f_z^A)^2}} = \frac{f_{xx}^B x' x'' + f_{yy}^B y' y'' + f_{zz}^B z' z''}{\sqrt{(f_x^B)^2 + (f_y^B)^2 + (f_z^B)^2}}. \tag{6.80}$$

The third equation is due to the fact that the curvature vector is perpendicular to the tangent vector, i.e.

$$x'x'' + y'y'' + z'z'' = 0 \,. \tag{6.81}$$

Solving the linear system (6.79), (6.80) and (6.81) at point P, we have $(x'', y'', z'') = (0, -\frac{320}{319}, 0)$, thus the curvature of the intersection curve at P is $\kappa = \frac{320}{319}$.

7. Distance Functions

7.1 Introduction

The computation of maximal and minimal distance between point sets in Euclidean space is a basic problem in computational geometry and geometric modeling. It is useful in surface intersections [210], numerical control machining, tolerance region and access space representation in solid modeling, robotics, inspection of manufactured objects [353, 297, 186, 3], and in feature recognition through the construction of medial axis transforms [298, 81, 450]. For this purpose, it is important to have computational methods which are efficient and reliable to compute extrema for the distance between two variable points where each of those variable points assumes all possible positions in a given set. In practical situations, this set can be a surface, a curve, or a single point.

In this chapter, we examine the computation of the stationary points of the squared distance function in five cases [461]:

1. Between a given point and a variable point on a 3-D space parametric curve.
2. Between a given point and a variable point on a parametric surface patch.
3. Between two variable points located on two given 3-D space parametric curves.
4. Between two variable points, one of which is located on a 3-D space parametric curve and the other is located on a parametric surface patch.
5. Between two variable points, located on two different given parametric surface patches.

We assume that the given curves and surfaces are rational polynomial parametric. However, the methods can be extended to the cases where the given curves and surfaces are represented by implicit polynomials using Lagrange multiplier methods (see Sects. 5.4.1 and 8.4). If the parametric curves and surfaces are in integral/rational B-spline form, the first step is to subdivide the integral/rational B-spline curve or surface patch into a number of integral/rational Bézier curves or patches. The problem thus becomes computing the distances between two point sets, which can be a space point, a integral/rational Bézier curve or a integral/rational Bézier surface patch. The squared distance functions expressed in Bernstein form are developed here for such point sets. This development is based on direct addition and multiplication of two Bernstein forms [106, 315] (see Sect. 1.3.2).

7.2 Problem formulation

7.2.1 Definition of the distances between two point sets

The definitions of the squared distance function for five cases described in Sect. 7.1 are given as follows:

1. The squared distance function between a point $\mathbf{p}_o = (x_o \ y_o \ z_o)^T$ and an arbitrary point on a parametric curve $\mathbf{r}(t)$ in three dimensional Cartesian space is defined by

$$D(t) = |\mathbf{p}_o - \mathbf{r}(t)|^2 = (\mathbf{p}_o - \mathbf{r}(t)) \cdot (\mathbf{p}_o - \mathbf{r}(t)) . \tag{7.1}$$

2. The squared distance function between a point $\mathbf{p}_o = (x_o \ y_o \ z_o)^T$ and a parametric surface patch $\mathbf{r} = \mathbf{r}(u,v)$ in three dimensional Cartesian space is defined by

$$D(u,v) = |\mathbf{p}_o - \mathbf{r}(u,v)|^2 = (\mathbf{p}_o - \mathbf{r}(u,v)) \cdot (\mathbf{p}_o - \mathbf{r}(u,v)) . \tag{7.2}$$

3. The squared distance function between two parametric curves $\mathbf{p} = \mathbf{p}(u)$ and $\mathbf{q} = \mathbf{q}(v)$ in three-dimensional Cartesian space is defined by

$$D(u,v) = |\mathbf{p}(u) - \mathbf{q}(v)|^2 = (\mathbf{p}(u) - \mathbf{q}(v)) \cdot (\mathbf{p}(u) - \mathbf{q}(v)) . \tag{7.3}$$

4. The squared distance function between a parametric curve $\mathbf{p} = \mathbf{p}(t)$ and a parametric surface patch $\mathbf{q} = \mathbf{q}(u,v)$ in three dimensional Cartesian space is defined by

$$D(t,u,v) = |\mathbf{p}(t) - \mathbf{q}(u,v)|^2 = (\mathbf{p}(t) - \mathbf{q}(u,v)) \cdot (\mathbf{p}(t) - \mathbf{q}(u,v)) . \tag{7.4}$$

5. The squared distance function between two parametric surface patches $\mathbf{p} = \mathbf{p}(\sigma,t)$ and $\mathbf{q} = \mathbf{q}(u,v)$ in three dimensional Cartesian space is defined by

$$D(\sigma,t,u,v) = |\mathbf{p}(\sigma,t) - \mathbf{q}(u,v)|^2 = (\mathbf{p}(\sigma,t) - \mathbf{q}(u,v)) \cdot (\mathbf{p}(\sigma,t) - \mathbf{q}(u,v)) . \tag{7.5}$$

The parameters σ, t, u, v in (7.1) - (7.5) satisfy the following inequalities $0 \leq \sigma, t, u, v \leq 1$. The general approach to locate local minima of the squared distance function is to search for zeros of the gradient vector field ∇D and then examine if at those zeros the squared distance function attains minima. For each of the five cases, the condition $\nabla D = \mathbf{0}$ becomes

1. The stationary points of the squared distance function between a point $\mathbf{p}_o = (x_o \; y_o \; z_o)^T$ and an arbitrary point on a parametric curve $\mathbf{r}(t)$ satisfy the following equation

$$D_t(t) = 0 , \qquad (7.6)$$

which can be rewritten using (7.1) as

$$(\mathbf{p}_o - \mathbf{r}(t)) \cdot \mathbf{r}_t(t) = 0 . \qquad (7.7)$$

2. The stationary points of the squared distance function between a point $\mathbf{p}_o = (x_o \; y_o \; z_o)^T$ and a parametric surface patch $\mathbf{r} = \mathbf{r}(u, v)$ satisfy the following two equations

$$D_u(u, v) = D_v(u, v) = 0 , \qquad (7.8)$$

which can be rewritten using (7.2) as

$$(\mathbf{p}_o - \mathbf{r}(u, v)) \cdot \mathbf{r}_u(u, v) = (\mathbf{p}_o - \mathbf{r}(u, v)) \cdot \mathbf{r}_v(u, v) = 0 . \qquad (7.9)$$

3. The stationary points of the squared distance function between two parametric curves $\mathbf{p} = \mathbf{p}(u)$ and $\mathbf{q} = \mathbf{q}(v)$ satisfy the following two equations

$$D_u(u, v) = D_v(u, v) = 0 , \qquad (7.10)$$

which can be rewritten using (7.3) as

$$(\mathbf{p}(u) - \mathbf{q}(v)) \cdot \mathbf{p}_u(v) = (\mathbf{p}(u) - \mathbf{q}(v)) \cdot \mathbf{q}_v(v) = 0 . \qquad (7.11)$$

4. The stationary points of the squared distance function between a parametric curve $\mathbf{p} = \mathbf{p}(t)$ and a parametric surface patch $\mathbf{q} = \mathbf{q}(u, v)$ satisfy the following three equations

$$D_t(t, u, v) = D_u(t, u, v) = D_v(t, u, v) = 0 , \qquad (7.12)$$

which can be rewritten using (7.4) as

$$\begin{aligned} (\mathbf{p}(t) - \mathbf{q}(u, v)) \cdot \mathbf{p}_t(t) &= (\mathbf{p}(t) - \mathbf{q}(u, v)) \cdot \mathbf{q}_u(u, v) \\ &= (\mathbf{p}(t) - \mathbf{q}(u, v)) \cdot \mathbf{q}_v(u, v) = 0 . \end{aligned} \qquad (7.13)$$

5. The stationary points of the squared distance function between two parametric surfaces $\mathbf{p} = \mathbf{p}(\sigma, t)$ and $\mathbf{q} = \mathbf{q}(u, v)$ satisfy the following four equations

$$D_\sigma(\sigma, t, u, v) = D_t(\sigma, t, u, v) = D_u(\sigma, t, u, v) = D_v(\sigma, t, u, v) = 0 , \qquad (7.14)$$

which can be rewritten using (7.5) as

$$\begin{aligned} (\mathbf{p}(\sigma, t) - \mathbf{q}(u, v)) \cdot \mathbf{p}_\sigma(\sigma, t) &= (\mathbf{p}(\sigma, t) - \mathbf{q}(u, v)) \cdot \mathbf{p}_t(\sigma, t) = 0 , \\ (\mathbf{p}(\sigma, t) - \mathbf{q}(u, v)) \cdot \mathbf{q}_u(u, v) &= (\mathbf{p}(\sigma, t) - \mathbf{q}(u, v)) \cdot \mathbf{q}_v(u, v) = 0 . \end{aligned}$$
$$(7.15)$$

When the parametric curves and surfaces are in rational form, the squared distance $D(\mathbf{u})$ between two point sets can be represented by:

$$D(\mathbf{u}) = \frac{P(\mathbf{u})}{Q(\mathbf{u})}, \qquad (7.16)$$

where $\mathbf{u}$ is the vector $(u_1 \ u_2 \ \dots \ u_n)^{\mathrm{T}}$. For the problems considered in this chapter, $n \in \{1, 2, 3, 4\}$. The gradient of $D(\mathbf{u})$ is given by

$$\nabla D = \frac{\nabla P(\mathbf{u})Q(\mathbf{u}) - P(\mathbf{u})\nabla Q(\mathbf{u})}{Q^2(\mathbf{u})}. \qquad (7.17)$$

If $\nabla D(\mathbf{u}) = \mathbf{0}$, the numerator of the above expression should be zero. Since the numerator in (7.17) is a polynomial vector field, our problem is reduced to finding the singular points of an n-dimensional polynomial vector field, i.e. the roots of a system of n nonlinear polynomial equations in n unknowns within a given n-dimensional box [392, 300]. The solution of this problem is discussed in Chap. 4.

Because the curves and surfaces involved in these distance problems are bounded, it is necessary to break up the distance computation problem into a number of subproblems which deal with interior and boundary points of the geometric objects separately. The major steps of the algorithm are outlined below:

1. Find the minimal and maximal distances in the interior domain of point sets.
2. Find the distances at four corner points of the surface if one point set is a surface.
3. Find the minimal and maximal distances along the four edges of the surface if one point set is a surface.
4. Find the distances at end points if one point set is a curve.
5. Compare the distances to get the absolute minimum and maximum.

7.2.2 Geometric interpretation of stationarity of distance function

By examining the derivation of the equations in the preceding section, we can interpret the stationary point condition $\nabla D = \mathbf{0}$ in terms of the concept of *collinear normal points*. This idea is used in [376, 209, 210] to detect closed loops in surface-surface intersections (see (5.100)). Simply stated, two points on two surfaces are said to be collinear normal points if their associated normal vectors lie on the same line.

It is possible to interpret the conditions in (7.15) geometrically. Notice that the line joining $\mathbf{p}(\sigma, t)$ and $\mathbf{q}(u, v)$ must be orthogonal to the two partial derivative vectors on $\mathbf{p}$ at (σ, t) and to the two partial derivative vectors on $\mathbf{q}$ at (u, v). As long as these derivatives do not degenerate, they span the tangent planes to the two surfaces. Therefore the normals to these two surfaces must lie on the line joining the two points, and $\mathbf{p}(\sigma, t)$ and $\mathbf{q}(u, v)$ are collinear normal points as long as (7.15) are satisfied. The difference between (5.100) and (7.15) is that when two surfaces transversally intersect, i.e. $\mathbf{p}(\sigma, t) = \mathbf{q}(u, v)$ and $(\mathbf{p}_\sigma \times \mathbf{p}_t) \times (\mathbf{q}_u \times \mathbf{q}_v) \neq \mathbf{0}$ at the points of intersection, the first set of equations of (5.100) are not satisfied, while (7.15) are automatically satisfied.

Similar geometrical interpretations may be made for other distance problems. In the point-surface distance problem, (7.9) states that for a point (u_0, v_0) to be a stationary point, both partial derivatives to the surface at that point must be orthogonal to the line joining $\mathbf{p}_0$ and $\mathbf{r}(u_0, v_0)$. If these derivatives do not degenerate, an equivalent statement is that the normal to the surface at (u, v) is collinear with the line joining the point and the surface. In other words point $\mathbf{r}(u_0, v_0)$ is an orthogonal projection of $\mathbf{p}_0$ onto a surface $\mathbf{r}(u, v)$. Pegna and Wolter [306] derived a set of differential equations to orthogonally project a space curve onto parametric surfaces as well as implicit surfaces and solved them efficiently as an initial value problem. These methods were also applied in [3] for inspection of sculptured surfaces.

7.3 More about stationary points

If it is only necessary to determine the absolute maximum or absolute minimum of the squared distance function between two geometric objects, then simply computing the maximum or minimum of the set of distances at the stationary points is sufficient, provided that this set is finite. However, it is often necessary to classify each stationary point as a local maximum, local minimum, or saddle point. Furthermore, the set of stationary points may not be finite (for example, consider the distance function between two concentric circles) and it may be required to trace out these infinite point sets. In this section we examine these questions.

7.3.1 Classification of stationary points

Let us first recall the definitions of local extrema at stationary points:

Definition 7.3.1. *Suppose that* $D : \mathbf{R}^n \to \mathbf{R}$ *is a scalar field on* $\mathbf{R}^n$. *Let* $\mathbf{u}$ *be a stationary point of* D, *that is* $\nabla D(\mathbf{u}) = \mathbf{0}$. *Then*

1. $\mathbf{u}$ *is a local maximum if there exists a neighborhood* U *of* $\mathbf{u}$ *such that for all* $\mathbf{v} \in U$, $D(\mathbf{v}) \leq D(\mathbf{u})$.
2. $\mathbf{u}$ *is a local minimum if there exists a neighborhood* U *of* $\mathbf{u}$ *such that for all* $\mathbf{v} \in U$, $D(\mathbf{v}) \geq D(\mathbf{u})$.

To illustrate the application of these definitions, let us begin with a simple one-parameter function $D(u)$ (which may arise, for example, in the point-curve distance problem). Suppose u_0 is a stationary point of $D(u)$, i.e. $\dot{D}(u_0) = 0$. Then a Taylor series expansion tells us that

$$D(u_0 + h) - D(u_0) = h\dot{D}(u_0) + \frac{1}{2}h^2\ddot{D}(u_0 + ht)$$

$$= \frac{1}{2}h^2\ddot{D}(u_0 + ht), \qquad (7.18)$$

where $t \in [0,1]$ is some specific value depending on h. From Definition 7.3.1, we see that u_0 is a local maximum if the right hand side of (7.18) is never positive for h close to 0 and a local minimum if the right hand side is nonnegative for h close to 0. Now, if $\ddot{D}(u_0) \neq 0$, then by continuity, there is some δ such that for $|h| < \delta$, $\ddot{D}(u_0)$ has the same sign as $\ddot{D}(u_0 + ht)$. Therefore, if $\ddot{D}(u_0) < 0$, then $h^2\ddot{D}(u_0 + ht) < 0$ for $|h| < \delta$ and hence u_0 is a local maximum. Similarly, if $\ddot{D}(u_0) > 0$, then u_0 is a local minimum. These results are familiar from calculus [166].

Unfortunately, the problem becomes more complicated if $\ddot{D}(u_0) = 0$. In this case we can not say anything about the sign of $\ddot{D}(u_0 + ht)$ and must therefore expand the Taylor series to higher derivatives. For example, if the second derivative is zero but the third derivative is nonzero, then we will have neither a maximum nor a minimum but a *point of inflection*. Consider the function $y = x^3$; in any neighborhood of the stationary point $x = 0$, the function takes on both positive and negative values and thus $x = 0$ is neither a maximum nor a minimum. If the third derivative is also zero, we have to look at the fourth derivative, and so on. Fortunately, these exceptions are rare, but a robust point classification algorithm should be able to handle such cases. In general, if $D^{(n)}(a)$ is the first derivative function that does not vanish, then the function $D(u)$ has [152]

- if n is odd; neither maximum nor minimum.
- if n is even;
  - if $D^{(n)}(a) < 0$; maximum,
  - if $D^{(n)}(a) > 0$; minimum.

Similar analysis can be performed with functions of more than one variable. Consider a function $D(u,v)$ of two variables, which might arise, for example, in a curve-curve distance problem. A Taylor expansion around the stationary point (u_0, v_0) to second order leads to

$$D(u_0 + h, v_0 + k) - D(u_0, v_0) = \frac{h^2}{2}D_{uu}(u_0 + ht, v_0 + kt)$$

$$+ hkD_{uv}(u_0 + ht, v_0 + kt)$$

$$+ \frac{k^2}{2}D_{vv}(u_0 + ht, v_0 + kt), \quad (7.19)$$

where $t \in [0,1]$. Instead of one second order term, this time we have three to deal with, which of course makes classification more difficult. However, by simply *completing the square* with the second order terms, we can easily formulate conditions for minima, maxima, and saddle points. To see this, let $a \equiv \frac{D_{uu}}{2}, b \equiv \frac{D_{uv}}{2}$, and $c \equiv \frac{D_{vv}}{2}$. Then we have

$$D(u_0 + h, v_0 + k) - D(u_0, v_0) = ah^2 + 2bhk + ck^2$$

$$= a(h^2 + \frac{2b}{a}hk + \frac{b^2}{a^2}k^2) + (c - \frac{b^2}{a})k^2$$

$$= a(h + \frac{b}{a}k)^2 + (c - \frac{b^2}{a})k^2 . \qquad (7.20)$$

Now h and k appear only within squared expressions. This new form enables us to ignore the signs of h and k and instead concentrate on the signs of the coefficients a and $(c - \frac{b^2}{a})$. To see this, notice that within any arbitrarily small neighborhood of (u_0, v_0), the ratio $\frac{h}{k}$ takes on all possible values between $-\infty$ and ∞, and therefore the first term of the right hand side of (7.20) can become extremely small or extremely large compared to the second term. However, the signs of each term are not so easily changed. These are determined solely by the signs of a and $(c - \frac{b^2}{a})$. If both coefficients are nonzero at (u_0, v_0), a continuity argument shows that the expressions $D_{uu}(u_0 + ht, v_0 + kt)$ and $D_{vv}(u_0 + ht, v_0 + kt) - \frac{D_{uv}^2(u_0+ht,v_0+kt)}{D_{uu}(u_0+ht,v_0+kt)}$ do not change sign as long as $(u_0 + ht, v_0 + kt)$ is sufficiently close to (u_0, v_0). Therefore, in order to determine the signs of these coefficients within a small neighborhood of (u_0, v_0), we can simply evaluate them at (u_0, v_0). If the coefficients are nonzero there, then their signs will enable us to classify the stationary point. For example, if both coefficients are positive at the stationary point, then clearly the right hand side of (7.20) is positive within some neighborhood of the stationary point, and hence (u_0, v_0) can be classified as a minimum. If the first coefficient is negative and the second is positive, then because either term may be made zero at different parts of any neighborhood, the right hand side of (7.20) may be positive or negative arbitrarily close to the stationary point, and thus (u_0, v_0) is neither a maximum nor a minimum. The other possibilities lead to the following theorem, which is proven more fully in [166]:

Theorem 7.3.1.

1. *If $D_{uu} < 0$ and $D_{uv}^2 < D_{uu}D_{vv}$ at the stationary point (u_0, v_0), then (u_0, v_0) is a local maximum.*
2. *If $D_{uu} > 0$ and $D_{uv}^2 < D_{uu}D_{vv}$, then (u_0, v_0) is a local minimum.*
3. *If $D_{uv}^2 > D_{uu}D_{vv}$ then (u_0, v_0) is a saddle point (neither a maximum nor a minimum).*
4. *If none of the above conditions apply, then it is necessary to examine higher-order derivatives.*

Notice that the third condition above applies even if $D_{uu} = 0$. In this case, the second order term reduces to $2bhk + ck^2$. As long as $b \neq 0$, specific choices of h and k within any arbitrarily small neighborhood of the stationary point can make this term either positive or negative, and hence (u_0, v_0) is neither a maximum nor a minimum.

It is possible to complete the square for functions of more than two variables; however, in order to develop general conditions for minima and maxima of such functions, we will employ the powerful theory of *quadratic forms* [410].

Definition 7.3.2. *A* quadratic form *is an expression of the form* $\mathbf{u}^T A \mathbf{u}$, *where* $\mathbf{u} = (u_1\ u_2\ \ldots\ u_n)^T$ *is an n-dimensional column vector of unknowns and A is an $n \times n$ matrix. A quadratic form $\mathbf{u}^T A \mathbf{u}$ is said to be*

1. positive definite *if $\mathbf{u}^T A \mathbf{u} > 0$ for all $\mathbf{u} \neq 0$.*
2. positive semidefinite *if $\mathbf{u}^T A \mathbf{u} \geq 0$ for all $\mathbf{u}$.*
3. negative definite (semidefinite) *if $-\mathbf{u}^T A \mathbf{u}$ is positive definite (semidefinite).*
4. indefinite *otherwise (i.e. the form takes on both positive and negative values).*

To see how these quadratic forms are used, let us take a Taylor expansion of a function of n variables $D(\mathbf{u})$ around the stationary point $\mathbf{u} = \mathbf{u}_0$:

$$D(\mathbf{u}_0 + \mathbf{h}) - D(\mathbf{u}_0) = \frac{1}{2}\mathbf{h}^T H \mathbf{h} + O(|\mathbf{h}|^3) \,. \tag{7.21}$$

Here H is the *Hessian matrix* of D; the element of the ith row, jth column, is given by

$$H_{ij} = \frac{\partial^2 D}{\partial u_i \partial u_j}(\mathbf{u}_0) \,. \tag{7.22}$$

Now clearly, if the quadratic form $\mathbf{h}^T H \mathbf{h}$ is positive definite, then within some neighborhood of the stationary point $\mathbf{u}_0$, the right hand side of (7.21) is nonnegative, and therefore $\mathbf{u}_0$ is a local minimum. Similarly, if the quadratic form is negative definite, then $\mathbf{u}_0$ is a local maximum.

At this point, we can use a familiar theorem of linear algebra whose proof is given in [410]:

Theorem 7.3.2. *Let $\mathbf{u}^T A \mathbf{u}$ be a quadratic form, let A be a real symmetric matrix, and for $i = 1, 2, \ldots, n$, let d_i be the determinant of the upper left $i \times i$ submatrix of A (which we will denote A_i). Then the quadratic form is positive definite if and only if $d_i > 0$ for all i.*

If the $>$ condition in the above theorem is relaxed to $\geq$, a corresponding theorem applies to positive semidefinite forms (that is, a form is positive semidefinite if and only if $d_i \geq 0$ for all i).

Theorem 7.3.2 immediately gives us a necessary and sufficient condition for negative definiteness as well. For, if we let $B \equiv -A$, the quadratic form $\mathbf{u}^T A \mathbf{u}$ is negative definite if and only if $\mathbf{u}^T B \mathbf{u}$ is positive definite if and only if $\det B_i > 0$ for all i. But $\det B_i = (-1)^i \det A_i$ and so we have the following corollary:

Corollary 7.3.1. *Let A and d_i be as before. Then the quadratic form $\mathbf{u}^T A \mathbf{u}$ is negative definite if and only if $(-1)^i d_i > 0$ for all i.*

Simply applying these conditions gives us the following theorem:

Theorem 7.3.3. *Let D be as before, and let H be its Hessian matrix, evaluated at the stationary point $\mathbf{u}_0$. Let d_i be the determinant of the upper left $i \times i$ submatrix H_i of H.*

1. *If $d_i > 0$ for all i, then $\mathbf{u}_0$ is a local minimum.*
2. *If $(-1)^i d_i > 0$ for all i, then $\mathbf{u}_0$ is a local maximum.*

If, on the other hand, the form $\mathbf{h}^T H \mathbf{h}$ is *indefinite*, we can conclude that $\mathbf{u}_0$ is neither a minimum nor a maximum (in two dimensions, such points are usually called *saddle points*), as the following theorem shows:

Theorem 7.3.4. *Let $H, D, \mathbf{u}_0$, and d_i be as before. Suppose that the d_i do not satisfy the conditions for definiteness or semidefiniteness. Then $\mathbf{u}_0$ is neither a local minimum nor a local maximum.*

Proof: Because $\mathbf{h}^T H \mathbf{h}$ is indefinite, there exist vectors $\mathbf{h}_1$ and $\mathbf{h}_2$ such that $\mathbf{h}_1^T H \mathbf{h}_1 < 0$ and $\mathbf{h}_2^T H \mathbf{h}_2 > 0$. Pick an arbitrary neighborhood U of $\mathbf{u}_0$. Then for any $\varepsilon > 0$ smaller than some positive number δ, $\mathbf{u}_0 + \varepsilon \mathbf{h}_1 \in U, \mathbf{u}_0 + \varepsilon \mathbf{h}_2 \in U, (\varepsilon \mathbf{h}_1)^T H (\varepsilon \mathbf{h}_1) < 0$, and $(\varepsilon \mathbf{h}_2)^T H (\varepsilon \mathbf{h}_2) > 0$.

Using (7.21), we obtain

$$D(\mathbf{u}_0 + \varepsilon \mathbf{h}_1) - D(\mathbf{u}_0) = \frac{1}{2}\varepsilon^2 \mathbf{h}_1^T H \mathbf{h}_1 + O(\varepsilon^3 |\mathbf{h}_1|^3), \qquad (7.23)$$

$$D(\mathbf{u}_0 + \varepsilon \mathbf{h}_2) - D(\mathbf{u}_0) = \frac{1}{2}\varepsilon^2 \mathbf{h}_2^T H \mathbf{h}_2 + O(\varepsilon^3 |\mathbf{h}_2|^3). \qquad (7.24)$$

Clearly, for sufficiently small ε, the right hand side of (7.23) will remain negative, while the right hand side of (7.24) will remain positive. Thus we have, for sufficiently small ε,

$$D(\mathbf{u}_0 + \varepsilon \mathbf{h}_1) < D(\mathbf{u}_0) < D(\mathbf{u}_0 + \varepsilon \mathbf{h}_2). \qquad (7.25)$$

Since U was an arbitrary neighborhood of $\mathbf{u}_0$, there is no neighborhood of $\mathbf{u}_0$ where $D(\mathbf{u}_0) \geq D(\mathbf{u})$ for all $u \in U$ or where $D(\mathbf{u}_0) \leq D(\mathbf{u})$ for all $u \in U$. Therefore $\mathbf{u}_0$ is neither a local minimum nor a local maximum. ∎

Unfortunately, trouble can occur if the Hessian is only semidefinite and not definite. In this case we need to look at higher order derivatives or use some other method to classify the point precisely. Although this is not too difficult for one-parameter functions, it becomes progressively more involved as the number of variables increases (see for example, [323, 129]).

7.3.2 Nonisolated stationary points

Because the equation $\nabla D = \mathbf{0}$ is equivalent to a system of n equations in n unknowns, we expect that in most cases the solution set of this system will consist of a few discrete, isolated points in $\mathbf{R}^n$. However, it is possible for the solution set to contain curves, surfaces, or hypersurfaces as well as points. For example, suppose $D(x, y) = x^2 y^2$; then the solution set of $\nabla D = \mathbf{0}$ consists of the two lines $x = 0$ and $y = 0$.

In this section, we will examine a marching method helpful in tracing curves of critical points. This type of degeneracy can occur if, for example, we are trying to find the minimum of the squared distance between two surfaces which happen to intersect. The method we use to trace out such curves involves setting up a system of differential equations which can be solved numerically using a standard ordinary differential equation system solver (see Sect. 5.8.1).

We begin our discussion by defining the concept of a *degenerate critical point* [323]:

Definition 7.3.3. *A critical point* $\mathbf{u}_0$ *of a function* $D : \mathbf{R}^n \to \mathbf{R}$ *(that is, a point at which* $\nabla D = \mathbf{0}$*) is called* degenerate *if the Hessian matrix of* D *evaluated at* $\mathbf{u}_0$ *is singular.*

The following well-known theorem of differential geometry relates the concepts of degeneracy and isolation.

Theorem 7.3.5. *Suppose a critical point* $\mathbf{u}_0$ *of* D *is nondegenerate, i.e. its Hessian is nonsingular. Then there exists a neighborhood* U *of* $\mathbf{u}_0$ *such that for all* $\mathbf{u} \in U - \{\mathbf{u}_0\}$, $\nabla D(\mathbf{u}) \neq \mathbf{0}$. *(A point* $\mathbf{u}_0$ *having this property is called an* isolated critical point.*)*

Proofs of this theorem are given in [323, 129]. Unfortunately, the converse is not true; all nondegenerate critical points are isolated, but some isolated critical points are degenerate. For example, the function $D(x, y) = x^3 - 3xy^2$ has an isolated critical point at the origin, but its Hessian matrix there is the zero matrix and therefore singular. Nevertheless, this theorem is practically helpful in eliminating most cases.

We will use this theorem to help us detect curve branches of nonisolated stationary points as follows:

1. Run the IPP algorithm described in Chap. 4 on our initial box of search, with a fairly coarse level of accuracy. For example, if we start with the search box $[0, 1]^n$, we may run our root-finding algorithm with a tolerance of 10^{-2} or 10^{-3}.
2. Check the remaining bounding boxes after this step. If we observe a number of boxes adjacent to one another, there may be a curve in the solution set.

3. Use Newton-Raphson iteration to find roots within these boxes to a high accuracy. Check the Hessian at these points; if it is singular, it is very likely that a curve exists. Accordingly, choose one of these roots as a starting point for our tracing technique.

This approach is by no means infallible; it is theoretically possible for there to be a number of degenerate isolated roots that will fool us into thinking that there is a curve involved. It is also possible that the solution set is a surface or even a higher dimensional entity. However, in practice these exceptions rarely occur.

Let us now assume that we have found a point on this curve, which we will call $\mathbf{r} : \mathbf{R} \to \mathbf{R}^n$. We will assume that $\mathbf{r}$ is a function of the parameter t, which takes on a value 0 at our starting point and that $\mathbf{r}$ has unit speed everywhere (i.e. $|\mathbf{r}(t)| = 1$ everywhere). This unit speed condition is imposed to ensure that $\mathbf{r}$ has an arc length parameterization.

Rewrite the equation $\nabla D = \mathbf{0}$ as a system of n equations in n unknowns:

$$D_{u_1}(u_1, \ldots, u_n) = 0 ,$$
$$\cdots$$
$$D_{u_n}(u_1, \ldots, u_n) = 0 . \tag{7.26}$$

Now, in order to stay on the curve $\mathbf{r}$ in moving a small distance from $\mathbf{r}(t)$ to $\mathbf{r}(t + dt)$, we need to know what increments have to be added to $u_1, \ldots, u_n$. Accordingly, we take a Taylor expansion of each equation around the critical point $\mathbf{u}_0$ to first order:

$$D_{u_1 u_1}(\mathbf{u}_0)du_1 + \ldots D_{u_1 u_n}(\mathbf{u}_0)du_n = 0 ,$$
$$\cdots$$
$$D_{u_n u_1}(\mathbf{u}_0)du_1 + \ldots D_{u_n u_n}(\mathbf{u}_0)du_n = 0 . \tag{7.27}$$

Or, rewriting this as a matrix equation, we have

$$H\mathbf{du} = \mathbf{0} , \tag{7.28}$$

where $\mathbf{du} = (du_1 \ \ldots \ du_n)^{\mathrm{T}}$ and H is the Hessian of $D(\mathbf{u})$ evaluated at $\mathbf{u}_0$.

Because $\mathbf{u}_0$ is degenerate, the rank of H should be less than n. In fact, we anticipate that it will be $n-1$, since any $\mathbf{du}$ satisfying (7.28) must be a vector tangent to $\mathbf{r}$ at $\mathbf{u}_0$, and hence the nullspace of H should have dimension 1. (It is possible on rare occasions that there is a curve passing through $\mathbf{u}_0$ but the rank of the Hessian is less than $n - 1$. This problem may occur if the first-order Taylor expansion in (7.27) contains insufficient information due to zero derivatives. Fortunately, these cases are extremely rare in practice; see [323] for more information.)

Because we need to find a tangent vector to $\mathbf{r}$ at $\mathbf{u}_0$ with unit length, we need to find some vector in the nullspace of H and then normalize it to length 1. The Singular Value Decomposition method [410] may be applied to H to generate such a vector in a stable manner. Then, after making the vector unit length, we pass it to our ordinary differential equation solver as the tangent to $\mathbf{r}$ at $\mathbf{u}_0$. We can trace backwards by changing the sign of the tangent vector.

Table 7.1. Curve and surface description (adapted from [461])

Case	Property	Degree(s)	Property	Degree(s)
P/C			rational	10
P/S			rational	8 and 8
C/C	integral	2	rational	2
C/S	rational	2	rational	4 and 3
S/S	rational	1 and 2	rational	1 and 2

7.4 Examples

In this section, we give examples of distance computations for the five cases: a space point and a variable point on a 3D space curve (P/C); a space point and a variable point on a surface (P/S); two variable points located on two 3D space curves (C/C); two variable points, one of which is located on a space curve and the other is located on a surface (C/S); two variable points located on two surfaces (S/S). The curves and surfaces involved are expressed as rational Bézier entities whose property and degree are listed in Table 7.1.

The curve in the P/C example is a high degree rational curve which gives rise to a system with 4 roots as shown in Fig. 7.1. The surface in the P/S example is a high degree rational surface which also generates a system with 4 roots (see Fig. 7.2). In the C/C example, one curve is a parabola and the other is a quarter circle represented as a rational Bézier curve; the resulting system of equations has three roots (see Fig. 7.3). In the C/S example, the curve is a parabola and the surface is a saddle-like rational Bézier surface; there is only one root, which occurs at the intersection point of the curve and the surface as illustrated in Fig. 7.4. In the S/S example, the two surfaces are ruled rational surfaces with one root (see Fig. 7.5).

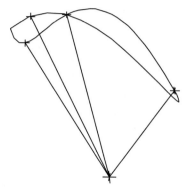

Fig. 7.1. Distances of a point and a high degree rational Bézier curve (adapted from [461])

Fig. 7.2. Distances of a point and a high order rational Bézier surface (adapted from [461])

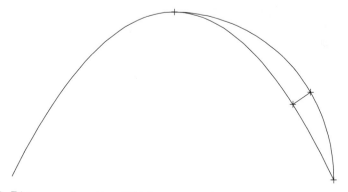

Fig. 7.3. Distances of a rational Bézier curve and an integral Bézier curve (adapted from [461])

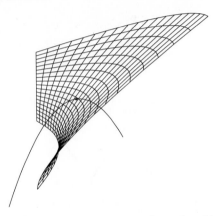

Fig. 7.4. Distances of a rational Bézier curve and a rational Bézier patch (adapted from [461])

Fig. 7.5. Distances of two linear-quadratic Bézier patches (adapted from [461])

8. Curve and Surface Interrogation

8.1 Classification of interrogation methods

Interrogation is the process of extraction of information from a geometric model. In this chapter we focus on free-form curve and surface interrogation. Free form surfaces, also called sculptured surfaces, are widely used in scientific and engineering applications. For example, the hydrodynamic shape of propeller blades has an important role in marine applications, and the aerodynamic shape of turbine blades determines the performance of aircraft engines. Free-form surfaces arise also in the bodies of the ships, automobiles and aircraft, which have both functionality and attractive shape requirements. Many electronic devices as well as consumer products are designed with aesthetic shapes, which involve free-form surfaces. During the last two decades, many curve and surface interrogation methods have been proposed (see references in [275, 61]), and we will introduce some of them in this chapter.

Propeller and turbine blades are manufactured by numerically controlled (NC) milling machines. When a ball-end mill cutter is used, the cutter radius must be smaller than the smallest concave radius of curvature of the surface to be machined to avoid *local overcut (gouging)* (see Sect. 11.1.2). Gouging is the one of the most critical problems in NC machining of free-form surfaces [184]. Therefore, we must determine the distribution of the principal curvatures of the surface, which are upper and lower bounds on the curvature at a given point, to select the cutter size [116, 94]. Visualization techniques of various curvature measures have been developed by Dill [74], Beck et al. [22], Munchmeyer [280, 279], Higashi and Kaneko [162], Pottmann and Optiz [328], Maekawa and Patrikalakis [255], Elber and Cohen [87] and Tuohy [423]. Higashi et al. [163] introduced the loci of points corresponding to extrema of curvature values of the design surface called *surface edges*, which show how the surface is waving and where the peaks of the wave exist. Kase et al. [189] presented local and global evaluation methods for shape errors of free-form surfaces which have been applied to the evaluation of sheet metal surfaces.

Developable surfaces [13, 222, 120, 32, 326, 252, 329] are surfaces which can be unfolded or developed onto a plane without stretching or tearing. They are of considerable importance to plate-metal-based industries as shipbuilding. For a developable surface the Gaussian curvature is zero everywhere [76]. Thus the manufacturer would profit from prior knowledge of the distribution of the Gaussian curvature of the metal plate. A line of curvature indicates a directional flow for the maximum or the minimum curvature across the surface [22, 279, 173, 257, 251, 98] which can be used for determining feeding directions to the rolling machine for the metal plate.

Fairing [365, 339, 144, 318, 148, 45, 317] is the process of eliminating shape irregularities in order to produce a smoother shape. *Reflection lines* [202, 190, 61] are a standard surface interrogation method to assess the fairness of design surfaces in the automobile industry. *Isophotes* are used for detection of surface irregularities [319, 144] and for continuity evaluation at the boundaries of adjacent patches [146]. Also *focal surfaces* [146, 145] are used to detect undesired curvature properties of a design surface. The set of curvature extrema of a fair surface should coincide with the designer's intention. Therefore, computation of all extrema of curvatures is desirable. The Gaussian, mean and principal curvatures are used for the detection of surface irregularities [280, 279, 255]. On the other hand Andersson [7] developed a method to specify curvature in a surface design method and Higashi et al. [164] proposed a method to generate a smooth surface by controlling the curvature distribution. Theisel and Farin [419] studied the curvature of characteristic curves on a surface, such as contour lines, lines of curvature, asymptotic lines, isophotes and reflection lines.

As we will see in this chapter, the governing equations for shape interrogation often result in n polynomial equations with n unknowns when the input curves and surfaces are in integral/rational Bézier form. If the parametric curves and surfaces are in integral/rational B-spline form, some shape interrogation methods involve a first step where the integral/rational B-spline curve or surface patch is subdivided into a number of integral/rational Bézier curves or rational Bézier patches. In the computer implementation we evaluate the coefficients of the governing nonlinear polynomial equations in multivariate Bernstein form starting from the given input Bézier curve or surface using the arithmetic operations in Bernstein form [106, 315] (see Sect. 1.3.2). The system of nonlinear polynomial equations can be solved robustly and accurately by the IPP algorithm presented in Chap. 4.

In this section we will classify the interrogation methods by the order of derivatives of the curve or surface position vector which are involved. An interrogation method is characterized as nth-order, if derivatives of the curve or surface position vector of order n are involved.

8.1.1 Zeroth-order interrogation methods

Wireframe. The wireframe of a surface patch is produced by displaying its boundary curves and a number of other iso-parametric curves in both parametric directions, as depicted in Fig. 8.1. Small-scale wireframes on raster screens provide only a rough idea of the underlying shape and are not very appropriate means to judge the nature of the surface and its smoothness and fairness. Furthermore, depending on the density of iso-parametric lines and

their actual shape, they may mislead the designer about the actual shape of
the surface.

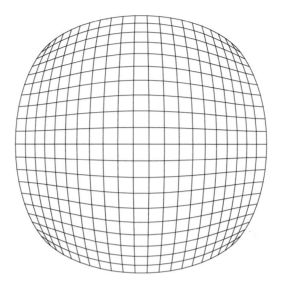

Fig. 8.1. Wireframe model of bicubic B-spline patch

Contouring. A better indication of the surface is normally obtained using
contour maps. There are different types of contour maps. The simplest contouring method is to intersect the surface with a family of user-defined non-
uniformly spaced parallel planes [22, 95] (see Fig. 5.1). Using $z = constant$
contour maps, maxima, minima and saddle points of the height $z(u, v)$ of
the surface can be identified. Contouring with planes is, for example, employed in ship hull and aircraft fuselage design. More complex contouring
methods involve intersections with a series of co-axial cylinders or cones of
non-uniformly spaced radii. Propeller and turbine blades are normally con-
toured using such methods. The intersection methods of Chap. 5 need to be
invoked to perform robust contouring.

8.1.2 First-order interrogation methods

Shading and ray tracing. A *shaded image* of a surface usually gives a more
realistic visual representation in comparison to a wireframe model [118]. A
simple illumination model, based on Lambert's cosine law and incorporating
specular reflection, gives the reflected intensity I as a function of the incident
intensity I_i from a point light source, the ambient intensity I_a, the diffuse

reflection constant k_d, the specular reflection constant k_s, the ambient diffuse reflection constant k_a, the angle θ between the unit surface normal $\mathbf{N}$ and a light direction vector, and of the angle α between a viewpoint direction vector and a vector in the direction of reflection:

$$I = I_a k_a + \frac{I_i}{r} (k_d \cos\theta + k_s \cos\alpha) , \tag{8.1}$$

for $0 \leq k_a, k_d, k_s \leq 1$ and $0 \leq \theta \leq \frac{\pi}{2}$ where r represents the distance from the perspective viewpoint to the point on the surface. More complicated shading models, which take into account the properties of the material, the angle of incidence, and the wavelength of the incidence light also exist [118].

The ray tracing technique [188, 118] gives more realistic images than a simple shaded image method, but is much slower. The intensity for each pixel is determined using a ray from the viewpoint through the pixel into the object. Ray tracing is in fact an intersection problem. The ray to surface intersection is computed for every ray and surface in a scene. Using the ray tracing technique, the hidden surface problem is solved during such computation and also other attributes, for example multiple reflections and shadowing, can be included in the model.

Isophotes. *Isophotes* are curves of constant light intensity on a surface, created by a point light source at infinity with direction $\mathbf{l}$ ($|\mathbf{l}| = 1$), specified by the user. These curves can be used for the detection of surface irregularities [319, 144]. An isophote is a curve for which the quantity

$$\mathbf{N}(u, v) \cdot \mathbf{l} = \cos\theta , \tag{8.2}$$

is constant and equal to c, for $0 \leq c \leq 1$ and $0 \leq \theta \leq 90°$ where $\mathbf{N}(u, v)$ is the unit surface normal vector. When the surface is locally planar (or flat) all normals are parallel and the isophotes do not generally exist. If the surface is C^M continuous then the isophote line will be C^{M-1} continuous. For rendering isophotes, the values of $\mathbf{N}(u, v) \cdot \mathbf{l}$ are computed on a lattice and a number of isophotes are generated by connecting points of equal value found by interpolation between straddling grid points [3]. Color Plate A.1 shows isophotes of the bicubic B-spline surface in Fig. 8.1 with $\mathbf{l} = \left(\frac{1}{2}, \frac{1}{2}, \frac{\sqrt{2}}{2} \right)^T$.

Reflection lines. *Reflection lines* are another first-order interrogation method used in the automotive industry to assess the fairness of a surface. Reflection lines simulate the mirror images of a number of parallel straight fluorescent lights on an automobile surface. In this method, deviations of the surface from a smooth shape can be detected by irregularities of the reflection lines. Originally a reflection line was defined as a reflected image of a linear light source on a surface by Klass [202]. Kaufmann and Klass [190] modified the above definition to reduce the computation as follows. A family of curves $\mathbf{q}_i(t)$, $i = 1 \ldots n$ on the surface, which are intersection curves of the surface with a specific family of planes parallel to a unit vector $\mathbf{v}$, are evaluated. For each intersection curve $\mathbf{q}_i(t)$, parameter t_i that satisfies

$$\frac{\dot{\mathbf{q}}_i(t)}{|\dot{\mathbf{q}}_i(t)|} \cdot \mathbf{v} = \cos\phi , \qquad (8.3)$$

is evaluated. Then points $\mathbf{q}_1(t_1)$, $\mathbf{q}_2(t_2)$, ..., $\mathbf{q}_n(t_n)$ are connected to form the reflection line. The procedure is repeated for different values of ϕ. If iso-parametric lines are used instead of intersection curves of the surface with a family of parallel planes, computational efficiency is further improved as in [3]. Choi and Lee [61] applied the Blinn-Newell type of reflection mapping [118], which uses simple and physically acceptable mapping algorithm, to generate reflection lines on a trimmed NURBS surface. Choi and Lee [61] also provide a thorough recent review of this topic.

Color Plate A.2 depicts reflection lines on the bicubic B-spline surface patch shown in Fig. 8.1, where values of $\frac{\dot{\mathbf{q}}(t)}{|\dot{\mathbf{q}}(t)|} \cdot \mathbf{v}$ with $\mathbf{v} = \left(\frac{1}{2}, \frac{1}{2}, \frac{\sqrt{2}}{2}\right)^T$ are evaluated along iso-parametirc curves and the equal value points found by interpolation between mesh points are connected to form the reflection lines.

Highlight lines. Beier and Chen [23] introduced the concept of a *highlight line* where a set of points on a surface are determined such that the distance between a linear light source and an extended surface normal at the highlight lines is zero (see Fig. 8.2). Let us denote the linear light source by

$$\mathbf{l}(t) = \mathbf{a} + \mathbf{b}t , \qquad (8.4)$$

where $\mathbf{a}$ is a point on the linear light source, $\mathbf{b}$ is a directional vector and t is a parameter. Also let us define the extended surface normal vector $\mathbf{e}$ at the surface point $\mathbf{q}$ by

$$\mathbf{e}(\tau) = \mathbf{q} + \mathbf{S}\tau , \qquad (8.5)$$

where $\mathbf{S}$ is the surface normal vector at $\mathbf{q}$ and τ is a parameter. The distance between the two lines $\mathbf{l}(t)$ and $\mathbf{e}(\tau)$ is given by

$$d = \frac{|(\mathbf{b} \times \mathbf{S}) \cdot (\mathbf{a} - \mathbf{q})|}{|\mathbf{b} \times \mathbf{S}|} . \qquad (8.6)$$

This distance d will vanish if point $\mathbf{q}$ is on the highlight line. If we avoid cases such that the linear light source and the surface normal become parallel, the denominator of (8.6) is nonzero and the governing equation for determining highlight lines reduces to

$$(\mathbf{b} \times \mathbf{S}(u,v)) \cdot (\mathbf{a} - \mathbf{q}(u,v)) = 0 . \qquad (8.7)$$

Equation (8.7) can be traced using the same technique that we introduced in Sect. 5.8.1. Sone and Chiyokura [401] developed a method to control a hightlight line directly using a NURBS boundary Gregory patch. Zhang and Cheng [460] studied a method to remove local irregularities of NURBS surface patches by modifying its highlight lines for real time interactive design.

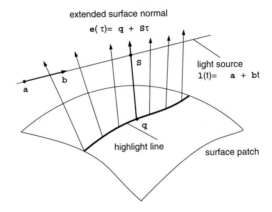

Fig. 8.2. Definition of highlight line (adapted from [23])

8.1.3 Second-order interrogation methods

Curvature plots. For a planar curve, an *inflection point* occurs at a point where the curvature changes sign (see Sect. 2.2). Note that a vanishing curvature does not necessarily imply inflections, e.g. consider the curve $y = x^4$ at $x = 0$. Using *curvature plots*, which consist of segments normal to the curve emanating from a number of points on the curve and whose lengths are proportional to the magnitude of the curvature given in (2.25) at the associated point, inflection points and the variation of curvature can be easily identified as illustrated in Fig. 8.3.

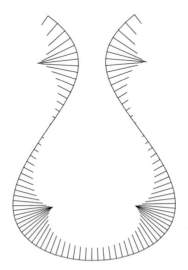

Fig. 8.3. Curvature plot of a planar curve with inflections

Zero curvature points. The signed curvature formula for a planar parametric curve is given in (2.25). Due to the regularity condition, a necessary condition to determine inflection points is

$$\dot{x}(t)\ddot{y}(t) - \dot{y}(t)\ddot{x}(t) = 0, \qquad t \in [t_1, t_2] \,. \tag{8.8}$$

An inflection point on a planar curve is shown in Fig. 8.7, marked by ×.

The curvature $\kappa(t)$ of a space curve is given in (2.26). The formula can be expressed as

$$\kappa(t) = \frac{\sqrt{(\dot{x}\ddot{y} - \dot{y}\ddot{x})^2 + (\dot{y}\ddot{z} - \dot{z}\ddot{y})^2 + (\dot{z}\ddot{x} - \dot{x}\ddot{z})^2}}{(\dot{x}^2 + \dot{y}^2 + \dot{z}^2)^{\frac{3}{2}}} \,. \tag{8.9}$$

Since we are assuming a regular curve, a condition to determine a point on a curve, where the curvature $\kappa(t)$ vanishes, is [57]

$$K_0(t) \equiv (\dot{x}\ddot{y} - \dot{y}\ddot{x})^2 + (\dot{y}\ddot{z} - \dot{z}\ddot{y})^2 + (\dot{z}\ddot{x} - \dot{x}\ddot{z})^2 = 0, \qquad t \in [t_1, t_2] \,, \tag{8.10}$$

or

$$\dot{x}\ddot{y} - \dot{y}\ddot{x} = \dot{y}\ddot{z} - \dot{z}\ddot{y} = \dot{z}\ddot{x} - \dot{x}\ddot{z} = 0, \qquad t \in [t_1, t_2] \,. \tag{8.11}$$

Curvature vanishing points on a space curve $\mathbf{r}(t)$ are shown in Fig. 8.8, marked by ×.

Radial curves. For a parametric space curve $\mathbf{r}(t)$, a *radial curve* is defined as

$$\mathbf{f}(t) = \mathbf{p} + \mathbf{n}(t)/\kappa(t) \,, \tag{8.12}$$

where $\mathbf{p}$ is a fixed reference point in space, $\mathbf{n}(t)$ is the unit principal normal vector and $\kappa(t)$ is a nonzero curvature of the curve [227]. Here the sign convention (a) (see Fig. 3.7 (a)) is adopted for the curvature. For sign convention (b) we simply replace the plus sign with a minus sign. Radial curves are a method for visualizing curvature in a manner decoupled from the shape of the curve as illustrated in Fig. 8.4. When an inflection point is involved, as in Fig. 8.4, spikes in opposite directions occur. By viewing the radial curve, a curvature measure is visualized. However, the user is left without a direct reference as to the relationship between the curve point and the curvature value. The radial curve also provides a method for viewing the range of variation of the curve's normal vector. This range of variation can be obtained from the angular sector described by all rays emanating from $\mathbf{p}$ and passing through all points on the radial curve. Radial curves are useful in accessibility and interference analyses.

Surface curvatures and curvature maps. Surface interrogation may be performed using different curvature measures of the surface [22, 280, 18, 92]. The Gaussian, mean, absolute [92], and root mean square (rms) [246] curvatures are the product (3.61), the average (3.62), the sum of the absolute values

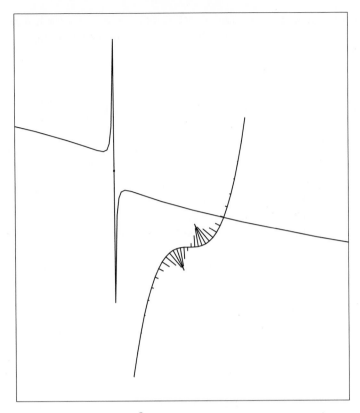

Fig. 8.4. A radial curve of $y = x^3$ with a fixed reference point $\mathbf{p} = (-2, 2, 0)$. The *thin curve* is the radial curve of the *thick curve* ($y = x^3$ with a curvature plot)

$$k_{abs} = \mid k_{max} \mid + \mid k_{min} \mid , \qquad (8.13)$$

and the square root of the sum of the squares of the principal curvatures

$$k_{rms} = \sqrt{k_{max}^2 + k_{min}^2} , \qquad (8.14)$$

respectively. These surface curvature functions are all scalar valued functions. The variation of any of these quantities can be displayed using a color-coded *curvature map* [74, 22]. Contour lines of constant curvature can also be used to display and visualize the variation of these curvature functions. Munchmeyer [280] calculates the curvature on a lattice and linearly interpolates the contour points. Maekawa and Patrikalakis [255] present a method which allows a robust and accurate computation of all stationary points and all contour lines for functions describing Gaussian, mean and principal curvatures of B-spline surfaces. This method allows us to divide the surface into regions of specific range of curvature. A summary with further applications can be found in [255].

A surface inflection exists on a surface at a point P if the surface crosses the tangent plane at P [172]. The point P is then called an *inflection point*. If the Gaussian curvature is positive at every point of a region of a surface then there are no inflections in that region. If the Gaussian curvature changes sign in a region of a surface then there is an inflection in that region. In areas where the surface has Gaussian curvature very close to or equal to zero the Gaussian curvature alone cannot provide adequate information about the shape of the surface. In such a case the surface has an inflection point in the region only if the mean curvature changes sign. The Gaussian and mean curvatures together provide sufficient information in order to identify surface inflections on a surface [279, 280].

The curvature maps of the principal curvatures can also help to select a spherical cutter of suitable radius in order to avoid gouging during machining of the surface [22, 255]. We will discuss further on how to construct contour lines of constant curvature in Sect. 8.5.

Focal curves and surfaces. For a parametric space curve $\mathbf{r}(t)$, a *focal curve* or an *evolute*, shown in Fig. 8.5, is defined as

$$\mathbf{f}(t) = \mathbf{r}(t) + \mathbf{n}(t)/\kappa(t) , \qquad (8.15)$$

where $\mathbf{n}(t)$ is a unit principal normal vector and $\kappa(t)$ is a nonzero curvature of the curve. Thus the focal curve or evolute of a curve is the locus of its centers of curvature. *Focal surfaces* [146, 145] can be constructed in a similar way by using the principal curvature functions of the given surface. For a given parametric surface patch $\mathbf{r}(u, v)$ the two associated focal surfaces are defined as

$$\mathbf{f}(u, v) = \mathbf{r}(u, v) + \mathbf{N}(u, v)/\kappa(u, v) , \qquad (8.16)$$

where $\mathbf{N}(u, v)$ is a unit surface normal and $\kappa(u, v)$ is a nonzero principal curvature ($\kappa_{min}(u, v)$ or $\kappa_{max}(u, v)$). Here sign convention (a) (see Fig. 3.7 (a)) is employed for both (8.15) and (8.16). When sign convention (b) is assumed, we simply replace the plus sign by the minus sign.

Focal curves and surfaces provide another method for visualizing curvature. An important application of focal curves is in testing the curvature continuity of surfaces across a common boundary. Curves on the focal surfaces of each surface, which are at the common boundary, $\mathbf{f}(t) = \mathbf{r}(u(t), v(t)) + \mathbf{N}(u(t), v(t))/\kappa(u(t), v(t))$, where $\mathbf{r} = \mathbf{r}(u(t), v(t))$ is the common boundary curve (or linkage curve) on the progenitor surface $\mathbf{r} = \mathbf{r}(u, v)$, can be compared to determine the curvature continuity of the surfaces. According to the *linkage curve theorem* by Pegna and Wolter [305] two surfaces joined with first-order or tangent plane continuity along a first-order continuous linkage curve can be shown to be second-order smooth on the linkage curve if the normal curvatures along the linkage curve on each surface agree in one direction other than the tangent direction to the linkage curve. Comparison of focal surfaces allows for a visual assessment of the accuracy for

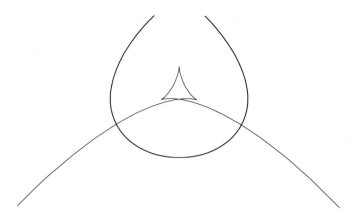

Fig. 8.5. Focal curve. The *thin curve* is the focal curve of the *thick curve*

the second-order contact of two patches along their linkage curve. Namely if the two focal curve point sets (defined via κ_{min} and κ_{max}) of one surface patch agree along the linkage curve with the two corresponding focal curve point sets of the adjacent patch then these surface patches have curvature continuous surface contact along the linkage curve. The deviation of the corresponding focal curves indicates the amount of discontinuity of curvature of both adjacent patches along the linkage curve.

Orthotomics. *Orthotomic curves and surfaces* are used to display the angle between the position vector of a curve (or surface) and the normal of the curve or surface respectively. Orthotomic curves and surfaces are useful for indicating the presence of inflection points [146]. A σ-orthotomic curve $\mathbf{y}_\sigma(t)$ of a planar curve $\mathbf{r}(t)$ with respect to a point $\mathbf{p}$, not on $\mathbf{r}(t)$ or any of its tangents, is defined as

$$\mathbf{y}_\sigma(t) = \mathbf{p} + \sigma[(\mathbf{r}(t) - \mathbf{p}) \cdot \mathbf{n}(t)]\mathbf{n}(t) , \qquad (8.17)$$

where $\mathbf{n}(t)$ is the unit normal vector of $\mathbf{r}(t)$ and σ is a scaling factor chosen for appropriate visualization. The tangent vector of an orthotomic curve is zero (and the orthotomic curve usually has a cusp-like singularity) at any parameter value of t at which the curve $\mathbf{r}(t)$ has an inflection point. An illustrative example is shown in Fig. 8.6.

A σ-orthotomic surface $\mathbf{y}_\sigma(u, v)$ of a surface $\mathbf{r}(u, v)$ with respect to a point $\mathbf{p}$, not on $\mathbf{r}(u, v)$ or any of its tangent planes, is defined as

$$\mathbf{y}_\sigma(u, v) = \mathbf{p} + \sigma[(\mathbf{r}(u, v) - \mathbf{p}) \cdot \mathbf{N}(u, v)]\mathbf{N}(u, v) , \qquad (8.18)$$

where $\mathbf{N}(u, v)$ is the unit normal vector of the surface $\mathbf{r}(u, v)$ and σ is a scaling factor. An orthotomic surface has a singularity, i.e. a degenerate tangent plane, at all values of u, v at which the Gaussian curvature of the surface $\mathbf{r}(u, v)$ vanishes or changes sign.

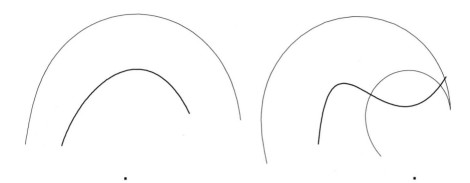

Fig. 8.6. Orthotomics of two curves (adapted from [3]). Note that the orthotomic on the right shows the inflection point of the curve. The *thin curves* are the orthotomics of the *thick curves*

Curvature lines. A *line of curvature* is a curve on a surface whose tangent at every point is aligned along a principal curvature direction. We have studied the basics of lines of curvature in Sect. 3.4 and will study them further in Chap. 9. The principal directions at a given point are those directions for which the normal curvature takes on minimum and maximum values. A line of curvature indicates a directional flow for the maximum or the minimum curvature across the surface [22]. Curvature lines provide some useful information about the surface. For example, when a plate is going to be shaped by rolling it is fed into the rolls using a principal direction and the forming rolls are adjusted according to the principal curvature [279]. The network of curvature lines may also be useful in idealization, meshing and structural analysis of shells with free-form surfaces as boundaries, because of their orthogonality property.

Geodesics. Geodesics are curves of zero geodesic curvature on a surface and provide candidates for arcs of minimum length between two points on a surface [412, 235, 136]. We will study the formulation of the governing equations of geodesics on a parametric surface as well as on an implicit surface and their solution methodology in Chap. 10.

8.1.4 Third-order interrogation methods

Torsion of space curves. As we discussed in Sect. 2.3 torsion describes the deviation of a space curve away from its osculating plane spanned by the curve's tangent and normal vectors. For planar curves, the torsion is always zero. For an arbitrary speed parametric space curve $\mathbf{r}(t)$, $t \in [t_1, t_2]$ the torsion at points with nonzero curvature is given by (2.48).

From (2.48), a condition for zero torsion at a nonzero curvature point is

$$T_0(t) \equiv \dddot{x}\,(\dot{y}\ddot{z} - \ddot{y}\dot{z}) + \dddot{y}\,(\dot{z}\ddot{x} - \ddot{z}\dot{x}) + \dddot{z}\,(\dot{x}\ddot{y} - \ddot{x}\dot{y}) = 0, \quad t \in [t_1, t_2] \,. \quad (8.19)$$

Figure 8.8 shows a point of zero torsion, marked by $\oplus$ at the midpoint of the curve. The sign of torsion has geometric significance [206]. If $\tau(t)$ changes its sign from $+(-)$ to $-(+)$ when passing a point its features change from a right-handed (left-handed) curve to a left-handed (right-handed) one, respectively.

Stationary points of curvature of planar and space curves. Modern CAD/CAM systems allow users to access specific application programs for performing several tasks, such as displaying objects on a graphic display, mass property calculation, finite element or boundary element meshing for analysis. These application programs often operate on piecewise linear approximation of the exact geometric definition. When a coarse approximation of good quality is required for 2-D and 3-D curves, stationary points of curvature play an important role in successful discretization [193, 97, 150, 57].

Using (2.25), the first derivative of the curvature function of a planar curve is given by

$$\dot{\kappa}(t) = \frac{(\dot{x}^2 + \dot{y}^2)(\dot{x}\,\dddot{y} - \dddot{x}\,\dot{y}) - 3(\dot{x}\ddot{x} + \dot{y}\ddot{y})(\dot{x}\ddot{y} - \ddot{x}\dot{y})}{(\dot{x}^2 + \dot{y}^2)^{\frac{5}{2}}} \,. \quad (8.20)$$

Since we are assuming a regular curve, the necessary condition to have a local maximum or minimum of curvature is given by

$$(\dot{x}^2 + \dot{y}^2)(\dot{x}\,\dddot{y} - \dddot{x}\,\dot{y}) - 3(\dot{x}\ddot{x} + \dot{y}\ddot{y})(\dot{x}\ddot{y} - \ddot{x}\dot{y}) = 0 \,. \quad (8.21)$$

Figure 8.7 shows the corresponding stationary points of curvature on $\mathbf{r}(t)$, labeled by $\circ$. Furthermore, comparing (8.8) and (8.21) we see both $\kappa(t)$ and $\dot{\kappa}(t)$ of a regular planar curve $\mathbf{r}(t)$ vanish for some t if and only if $\dot{x}\ddot{y} = \ddot{x}\dot{y}$ and $\dot{x}\,\dddot{y} = \dddot{x}\,\dot{y}$, simultaneously [57].

Similarly for a regular space curve, the necessary condition to have stationary points of curvature $\kappa(t)$ at nonzero curvature points is given by

$$K_1(t) \equiv G_1(t)G_2(t) - 3G_3(t)K_0(t) = 0 \,, \quad (8.22)$$

where

$$G_1(t) = \dot{x}^2 + \dot{y}^2 + \dot{z}^2 \,, \quad (8.23)$$

$$G_2(t) = (\dot{x}\ddot{y} - \ddot{x}\dot{y})(\dot{x}\,\dddot{y} - \dddot{x}\,\dot{y}) + (\dot{y}\ddot{z} - \ddot{y}\dot{z})(\dot{y}\,\dddot{z} - \dddot{y}\,\dot{z}) \quad (8.24)$$

$$+ (\dot{z}\ddot{x} - \ddot{z}\dot{x})(\dot{z}\,\dddot{x} - \dddot{z}\,\dot{x}) \,,$$

$$G_3(t) = \dot{x}\ddot{x} + \dot{y}\ddot{y} + \dot{z}\ddot{z} \,, \quad (8.25)$$

and $K_0(t)$ is defined in (8.10). Note that at points $\kappa(t) = 0$, the necessary condition (8.22) is automatically satisfied (see (8.11)), thus at those points $\dot{\kappa}(t) = 0$ is equivalent to

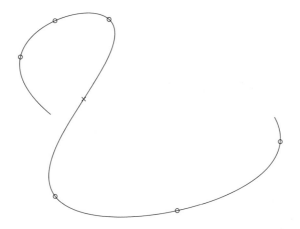

Fig. 8.7. Significant points on a planar Bézier curve of degree 5 (adapted from [57])

$$\dot{x}\,\overset{...}{y} - \overset{...}{x}\,\dot{y} = \dot{y}\,\overset{...}{z} - \overset{...}{y}\,\dot{z} = \dot{z}\,\overset{...}{x} - \overset{...}{z}\,\dot{x} = 0 \ . \tag{8.26}$$

Three points in Fig. 8.8 satisfy (8.22). Two of them are marked by o's and one is marked by ⊕ at the midpoint of the curve.

Once all the stationary points are identified, we may use the extrema theory of functions for a single variable given in Sect. 7.3.1 to classify stationary points of curvature.

Stationary points of curvature of parametric surfaces. Stationary points of surface curvature are important in methods for the correct topological decomposition of the surface on the basis of curvature [255]. Let the curvature in question of a parametric surface $\mathbf{r} = \mathbf{r}(u,v)$ defined over $(u,v) \in [0,1]^2$ be denoted by a scalar function $C(u,v)$, then the following (see also Sect. 7.2.1), need to be evaluated to locate all the stationary points of curvature and to find the global maximum and minimum values of $C(u,v)$ to provide a correct topological decomposition of the surface [255].

1. The four values of curvature at the parameter domain corners

$$C(0,0), \ C(0,1), \ C(1,0), \ C(1,1) \ . \tag{8.27}$$

2. Stationary points along parameter domain boundaries (roots of the four equations)

$$\begin{aligned} C_u(u,0) &= 0, \ C_u(u,1) = 0, \ 0 \le u \le 1 \ , \\ C_v(0,v) &= 0, \ C_v(1,v) = 0, \ 0 \le v \le 1 \ . \end{aligned} \tag{8.28}$$

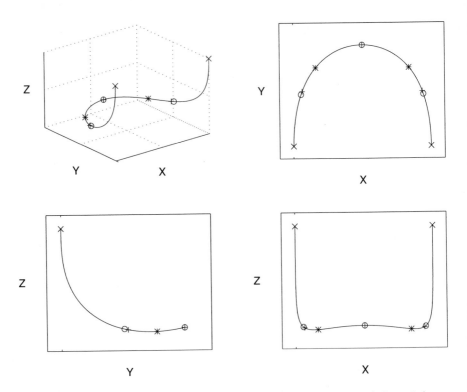

Fig. 8.8. Significant points on a space curve and its projections (adapted from [57])

3. Stationary points within the parameter domain (roots of the two simultaneous equations)

   $$C_u(u,v) = 0,\ C_v(u,v) = 0,\ 0 \le u,\ v \le 1 . \qquad (8.29)$$

The curvature values at the parameter domain corners are readily computed.
The classification of stationary points of functions of two variables is given in
Theorem 7.3.1. The formulations for the stationary points of the Gaussian,
mean and principal curvatures for parametric surfaces are derived in Sect.
8.2.

8.1.5 Fourth-order interrogation methods

Stationary points of torsion of space curve. At nonzero curvature point
on a regular space curve where the twist out of its osculating plane attains a
local maximum or *minimum*, namely $\dot{\tau}(t) = 0$, leads to the necessary condition [57]

$$T_1(t) \equiv G_4(t)K_0(t) - 2G_2(t)T_0(t) = 0, \qquad t \in [t_1, t_2] , \tag{8.30}$$

where

$$G_4(t) = x^{(4)}(\dot{y}\ddot{z} - \ddot{y}\dot{z}) + y^{(4)}(\dot{z}\ddot{x} - \ddot{z}\dot{x}) + z^{(4)}(\dot{x}\ddot{y} - \ddot{x}\dot{y}) , \tag{8.31}$$

and $K_0(t)$, $G_2(t)$ and $T_0(t)$ are defined in (8.10), (8.24) and (8.19), respectively.

Figure 8.8 shows the corresponding significant points, marked by ∗'s. Moreover, comparing (8.19) and (8.30), both $\tau(t)$ and $\dot{\tau}(t)$ of a regular curve vanish at nonzero curvature points if and only if $T_0(t) = G_4(t) = 0$.

Stationary points of total curvature. We now consider the notion of the *Darboux* vector $\boldsymbol{\Omega}(t)$ defined by [206],

$$\boldsymbol{\Omega}(t) = \tau(t)\mathbf{t}(t) + \kappa(t)\mathbf{b}(t) , \tag{8.32}$$

where $\mathbf{t}(t)$ and $\mathbf{b}(t)$ are the unit tangent and binormal vectors, respectively. The *Darboux* vector turns out to be a *rotation* vector of the *Frenet frame* while moving along the curve and therefore, its Euclidean norm $|\boldsymbol{\Omega}(t)|$ indicates the angular speed $\omega(t)$ of the moving local frame. The angular speed $\omega(t)$ is sometimes called *total curvature* of a curve and defined by

$$\omega(t) = \sqrt{\kappa^2(t) + \tau^2(t)} . \tag{8.33}$$

In a planar curve, $\omega(t)$ reduces to $\kappa(t)$ and the binormal vector becomes the axis of rotation. We notice that the total curvature $\omega(t)$ captures the *coupled* effect of both intrinsic features of a *space* curve, and hence, we may consider it as a criterion function for detecting a significant point on a space curve [57].

At a nonzero curvature point on a regular space curve, where the moving frame has its locally highest or lowest angular speed, satisfies the equation $\dot{\omega}(t) = 0$, i.e.

$$K_0^3(t)K_1(t) + G_1^4(t)T_0(t)T_1(t) = 0, \qquad t \in [t_1, t_2] , \tag{8.34}$$

where $K_0(t)$, $K_1(t)$, $G_1(t)$, $T_0(t)$ and $T_1(t)$ are defined in (8.10), (8.22), (8.23), (8.19) and (8.30), respectively. Comparing each function we can roughly see each contribution of $\kappa(t)$, $\dot{\kappa}(t)$, $\tau(t)$ and $\dot{\tau}(t)$ to (8.34). For a special example, if $\dot{\kappa}(t)$ and one of $\tau(t)$ or $\dot{\tau}(t)$ vanish at some t, (8.34) is also satisfied there. We note $K_0(t)$ and $G_1(t)$ are always *positive* at a nonzero curvature point. Three points in Fig. 8.8 satisfy (8.34). Two points, marked by +'s, are located close to the points of curvature extrema, and the other point, marked by ⊕, is located at the midpoint of the curve where $\dot{\kappa}(t)$ and $\tau(t)$ also vanish.

Finally, for the special case where $\kappa(t)$, $\tau(t)$ and consequently $\omega(t)$ are constant, the curve is a *circular helix*.

8.2 Stationary points of curvature of free-form parametric surfaces

For simplicity, the underlying surface is assumed to be an integral Bézier patch (polynomial patch) $\mathbf{r} = \mathbf{r}(u, v)$ defined over $(u, v) \in [0, 1]^2$ of degrees m and n in u and v, respectively. Extension to rational Bézier patch and rational B-spline patch, although tedious does not present conceptual difficulties. We also assume that the surface is regular. In the rest of this chapter we employ convention (b) (see Fig. 3.7 (b) and Table 3.2) such that the normal curvature κ of a surface at point P is positive when the center of curvature is on the opposite direction of the unit normal vector $\mathbf{N}$ of the surface. More details for obtaining stationary points of curvature of free-form parametric surfaces are given in [255].

8.2.1 Gaussian curvature

To formulate the governing equations for computing the stationary points of Gaussian curvature $K(u, v)$ within the domain, we substitute (3.46) into (8.29) which yields [255]

$$K_u(u,v) = \frac{\check{A}(u,v)}{S^6(u,v)} = 0, \quad K_v(u,v) = \frac{\bar{A}(u,v)}{S^6(u,v)} = 0, \quad 0 \le u, v \le 1 ,$$

$$(8.35)$$

where

$$\mathbf{S} = \mathbf{r}_u \times \mathbf{r}_v , \tag{8.36}$$
$$S(u,v) = |\mathbf{S}| = |\mathbf{r}_u \times \mathbf{r}_v| , \tag{8.37}$$
$$\check{A}(u,v) = A_u S^2 - 4(\mathbf{S} \cdot \mathbf{S}_u)A , \tag{8.38}$$
$$\bar{A}(u,v) = A_v S^2 - 4(\mathbf{S} \cdot \mathbf{S}_v)A . \tag{8.39}$$

Polynomial A and its partial derivatives are given by

$$A = \tilde{L}\tilde{N} - \tilde{M}^2, \ A_u = \tilde{L}_u\tilde{N} + \tilde{L}\tilde{N}_u - 2\tilde{M}\tilde{M}_u, \ A_v = \tilde{L}_v\tilde{N} + \tilde{L}\tilde{N}_v - 2\tilde{M}\tilde{M}_v,$$

$$(8.40)$$

where

$$\tilde{L} = SL = \mathbf{S} \cdot \mathbf{r}_{uu}, \ \tilde{L}_u = \mathbf{S}_u \cdot \mathbf{r}_{uu} + \mathbf{S} \cdot \mathbf{r}_{uuu}, \ \tilde{L}_v = \mathbf{S}_v \cdot \mathbf{r}_{uu} + \mathbf{S} \cdot \mathbf{r}_{uuv} ,$$

$$(8.41)$$

$$\tilde{M} = SM = \mathbf{S} \cdot \mathbf{r}_{uv}, \ \tilde{M}_u = \mathbf{S}_u \cdot \mathbf{r}_{uv} + \mathbf{S} \cdot \mathbf{r}_{uuv}, \ \tilde{M}_v = \mathbf{S}_v \cdot \mathbf{r}_{uv} + \mathbf{S} \cdot \mathbf{r}_{uvv} ,$$

$$(8.42)$$

$$\tilde{N} = SN = \mathbf{S} \cdot \mathbf{r}_{vv}, \ \tilde{N}_u = \mathbf{S}_u \cdot \mathbf{r}_{vv} + \mathbf{S} \cdot \mathbf{r}_{uvv}, \ \tilde{N}_v = \mathbf{S}_v \cdot \mathbf{r}_{vv} + \mathbf{S} \cdot \mathbf{r}_{vvv} ,$$

$$(8.43)$$

and

$$\mathbf{S}_u = \mathbf{r}_{uu} \times \mathbf{r}_v + \mathbf{r}_u \times \mathbf{r}_{uv}, \quad \mathbf{S}_v = \mathbf{r}_{uv} \times \mathbf{r}_v + \mathbf{r}_u \times \mathbf{r}_{vv} . \tag{8.44}$$

As $S \neq 0$, (8.35) are satisfied if

$$\check{A}(u,v) = 0, \quad \bar{A}(u,v) = 0, \quad 0 \le u, v \le 1 , \tag{8.45}$$

which are two simultaneous bivariate polynomial equations of degree $(10m - 7, 10n - 6)$, $(10m - 6, 10n - 7)$ in u and v, respectively. For example, if the input surface is a bicubic Bézier patch, the degrees of the two simultaneous bivariate polynomial equations become (23, 24) and (24, 23) in u and v. System (8.45) can be solved robustly with the IPP algorithm described in Chap. 4 (see [255] for details).

The stationary points along the four boundary edges are easily obtained by solving the four univariate polynomial equations,

$$\check{A}(u,0) = 0, \quad \check{A}(u,1) = 0, \quad 0 \le u \le 1 , \tag{8.46}$$
$$\bar{A}(0,v) = 0, \quad \bar{A}(1,v) = 0, \quad 0 \le v \le 1 , \tag{8.47}$$

using the IPP algorithm described in Chap. 4 (see [255] for details).

Example 8.2.1. A hyperbolic paraboloid $\mathbf{r}(u,v) = (u, v, uv)^T$, $(u,v) \in [0,1]^2$ (bilinear surface), illustrated in Fig. 3.4, can be expressed in a Bézier form as

$$\mathbf{r} = \sum_{i=0}^{1} \sum_{j=0}^{1} \mathbf{b}_{ij} B_{i,1}(u) B_{j,1}(v) ,$$

where $\mathbf{b}_{00} = (0,0,0)^T$, $\mathbf{b}_{01} = (0,1,0)^T$, $\mathbf{b}_{10} = (1,0,0)^T$ and $\mathbf{b}_{11} = (1,1,1)^T$. The first and second fundamental form coefficients are readily computed using (3.63) (3.64) (3.65)

$$E = \mathbf{r}_u \cdot \mathbf{r}_u = 1 + v^2, \quad F = \mathbf{r}_u \cdot \mathbf{r}_v = uv, \quad G = \mathbf{r}_v \cdot \mathbf{r}_v = 1 + u^2 ,$$
$$\mathbf{N} = \frac{(-v, -u, 1)^T}{\sqrt{u^2 + v^2 + 1}} ,$$
$$L = \mathbf{r}_{uu} \cdot \mathbf{N} = 0, \quad M = \mathbf{r}_{uv} \cdot \mathbf{N} = \frac{1}{\sqrt{u^2 + v^2 + 1}}, \quad N = \mathbf{r}_{vv} \cdot \mathbf{N} = 0 .$$

Thus the Gaussian curvature is given by

$$K = \frac{LN - M^2}{EG - F^2} = -\frac{1}{u^2 + v^2 + 1} .$$

The Gaussian curvature is always negative and hence all the points are hyperbolic.

The stationary points within the parameter domain are given by finding the roots of $K_u = K_v = 0$. Since

$$\mathbf{S} = \mathbf{r}_u \times \mathbf{r}_v = (-v, -u, 1)^T, \quad S = |\mathbf{S}| = \sqrt{u^2 + v^2 + 1},$$
$$\mathbf{S}_u = (0, -1, 0)^T, \quad \mathbf{S}_v = (-1, 0, 0)^T,$$

and from (8.40) through (8.43)

$$\tilde{L} = SL = 0, \quad \tilde{M} = SM = 1, \quad \tilde{N} = SN = 0,$$
$$A = \tilde{L}\tilde{N} - \tilde{M}^2 = -1,$$
$$A_u = \tilde{L}_u\tilde{N} + \tilde{L}\tilde{N}_u - 2\tilde{M}\tilde{M}_u = 0, \quad A_v = \tilde{L}_v\tilde{N} + \tilde{L}\tilde{N}_v - 2\tilde{M}\tilde{M}_v = 0,$$

and from (8.38) and (8.39)

$$\check{A} = A_u S^2 - 4(\mathbf{S} \cdot \mathbf{S}_u)A = 4u, \quad \bar{A} = A_v S^2 - 4(\mathbf{S} \cdot \mathbf{S}_v)A = 4v,$$

hence we have

$$K_u = \frac{4u}{(u^2 + v^2 + 1)^3}, \quad K_v = \frac{4v}{(u^2 + v^2 + 1)^3}.$$

Thus the stationary point within the domain is given by $(u, v) = (0, 0)$, which coincides with one of the corner points of the bilinear surface patch. Using Theorem 7.3.1, we find that $K(0, 0) = -1$ is a minimum.

At the remaining of corner points of the patch, we can readily compute

$$K(0, 1) = -\frac{1}{4}, \quad K(1, 0) = -\frac{1}{4}, \quad K(1, 1) = -\frac{1}{9}.$$

Stationary points along parameter domain boundaries are obtained by solving the following four univariate equations

$$K_u(u, 0) = \frac{4u}{(u^2 + 1)^3} = 0,$$

$$K_u(u, 1) = \frac{4u}{(u^2 + 2)^3} = 0,$$

$$K_v(0, v) = \frac{4v}{(v^2 + 1)^3} = 0,$$

$$K_v(1, v) = \frac{4v}{(v^2 + 2)^3} = 0.$$

The roots are $u = 0$, $u = 0$, $v = 0$ and $v = 0$, thus they coincide with the corner points $(0, 0)$, $(0, 1)$, $(0, 0)$ and $(1, 0)$. Therefore, the range of the Gaussian curvature is $-1 \le K \le -\frac{1}{9}$.

8.2.2 Mean curvature

Similarly to the Gaussian curvature, we have the following equations to evaluate the stationary points of mean curvature H given by (3.47) (convention (b) (see Fig. 3.7 (b) and Table 3.2) is employed) within the domain [255]:

$$H_u(u,v) = \frac{\check{B}(u,v)}{2S^5(u,v)} = 0, \quad H_v(u,v) = \frac{\bar{B}(u,v)}{2S^5(u,v)} = 0, \quad 0 \le u,v \le 1 \,,$$

$$(8.48)$$

where

$$\check{B}(u,v) = B_u S^2 - 3(\mathbf{S} \cdot \mathbf{S}_u)B \,, \qquad (8.49)$$
$$\bar{B}(u,v) = B_v S^2 - 3(\mathbf{S} \cdot \mathbf{S}_v)B \,. \qquad (8.50)$$

Polynomial B and its partial derivatives are given by

$$B = 2F\tilde{M} - E\tilde{N} - G\tilde{L} \,, \qquad (8.51)$$
$$B_u = 2(F\tilde{M}_u + F_u\tilde{M}) - (E_u\tilde{N} + E\tilde{N}_u) - (G_u\tilde{L} + G\tilde{L}_u) \,, \quad (8.52)$$
$$B_v = 2(F\tilde{M}_v + F_v\tilde{M}) - (E_v\tilde{N} + E\tilde{N}_v) - (G_v\tilde{L} + G\tilde{L}_v) \,, \quad (8.53)$$

where

$$E_u = 2\mathbf{r}_u \cdot \mathbf{r}_{uu}, \quad E_v = 2\mathbf{r}_u \cdot \mathbf{r}_{uv} \,, \qquad (8.54)$$
$$F_u = \mathbf{r}_{uu} \cdot \mathbf{r}_v + \mathbf{r}_u \cdot \mathbf{r}_{uv}, \quad F_v = \mathbf{r}_{uv} \cdot \mathbf{r}_v + \mathbf{r}_u \cdot \mathbf{r}_{vv} \,, \qquad (8.55)$$
$$G_u = 2\mathbf{r}_v \cdot \mathbf{r}_{uv}, \quad G_v = 2\mathbf{r}_v \cdot \mathbf{r}_{vv} \,. \qquad (8.56)$$

Since $S \ne 0$, (8.48) reduce to two simultaneous bivariate polynomial equations

$$\check{B}(u,v) = 0, \quad \bar{B}(u,v) = 0, \quad 0 \le u,v \le 1 \,, \qquad (8.57)$$

of degree $(9m - 6, 9n - 5)$ and $(9m - 5, 9n - 6)$ in u and v. For a bicubic Bézier patch input, the degrees of the governing equations are $(21, 22)$ and $(22, 21)$ in u and v, respectively. The system (8.57) can be solved robustly with the IPP algorithm described in Chap. 4 (see [255] for details).

The stationary points of mean curvature along the domain boundary can be obtained by solving the following four univariate polynomial equations:

$$\check{B}(u,0) = 0, \quad \check{B}(u,1) = 0 \,, \quad 0 \le u \le 1 \,, \qquad (8.58)$$
$$\bar{B}(0,v) = 0, \quad \bar{B}(1,v) = 0 \,. \quad 0 \le v \le 1 \,. \qquad (8.59)$$

These equations can be solved robustly with the IPP algorithm described in Chap. 4 (see [255] for details).

8.2.3 Principal curvatures

For obtaining the stationary points of principal curvature κ within the domain, the simultaneous bivariate equations (8.29) become [255]

$$\kappa_u(u,v) = \frac{f_1(u,v) \pm f_2(u,v)\sqrt{f_3(u,v)}}{2S^5(u,v)\sqrt{f_3(u,v)}} = 0, \quad 0 \le u,v \le 1, \quad (8.60)$$

$$\kappa_v(u,v) = \frac{g_1(u,v) \pm g_2(u,v)\sqrt{f_3(u,v)}}{2S^5(u,v)\sqrt{f_3(u,v)}} = 0, \quad 0 \le u,v \le 1.$$

The plus and minus signs correspond to the maximum and minimum principal curvatures, and $f_1(u,v)$, $f_2(u,v)$, $f_3(u,v)$, $g_1(u,v)$ and $g_2(u,v)$ are polynomials of degree $(14m - 9, 14n - 8)$, $(9m - 6, 9n - 5)$, $(10m - 6, 10n - 6)$, $(14m - 8, 14n - 9)$, $(9m - 5, 9n - 6)$ in u and v parameters and are given by

$$f_1(u,v) = (BB_u - 2A_u S^2)S^2 + (8AS^2 - 3B^2)(\mathbf{S} \cdot \mathbf{S}_u), \quad (8.61)$$
$$f_2(u,v) = B_u S^2 - 3(\mathbf{S} \cdot \mathbf{S}_u)B, \quad (8.62)$$
$$f_3(u,v) = B^2 - 4AS^2, \quad (8.63)$$
$$g_1(u,v) = (BB_v - 2A_v S^2)S^2 + (8AS^2 - 3B^2)(\mathbf{S} \cdot \mathbf{S}_v), \quad (8.64)$$
$$g_2(u,v) = B_v S^2 - 3(\mathbf{S} \cdot \mathbf{S}_v)B. \quad (8.65)$$

Assuming $f_3 \neq 0$ and $S \neq 0$, we obtain

$$f_1(u,v) \pm f_2(u,v)\sqrt{f_3(u,v)} = 0, \quad 0 \le u,v \le 1, \quad (8.66)$$
$$g_1(u,v) \pm g_2(u,v)\sqrt{f_3(u,v)} = 0, \quad 0 \le u,v \le 1.$$

These are two simultaneous bivariate irrational equations involving polynomials and square roots of polynomials which arise from the analytical expressions of principal curvatures. We can introduce an auxiliary variable τ such that $\tau^2 = f_3$ to remove the radical and transform the problem into a system of three trivariate polynomial equations of degree $(14m - 9, 14n - 8, 1)$, $(14m - 8, 14n - 9, 1)$, $(10m - 6, 10n - 6, 2)$ in u, v and τ. For a bicubic Bézier patch the degrees of the trivariate polynomial equations are $(33,34,1)$, $(34,33,1)$ and $(24,24,2)$. The resulting system can be solved with the IPP algorithm of Chap. 4 (see [255] for details).

For the stationary points of principal curvatures along the boundary, we need to solve the following four univariate irrational equations involving polynomials and square roots of polynomials

$$f_1(u,0) \pm f_2(u,0)\sqrt{f_3(u,0)} = 0, \quad f_1(u,1) \pm f_2(u,1)\sqrt{f_3(u,1)} = 0, \quad 0 \le u \le 1,$$
$$(8.67)$$

$$g_1(0,v) \pm g_2(0,v)\sqrt{f_3(0,v)} = 0, \quad g_1(1,v) \pm g_2(1,v)\sqrt{f_3(1,v)} = 0, \quad 0 \le v \le 1,$$
$$(8.68)$$

which can be solved by the same auxiliary variable method as before.

When $f_3 = 0$ (or equivalently $H^2 - K = 0$ if $S \neq 0$), (8.60) become singular. This condition is equivalent to the point where the two principal curvatures are identical, i.e. an umbilical point. If the umbilical point coincides with a local maximum or minimum of the curvature, we cannot use (8.66) to locate such a point. In such case we need to locate the umbilical point first by finding the roots of the equation

$$H^2(u,v) - K(u,v) = \frac{f_3(u,v)}{4S^6(u,v)} = 0 , \tag{8.69}$$

which reduces to solving $f_3(u,v) = 0$, since $S \neq 0$. Because $W(u,v) \equiv \frac{f_3(u,v)}{4S^6(u,v)}$ is a non-negative function, $W(u,v)$ has a global minimum at the umbilic [257]. The condition for global minimum at the umbilic implies that $\nabla W = \mathbf{0}$ or equivalently (given that $f_3(u,v) = 0$)

$$W_u = \frac{\frac{\partial f_3}{\partial u}}{4S^6} = 0, \quad W_v = \frac{\frac{\partial f_3}{\partial v}}{4S^6} = 0 . \tag{8.70}$$

Therefore, assuming $S \neq 0$, the umbilics are the solutions of the following three simultaneous equations (see also Sect. 9.3):

$$\frac{\partial f_3(u,v)}{\partial u} = 0, \quad \frac{\partial f_3(u,v)}{\partial v} = 0, \quad f_3(u,v) = 0, \quad 0 \le u, v \le 1 . \tag{8.71}$$

These equations can be reduced to [257]:

$$BB_u - 2A_u S^2 - 4A(\mathbf{S} \cdot \mathbf{S}_u) = 0, \quad BB_v - 2A_v S^2 - 4A(\mathbf{S} \cdot \mathbf{S}_v) = 0 ,$$
$$B^2 - 4AS^2 = 0, \quad 0 \le u, v \le 1 . \tag{8.72}$$

Since $f_3(u,v) = 0$ at the umbilics, (8.66) reduce to $f_1(u,v) = 0$, $g_1(u,v) = 0$. If we substitute the first equation of (8.72) into (8.61) and use the fact $f_3 = B^2 - 4AS^2 = 0$, we obtain $f_1(u,v) = 0$. Similarly by substituting the second equation of (8.72) into (8.64), we obtain $g_1(u,v) = 0$. Consequently the solutions of (8.66) include not only the locations of extrema of principal curvatures but also the locations of the umbilical points. Then we use Theorem 9.5.1 at the umbilical points to check if the umbilical point is a local extremum of principal curvatures.

A cusp is an isolated singular point on the surface where the surface tangent plane is undefined, i.e. $\mathbf{r}_u \times \mathbf{r}_v = \mathbf{0}$. Cusps of an offset surface correspond to points on the progenitor where both of the principal curvatures are equal to $-\frac{1}{d}$ where d is the offset distance [94] (see also Sect. 11.3.2). In this manner, cusps on an offset surface are associated with umbilics of the progenitor. Hence we can locate the cusps on an offset surface using (8.71).

8.3 Stationary points of curvature of explicit surfaces

We can apply the procedures discussed in Sect. 8.2 to obtain the stationary points of curvature of explicit surfaces [248]. Locally any surface can be expressed as a graph of a differentiable function [76]. Given a point P on the parametric surface S, we can set an orthogonal Cartesian coordinate system xyz such that xy-plane coincides with the tangent plane of S at P and z-axis is along the normal at P. It follows that in the neighborhood of P any parametric surface S can be represented in the form $\mathbf{r}(x,y) = [x, y, h(x,y)]^T$, where h is a differentiable function with $h(0,0) = h_x(0,0) = h_y(0,0) = 0$.

We can Taylor expand $h(x,y)$ about $(0,0)$ as follows

$$h(x,y) = h(0,0) + [xh_x(0,0) + yh_y(0,0)] \tag{8.73}$$
$$+ \frac{1}{2!}[x^2 h_{xx}(0,0) + 2xy h_{xy}(0,0) + y^2 h_{yy}(0,0)]$$
$$+ \frac{1}{3!}[x^3 h_{xxx}(0,0) + 3x^2 y h_{xxy}(0,0) + 3xy^2 h_{xyy}(0,0) + y^3 h_{yyy}(0,0)]$$
$$+ R(x,y)(|x,y|^3) ,$$

where $R(x,y)$ is a remainder term with $\lim_{x\to 0, y\to 0} R(x,y) = 0$ and $|x,y| = \sqrt{x^2 + y^2}$. If we take into account that $h(0,0) = h_x(0,0) = h_y(0,0) = 0$, we can consider

$$h(x,y) = \frac{1}{2}[h_{xx}(0,0)x^2 + 2h_{xy}(0,0)xy + h_{yy}(0,0)y^2] , \tag{8.74}$$

as the *second order approximation* of $h(x,y)$.

If we denote E, F, G and L, M, N as coefficients of the first and second fundamental forms of the surface, and assume further that x and y axes are directed along the principal directions at P, assuming P is not an umbilic, then $F = M = 0$ [76]. It follows that $h_{xy}(0,0) = 0$, since $M = h_{xy}/\sqrt{1 + h_x^2 + h_y^2}$ (see Equation (3.65)). Although we have assumed P is not an umbilic, we can show that $h_{xy}(0,0)$ will also vanish when the point is an umbilic (see Sect. 9.2). Also the principal curvatures at P when $F = M = 0$ can be expressed as follows:

- If $h_{xx}(0,0) > h_{yy}(0,0)$; $\qquad\qquad\qquad\qquad\qquad\qquad$ (8.75)
$$\kappa_{min} = -\frac{L}{E} = -h_{xx}(0,0), \quad \kappa_{max} = -\frac{N}{G} = -h_{yy}(0,0) ,$$
- If $h_{xx}(0,0) < h_{yy}(0,0)$; $\qquad\qquad\qquad\qquad\qquad\qquad$ (8.76)
$$\kappa_{max} = -\frac{L}{E} = -h_{xx}(0,0), \quad \kappa_{min} = -\frac{N}{G} = -h_{yy}(0,0) ,$$

where the minus signs are due to convention (b) of the normal curvature (see Fig. 3.7 (b)).

If we set $\alpha = h_{xx}(0,0)$ and $\beta = h_{yy}(0,0)$ and assuming that P is a nonplanar point, the surface can be written locally as a second order approximation in the nonparametric form given by

$$z = \frac{1}{2}(\alpha x^2 + \beta y^2) . \tag{8.77}$$

Its corresponding parametric form is

$$\mathbf{r}(x,y) = [x, y, \frac{1}{2}(\alpha x^2 + \beta y^2)]^T .\qquad(8.78)$$

Equation (8.77) or (8.78) represents an *explicit quadratic surfaces* which can be categorized into four types according to combinations of α and β as listed in Table 8.1. The four types of explicit quadratic surfaces are depicted in Fig. 8.9.

Table 8.1. Four types of explicit quadratic surfaces according to α and β (adapted from [248])

Signs of α and β	Types of surfaces	Types of points at P
$\alpha\beta < 0$	Hyperbolic paraboloid	Hyperbolic point
$\alpha\beta > 0$ and $\alpha \neq \beta$	Elliptic paraboloid	Elliptic point
$\alpha = \beta$	Paraboloid of revolution	Umbilical point
$\alpha = 0$ or $\beta = 0$	Parabolic cylinder	Parabolic point

Since any regular surface can be locally approximated in the neighborhood of a point P by an explicit quadratic surface to the second order, we examine the stationary points of curvatures of explicit quadratic surfaces as representatives of explicit surfaces. We can apply the procedures in Sect. 8.2 to evaluate the stationary points of curvatures of explicit quadratic surfaces. We will only examine the stationary points of principal curvatures, since those of Gaussian and mean curvatures can be found in a similar way. In the sequel we assume that $\beta > 0$ and $\alpha \leq \beta$ without loss of generality. It follows that at (0,0,0) (8.76) holds and the x-axis will be the direction of maximum principal curvature and y-axis will be the direction for the minimum principal curvature. The surface is a hyperbolic paraboloid when $\alpha < 0$, an elliptic paraboloid when $0 < \alpha < \beta$, a paraboloid of revolution when $0 < \alpha = \beta$, and a parabolic cylinder when $\alpha = 0$. The paraboloid and the parabolic cylinder can be considered as degenerate cases of the elliptic paraboloid.

Gaussian curvature (see (3.66)) and mean curvature (see (3.67) in Table 3.2) are readily evaluated

$$K(x,y) = \frac{\alpha\beta}{(1 + \alpha^2 x^2 + \beta^2 y^2)^2}, \quad H(x,y) = -\frac{\alpha + \beta + \alpha\beta(\alpha x^2 + \beta y^2)}{2(1 + \alpha^2 x^2 + \beta^2 y^2)^{\frac{3}{2}}},$$

$$(8.79)$$

and hence the principal curvatures become

$$\kappa_{max}(x,y) = \frac{-(1 + \alpha^2 x^2)\beta - (1 + \beta^2 y^2)\alpha}{2(1 + \alpha^2 x^2 + \beta^2 y^2)^{\frac{3}{2}}}\qquad(8.80)$$

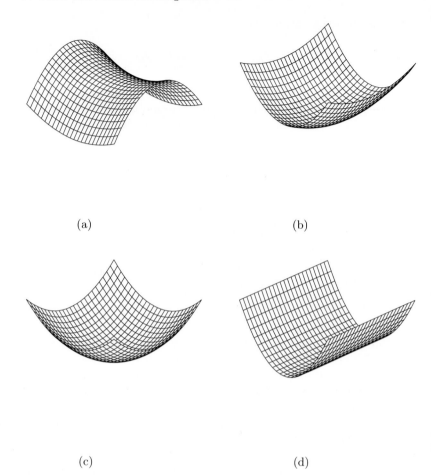

Fig. 8.9. Explicit quadratic surfaces $z = \frac{1}{2}(\alpha x^2 + \beta y^2)$ (adapted from [248]): (a) hyperbolic paraboloid ($\alpha = -3$, $\beta = 1$), (b) elliptic paraboloid ($\alpha = 1$, $\beta = 3$), (c) paraboloid of revolution($\alpha = \beta = 3$), (d) parabolic cylinder ($\alpha = 0$, $\beta = 3$)

$$\kappa_{min}(x,y) = \frac{-(1+\alpha^2 x^2)\beta - (1+\beta^2 y^2)\alpha}{2(1+\alpha^2 x^2 + \beta^2 y^2)^{\frac{3}{2}}}$$

$$+\frac{\sqrt{(\alpha-\beta)^2 - 2\alpha\beta(\alpha-\beta)(\alpha x^2 - \beta y^2) + \alpha^2\beta^2(\alpha x^2 + \beta y^2)^2}}{2(1+\alpha^2 x^2 + \beta^2 y^2)^{\frac{3}{2}}},$$

$$-\frac{\sqrt{(\alpha-\beta)^2 - 2\alpha\beta(\alpha-\beta)(\alpha x^2 - \beta y^2) + \alpha^2\beta^2(\alpha x^2 + \beta y^2)^2}}{2(1+\alpha^2 x^2 + \beta^2 y^2)^{\frac{3}{2}}}. \tag{8.81}$$

The stationary points of principal curvature in the domain must satisfy the simultaneous equations (8.66). The equations not only find the stationary points of principal curvatures but also find the locations of the umbilics (see Sect. 8.2.3 and [255]). For explicit quadratic surface f_1, f_2, f_3, g_1, g_2 (see

(8.61) - (8.65)) are given by

$$f_1(x,y) = -[\alpha^2\beta x(1+\alpha^2x^2) - 2\alpha^2\beta^3xy^2 + 3\alpha^3x(1+\beta^2y^2)][(1+\alpha^2x^2)\beta$$
$$+(1+\beta^2y^2)\alpha)] + 8\alpha^3\beta x(1+\alpha^2x^2+\beta^2y^2) , \qquad (8.82)$$
$$f_2(x,y) = \alpha^2\beta x(1+\alpha^2x^2) - 2\alpha^2\beta^3xy^2 + 3\alpha^3x(1+\beta^2y^2) , \qquad (8.83)$$
$$f_3(x,y) = (\alpha-\beta)^2 - 2\alpha\beta(\alpha-\beta)(\alpha x^2 - \beta y^2) + \alpha^2\beta^2(\alpha x^2+\beta y^2)^2 , \qquad (8.84)$$
$$g_1(x,y) = -[\alpha^2\beta y(1+\beta^2y^2) - 2\alpha^3\beta^2x^2y + 3\beta^3y(1+\alpha^2x^2)][(1+\alpha^2x^2)\beta$$
$$+(1+\beta^2y^2)\alpha] + 8\alpha\beta^3y(1+\alpha^2x^2+\beta^2y^2) , \qquad (8.85)$$
$$g_2(x,y) = \alpha\beta^2y(1+\beta^2y^2) - 2\alpha^3\beta^2x^2y + 3\beta^3y(1+\alpha^2x^2) . \qquad (8.86)$$

We can reduce the two simultaneous bivariate irrational equations (8.66) involving polynomials and square roots of polynomials into a system of three nonlinear polynomial equations in three variables through the introduction of *auxiliary variables* [255, 254] (see Sect. 4.5). As this is a system of low degree polynomial equations, we can solve these equations analytically using a symbolic manipulation program such as MATHEMATICA [446], MAPLE [51].

Since the maximum principal curvature can be obtained in a similar manner, we will only focus on the minimum principal curvature. Provided we find all the real roots, we check first if the roots (x,y) are umbilics or not by substituting the roots (x,y) into $\kappa_{min}(x,y) - H(x,y) = 0$. If the roots do not satisfy this equation, the points are not umbilics and we can use Theorem 7.3.1 to classify the stationary points of minimum principal curvature. To apply the extrema theory of functions to the minimum principal curvature function, the second order partial derivatives of the minimum principal curvatures are required; however, we avoid to present these here, since they are extremely lengthy. If the points are umbilics, we need to use a specialized *criterion* [257] (see Theorem 9.5.1), to check if the point is a local extremum of the principal curvature, since the curvature function κ_{min} is not differentiable at an umbilic.

All the real roots of (8.66) and their classifications are listed in Table 8.2. The hyperbolic paraboloid has only one real root $(x,y) = (0,0)$, which corresponds to point (0,0,0) on the surface, and gives a minimum of the minimum principal curvature function with $\kappa_{min}(0,0) = -\beta$ according to the extrema theory. Since there is no other extremum, this minimum is also a global minimum. The elliptic paraboloid has three real roots $(x,y) = (0,0)$ and $\left(0, \pm\frac{1}{\beta}\sqrt{\frac{\beta-\alpha}{\alpha}}\right)$. The first root corresponds to a non-umbilical point (0,0,0) on the surface, while the other two real roots correspond to generic lemon type umbilical points $\left(0, \pm\frac{1}{\beta}\sqrt{\frac{\beta-\alpha}{\alpha}}, \frac{\beta-\alpha}{2\alpha\beta}\right)$ on the surface as shown in Fig. 9.1. Using the extrema theory of functions, it follows that the root $(0,0)$ gives a minimum of the minimum principal curvature function with $\kappa_{min}(0,0) = -\beta$. Using the criterion in Theorem 9.5.1, we can find that the roots corresponding to the two lemon type umbilics do not provide extrema of the minimum

Table 8.2. Classification of roots according to types of explicit quadratic surfaces

Types of surfaces	Hyperbolic paraboloid	Elliptic paraboloid		
Signs of α and β	$\alpha < 0 < \beta$	$0 < \alpha < \beta$		
# of real roots	1	3		
Roots	$(0,0)$	$(0,0)$	$\left(0, \pm\frac{1}{\beta}\sqrt{\frac{\beta-\alpha}{\alpha}}\right)$	
Classification	Minimum	Minimum	Lemon type umbilics	
$\kappa_{min}(x,y)$ at roots	$-\beta$	$-\beta$	$-\alpha\sqrt{\frac{\alpha}{\beta}}$	
Types of surfaces	Paraboloid of revolution	Parabolic cylinder		
Signs of α and β	$0 < \alpha = \beta$	$0 = \alpha < \beta$		
# of real roots	1	∞		
Roots	$(0,0)$	Along x-axis		
Classification	Minimum	Minima		
$\kappa_{min}(x,y)$ at roots	$-\beta$	$-\beta$		

principal curvature function. Consequently the minimum $-\beta$ is a global minimum.

As α approaches β, two generic umbilics merge to one non-generic umbilic at $(0,0,0)$, and the surface reduces to a paraboloid of revolution as illustrated in Fig. 9.1. The paraboloid of revolution has only one real root $(x,y) = (0,0)$, which corresponds to point $(0,0,0)$ on the surface, and is a non-generic umbilical point as mentioned above. Since the root corresponds to an umbilical point, we cannot use the extrema theory of functions, nor can we use the criterion in Theorem 9.5.1, since all the second derivatives of W (see (9.25)) vanish. But it is apparent that the paraboloid of revolution has a global minimum of the minimum principal curvature function at the umbilic with $\kappa_{min}(0,0) = -\beta$, since the paraboloid of revolution can be constructed by rotating a parabola, which has a global minimum of its curvature at the origin, around the z-axis. Here we are employing the sign convention (b) (see Fig. 3.7 (b) and Table 3.2) of the curvature.

In the case of a parabolic cylinder, the minimum principal curvature reduces to a simple univariate function

$$\kappa_{min}(y) = -\beta(1 + \beta^2 y^2)^{-\frac{3}{2}} . \tag{8.87}$$

It is easy to show that (8.87) has a global minimum at $y = 0$. Therefore the parabolic cylinder has global minima with value $\kappa_{min} = -\beta$ along the x-axis. These discussions lead to the following lemma [248]:

Lemma 8.3.1. *The minimum principal curvature function of explicit quadratic surfaces, except for the parabolic cylinder, has only one extremum at $(0,0)$ with value $-\beta$, which corresponds to the point $(0,0,0)$ on the surface, and it is a global minimum. The parabolic cylinder has global minima $-\beta$ at $y = 0$, which corresponds to the x-axis on the surface, and has no other extrema.*

Similarly we can deduce the following lemma [248].

Lemma 8.3.2. *The maximum principal curvature function of explicit qua-dratic surfaces, except for the parabolic cylinder has only one extremum at $(0,0)$ with value $-\alpha$, which corresponds to the point $(0,0,0)$ on the surface, and it is a global minimum for elliptic paraboloid and paraboloid of revolution, while it is a global maximum for hyperbolic paraboloid (note that for hyperbo-lic paraboloid α is negative). The maximum principal curvature function of parabolic cylinders is zero everywhere.*

Note that Lemmata 8.3.1 and 8.3.2 are based on convention (b) of the normal curvature (see Fig. 3.7 (b)), and will be used in Sect. 11.3.4.

8.4 Stationary points of curvature of implicit surfaces

Now we provide a method to evaluate the stationary points of a curvature function $C(x, y, z)$ of implicit surfaces

$$f(x, y, z) = 0 \ . \tag{8.88}$$

We want to maximize or minimize the function $C(x, y, z)$ subject to a cons-traint (8.88). Introducing the *Lagrange multiplier* λ and the auxiliary function [166]

$$\phi(x, y, z) = C(x, y, z) + \lambda f(x, y, z) \ , \tag{8.89}$$

the necessary conditions for the auxiliary function $\phi(x, y, z)$ to attain a local maximum or minimum, when no constraints are imposed, are

$$\phi_x = 0, \quad \phi_y = 0, \quad \phi_z = 0 \ . \tag{8.90}$$

Equations (8.90) together with (8.88) form four equations with four unknowns x, y, z and λ. The curvature functions $C(x, y, z)$ including Gaussian, mean and principal curvatures are evaluated using the procedure described in Sect. 3.5.2. In a manner similar to parametric surfaces, the denominator and the numerator of the curvature functions consist of polynomials and square root of polynomials if $f(x, y, z)$ is a polynomial.

As an illustrative example, we will examine the stationary points of the minimum principal curvature $\kappa_{min}(x, y, z)$ of an ellipsoid (3.83) [249] where we assume $a \leq b \leq c$. The auxiliary function becomes

$$\phi = \kappa_{min}(x, y, z) + \lambda \left(\frac{x^2}{a^2} + \frac{y^2}{b^2} + \frac{z^2}{c^2} - 1 \right) \ , \tag{8.91}$$

and hence the system of equations to obtain the stationary points of minimum principal curvature of an ellipsoid reduce to

$$\phi_x(x, y, z) = \kappa_x(x, y, z) + \frac{2x\lambda}{a^2} = 0 , \qquad (8.92)$$

$$\phi_y(x, y, z) = \kappa_y(x, y, z) + \frac{2y\lambda}{b^2} = 0 , \qquad (8.93)$$

$$\phi_z(x, y, z) = \kappa_z(x, y, z) + \frac{2z\lambda}{c^2} = 0 , \qquad (8.94)$$

$$\frac{x^2}{a^2} + \frac{y^2}{b^2} + \frac{z^2}{c^2} - 1 = 0 , \qquad (8.95)$$

where for simplicity we have set $\kappa = \kappa_{min}$.

After some algebraic manipulation, we obtain

$$x(g_1 + g_2 t + 4\lambda a^4 b^2 c^2 f_2^2 \sigma t) = 0 , \qquad (8.96)$$

$$y(h_1 + h_2 t + 4\lambda a^2 b^4 c^2 f_2^2 \sigma t) = 0 , \qquad (8.97)$$

$$z(p_1 + p_2 t + 4\lambda a^2 b^2 c^4 f_2^2 \sigma t) = 0 , \qquad (8.98)$$

$$\frac{x^2}{a^2} + \frac{y^2}{b^2} + \frac{z^2}{c^2} - 1 = 0 , \qquad (8.99)$$

$$t^2 - g_3 = 0 , \qquad (8.100)$$

$$\sigma^2 - f_2 = 0 , \qquad (8.101)$$

where the last two equations are added through the introduction of auxiliary variables t and σ to remove the square roots of polynomials that appear in the denominator and numerator of the expression in (3.83), and

$$f_1 = x^2 + y^2 + z^2 - a^2 - b^2 - c^2 , \qquad (8.102)$$

$$f_2 = \frac{x^2}{a^4} + \frac{y^2}{b^4} + \frac{z^2}{c^4} , \qquad (8.103)$$

$$g_1 = -2f_1 f_2 a^4 - 8a^2 b^2 c^2 f_2 + 3f_1^2 , \qquad (8.104)$$

$$g_2 = 2a^4 f_2 - 3f_1 , \qquad (8.105)$$

$$g_3 = f_1^2 - 4a^2 b^2 c^2 f_2 , \qquad (8.106)$$

$$h_1 = -2f_1 f_2 b^4 - 8a^2 b^2 c^2 f_2 + 3f_1^2 , \qquad (8.107)$$

$$h_2 = 2b^4 f_2 - 3f_1 , \qquad (8.108)$$

$$p_1 = -2f_1 f_2 c^4 - 8a^2 b^2 c^2 f_2 + 3f_1^2 , \qquad (8.109)$$

$$p_2 = 2c^4 f_2 - 3f_1 . \qquad (8.110)$$

Now the system consists of six equations with six unknowns x, y, z, λ, t and σ. Since the degree of the polynomials is low we can solve the system by a symbolic manipulation program such as MATHEMATICA [446], MAPLE [51], which gives a global minimum of κ_{min} equal to $\frac{a}{c^2}$ at $(\pm a, 0, 0)$, a local minimum $\frac{b}{c^2}$ at $(0, \pm b, 0)$ and a global maximum $\frac{c}{b^2}$ at $(0, 0, \pm c)$.

8.5 Contouring constant curvature

8.5.1 Contouring levels

The variation of curvature can be displayed using a color coded map. Color coded maps provide a rough idea of the differential properties of surfaces but are not sufficient to provide detailed machining information nor permit automation of the machining process or of fairing algorithms. Iso-curvature curves can also be used to display and visualize the variation of curvature by computing the curvature on a lattice and linearly interpolating the contour points. The iso-curvature curves divide the surface into regions of specific range of curvature. However discrete color coded maps of curvature and lattice methods for curvature contouring do not guarantee to locate all the stationary points (local extrema and saddle points) of curvature, and hence may fail to provide the correct topological decomposition of the surface on the basis of curvature to the manufacturer or to a fairing process. A robust procedure for contouring curvature of a free-form parametric polynomial surface can be found in [255]. The contouring levels should be determined to faithfully represent the curvature distribution. To do this, we need to evaluate (8.27), (8.28) and (8.29).

Example 8.5.1. In Example 8.2.1 we have examined the range of Gaussian curvature of a bilinear surface. The contour lines of Gaussian curvature can be evaluated by setting $K = C_L$ with C_L a constant satisfying $-1 \leq C_L \leq -\frac{1}{9}$

$$K = -\frac{1}{(u^2 + v^2 + 1)^2} = C_L \,,$$

which can be rewritten as

$$u^2 + v^2 = \sqrt{-\frac{1}{C_L}} - 1 \,.$$

Since there is no local maxima, minima, nor saddle points of Gaussian curvature in the domain, the constant curvature lines are concentric circles in the parameter space with center at (0,0) and radius $\sqrt{\sqrt{-\frac{1}{C_L}} - 1}$.

8.5.2 Finding starting points

Contour lines in the parameter space of a bivariate function can be separated into three categories:

- Local maxima and minima of the function are encircled by closed contour curves [210].

- At the precise level of a saddle point, the contour curves cross (self-intersect) or exhibit more complex behavior (eg. $z = constant$ contour lines of monkey saddle $z = x^3 - 3xy^2$, dog saddle $z = 4x^3y - 4xy^3$ etc, [205]).
- Contour curves start from a domain boundary point and end at a domain boundary point.

If the surface is subdivided along the iso-parametric lines which contain the local maxima and minima of curvature inside the domain and the contouring levels of curvature are chosen such that the contour curves avoid saddle points, as shown in Figs. A.3, A.4, A.5, A.6, each sub-patch will contain simple contour branches without loops or singularities. Therefore we can find all the starting points of the various levels of contour curves along the parameter domain boundary of each sub-patch by finding the roots of the following equations.

Starting with the Gaussian curvature

$$K(u,0) = \frac{A(u,0)}{S^4(u,0)} = C_K,\ K(u,1) = \frac{A(u,1)}{S^4(u,1)} = C_K,\ 0 \le u \le 1\,,\ (8.111)$$

$$K(0,v) = \frac{A(0,v)}{S^4(0,v)} = C_K,\ K(1,v) = \frac{A(1,v)}{S^4(1,v)} = C_K,\ 0 \le v \le 1\,,\ (8.112)$$

where C_K is the constant Gaussian curvature value. These equations can be rewritten as follows:

$$C_K S^4(u,0) - A(u,0) = 0,\ C_K S^4(u,1) - A(u,1) = 0,\ 0 \le u \le 1\,,\ (8.113)$$
$$C_K S^4(0,v) - A(0,v) = 0,\ \ C_K S^4(1,v) - A(1,v) = 0,\ \ 0 \le v \le 1\,.\ (8.114)$$

Equations (8.113), (8.114) are univariate polynomials of degree 8m-4 and 8n-4, respectively.

Similarly for the mean curvature

$$H(u,0) = \frac{B(u,0)}{2S^3(u,0)} = C_H,\ H(u,1) = \frac{B(u,1)}{2S^3(u,1)} = C_H,\ 0 \le u \le 1\,,(8.115)$$

$$H(0,v) = \frac{B(0,v)}{2S^3(0,v)} = C_H,\ H(1,v) = \frac{B(1,v)}{2S^3(1,v)} = C_H,\ 0 \le v \le 1\,,\ (8.116)$$

where C_H is the constant mean curvature value. These equations can be rewritten as follows

$$B(u,0) - 2C_H\sqrt{S^2(u,0)}S^2(u,0) = 0,\quad B(u,1) - 2C_H\sqrt{S^2(u,1)}S^2(u,1) = 0\,,$$
$$(8.117)$$

$$B(0,v) - 2C_H\sqrt{S^2(0,v)}S^2(0,v) = 0,\quad B(1,v) - 2C_H\sqrt{S^2(1,v)}S^2(1,v) = 0\,,$$
$$(8.118)$$

where $0 \le u, v \le 1$. Equations (8.117), (8.118) are the univariate irrational functions involving polynomials and square roots of polynomials which come from the normalization of the normal vector of the surface (see (3.3)). $B(u,v)$ is a polynomial of degree $(5m-3, 5n-3)$ and $S^2(u,v)$ is a polynomial of degree $(4m-2, 4n-2)$.

Finally for the principal curvatures

$$\kappa(u,0) = \frac{B(u,0) \pm \sqrt{f_3(u,0)}}{2S^3(u,0)} = C_\kappa , \qquad (8.119)$$

$$\kappa(u,1) = \frac{B(u,1) \pm \sqrt{f_3(u,1)}}{2S^3(u,1)} = C_\kappa , \qquad (8.120)$$

$$\kappa(0,v) = \frac{B(0,v) \pm \sqrt{f_3(0,v)}}{2S^3(0,v)} = C_\kappa , \qquad (8.121)$$

$$\kappa(1,v) = \frac{B(1,v) \pm \sqrt{f_3(1,v)}}{2S^3(1,v)} = C_\kappa , \qquad (8.122)$$

where $0 \leq u, v \leq 1$, the $\pm$ signs correspond to the maximum and minimum principal curvatures, and C_κ is the constant value of principal curvature and $f_3(u,v)$ is a polynomial function defined in (8.63). Equations (8.119) through (8.122) can be rewritten as follows:

$$B(u,0) \pm \sqrt{f_3(u,0)} - 2C_\kappa S^2(u,0)\sqrt{S^2(u,0)} = 0, \quad 0 \leq u \leq 1, \quad (8.123)$$

$$B(u,1) \pm \sqrt{f_3(u,1)} - 2C_\kappa S^2(u,1)\sqrt{S^2(u,1)} = 0, \quad 0 \leq u \leq 1, \quad (8.124)$$

$$B(0,v) \pm \sqrt{f_3(0,v)} - 2C_\kappa S^2(0,v)\sqrt{S^2(0,v)} = 0, \quad 0 \leq v \leq 1, \quad (8.125)$$

$$B(1,v) \pm \sqrt{f_3(1,v)} - 2C_\kappa S^2(1,v)\sqrt{S^2(1,v)} = 0, \quad 0 \leq v \leq 1. \quad (8.126)$$

Equations (8.123) through (8.126) are the univariate irrational functions involving polynomials and two square roots of polynomials which come from the analytical expression of the principal curvature and normalization of the normal vector of the surface. The starting points for contour curves of curvature occur in pairs, since non-loop contour lines must start from a domain boundary and must end at a domain boundary point.

Cuspidal edges are loci of points on the surface which have tangent discontinuity and appear as sharp edges on the surface. Cuspidal edges of an offset surface correspond to points on the progenitor where one of the principal curvatures is equal to $-\frac{1}{d}$ (d is the offset distance) [94]. Therefore we can use (8.123) through (8.126) to compute the starting points for tracing cuspidal edges on the offset, if the surface is subdivided along the iso-parametric lines which contain the local maxima and minima of principal curvatures, such that each sub-patch will not contain loops of cuspidal edges in its interior.

8.5.3 Mathematical formulation of contouring

Contour curves for constant curvature satisfy the following equation

$$C(u,v) = constant , \qquad (8.127)$$

where $C(u, v)$ is a curvature at the given point (u, v). The procedure for tracing contour curves is same as that of tracing method in the intersection problem that we discussed in Sect. 5.8.1. We now consider a space curve which lies on the surface represented by the parametric form $\mathbf{r}(t) = \mathbf{r}[u(t), v(t)]$. Differentiating (8.127) with respect to t yields

$$C_u \dot{u} + C_v \dot{v} = 0 , \tag{8.128}$$

where $\dot{u}$, $\dot{v}$ are the first derivatives with respect to t, and $(\dot{u}, \dot{v})$ gives the direction of the contour line in parameter space. The solutions to (8.128) can be written as

$$\dot{u} = \xi C_v, \quad \dot{v} = -\xi C_u , \tag{8.129}$$

where ξ is an arbitrary nonzero factor. When the curvature map is constructed in the uv parameter space, ξ can be chosen to provide arc-length parametrization in the parameter domain as follows

$$\xi = \pm \frac{1}{\sqrt{C_u^2 + C_v^2}} , \tag{8.130}$$

or when it is constructed on the surface itself, ξ can be chosen to provide arc length parametrization using first fundamental form (3.13) of the surface as a normalization condition as in (5.91)

$$\xi = \pm \frac{1}{\sqrt{EC_v^2 - 2FC_uC_v + GC_u^2}} , \tag{8.131}$$

where C_u and C_v are given in (8.35), (8.48) and (8.60) for Gaussian, mean and principal curvatures respectively.

The points of the contour curves are computed successively by integrating the initial value problem for a system of coupled nonlinear differential equations (8.129) using, for example the Runge-Kutta method or a more sophisticated variable stepsize and variable order Adams method [69]. Starting points are computed by the method described in Sect. 8.5.2. Accuracy of the contour curve depends on the number of points used to represent the contour curve by straight line segments. Note that for principal curvatures, C_u and C_v become singular at an umbilical point, therefore we avoid the contour level which is equivalent to the curvature value at the umbilics.

To generate a color curvature map, graphics packages require closed polygonal geometry. Therefore, for display purposes and also for theoretical interest, it is necessary to decompose a parameter sub-domain further into closed polygonal regions. We may use the *trip algorithm* introduced by Preusser [337] to polygonize the area between contour curves.

A closed polygon is defined by contour curves and the border lines of the domain as illustrated in Fig. 8.10. Let S_i^s and S_i^e be the starting and ending point of a contour curve respectively, and V_i be the vertices of the rectangular sub-domain. The algorithm is given as follows:

Trip algorithm [337]

1. Start the first round trip from S_1^s.
2. At S_1^e continue in a counter-clockwise direction on the rectangular boundary to the next starting point S_2^s or vertex.
3. Search a way back to the starting point S_1^s by following a contour curve from S_2^s, or a rectangular boundary from a vertex. As a result, the first polygon is given by S_1^s-S_1^e-S_2^s-S_2^e-V_1-S_1^s.
4. A second polygon is started at the next unused starting point in counter-clockwise direction and is completed in the same way which results in S_3^s-S_3^e-S_4^s-S_4^e-V_2-S_3^s.
5. When there are no unused points S, we start the remaining polygons at points S that have served only as the end of a contour curve, i.e. S_3^e and S_5^e.

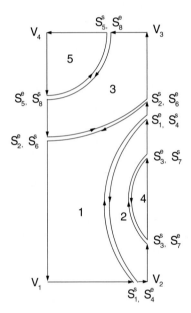

Fig. 8.10. Trip algorithm

8.5.4 Examples

To illustrate constant curvature contouring, we used a wave-like bicubic integral Bézier surface patch (see Fig. 8.11). The boundary twelve control points are coplanar so that the boundary curves form a square. The remaining four interior control points are not on the same plane. The control points are given as follows:

$$\begin{pmatrix} \mathbf{P}_{00} & \mathbf{P}_{01} & \mathbf{P}_{02} & \mathbf{P}_{03} \\ \mathbf{P}_{10} & \mathbf{P}_{11} & \mathbf{P}_{12} & \mathbf{P}_{13} \\ \mathbf{P}_{20} & \mathbf{P}_{21} & \mathbf{P}_{22} & \mathbf{P}_{23} \\ \mathbf{P}_{30} & \mathbf{P}_{31} & \mathbf{P}_{32} & \mathbf{P}_{33} \end{pmatrix} = \begin{pmatrix} (0, 0, 0) & (0, \frac{1}{3}, 0) & (0, \frac{2}{3}, 0) & (0, 1, 0) \\ (\frac{1}{3}, 0, 0) & (\frac{1}{3}, \frac{1}{3}, 1) & (\frac{1}{3}, \frac{2}{3}, \frac{3}{5}) & (\frac{1}{3}, 1, 0) \\ (\frac{2}{3}, 0, 0) & (\frac{2}{3}, \frac{1}{3}, -1) & (\frac{2}{3}, \frac{2}{3}, -\frac{3}{5}) & (\frac{2}{3}, 1, 0) \\ (1, 0, 0) & (1, \frac{1}{3}, 0) & (1, \frac{2}{3}, 0) & (1, 1, 0) \end{pmatrix}.$$

Therefore the surface is anti-symmetric with respect to $u = 0.5$. The wave-like surface can also be expressed in a graph form as $[x, y, h(x, y)]^T$ with

$$h(x, y) = 7.2x^3y^3 - 25.2x^3y^2 + 18x^3y - 10.8x^2y^3 + 37.8x^2y^2 - 27x^2y$$
$$+3.6xy^3 - 12.6xy^2 + 9xy , \qquad (8.132)$$

where $0 \leq x, y \leq 1$ and $x = u$, $y = v$. Although the wireframe of the surface looks simple (see Fig. 8.11), the surface is rich in its variety of differential geometry properties. The wave-like surface has four spherical umbilics at $(0.211, 0.052)$, $(0.211, 0.984)$, $(0.789, 0.052)$, $(0.789, 0.984)$ with principal curvature values 1.197, 0.267, -1.197, -0.267, and one flat umbilic at $(0.5, 0.440)$. None of them are local extrema according to the criterion 9.5.1.

To display the curvature of the subdivided surface clearly, we assigned discrete color to each closed region based on curvature level. The level was determined by taking the average value of the curvature values of the contour curves excluding the boundary lines which form the closed region. We assigned R (red), G (green) and B (blue) to the minimum, zero and maximum curvature values of the whole domain. The color of the curvature values in between is linearly interpolated.

Gaussian curvature. Color Plate A.3 shows a color map of the Gaussian curvature K. Since the surface is anti-symmetric with respect to $u = 0.5$, the Gaussian curvature which is the product of maximum and minimum principal curvatures is symmetric with respect to $u = 0.5$. The range of the curvature is $-81 \leq K \leq 10.297$. The global maximum Gaussian curvature $K = 10.297$ occurs at two stationary points within the domain $(0.195, 0.374)$, $(0.805, 0.374)$. The global minimum curvature $K = -81$ is located at two corners $(0, 0)$ and $(1, 0)$. There is also a saddle point inside the domain at $(0.5, 0.440)$, with value $K = 0$, which is a flat point of the surface. There are six local maxima and two local minima along the domain boundaries. Local maxima at $(0.211, 0)$, $(0.789, 0)$, $(0.211, 1)$, $(0.789, 1)$, $(0, 0.440)$, $(1, 0.440)$ with all values $K = 0$, and local minima at $(0.5, 0)$ with $K = -20.25$ and at $(0.5, 1)$ with $K = -7.29$. Since the two local maxima inside the domain have the same v coordinate, we subdivide the surface into two sub-domains along the iso-parametric line $v = 0.374$. In this picture, we avoid the curvature level $K = 0$ so as not to deal with the self-intersecting contour at the saddle point.

Mean curvature. Color Plate A.4 shows a color map of the mean curvature H. Because of anti-symmetry with respect to $u = 0.5$, mean curvature has $H = 0$ contour line at $u = 0.5$. Mean curvature varies from -4.056 to 4.056. Both global maximum and minimum curvature are located inside the domain at $(0.190, 0.414)$ and $(0.810, 0.414)$ respectively. There are seven local maxima and seven local minima along the boundary. The local maxima are located at $(0.116, 0)$, $(0.319, 0)$, $(0.789, 0)$, $(0.211, 1)$, $(0, 0.089)$, $(0, 0.861)$, $(1, 0.440)$ with $H = 0.539$, 0.539, -0.524, 0.121, 1.155, 1.155, -0.607. The local minima are located at $(0.211, 0)$, $(0.681, 0)$, $(0.884, 0)$, $(0.789, 1)$, $(0, 0.440)$, $(1, 0.089)$,

$(1, 0.861)$ with $H = 0.524, -0.539, -0.539, -0.121, 0.607, -1.155, -1.155$. Since the local maximum and minimum inside the domain are on the same iso-parametric line $v = 0.414$, we subdivide into two sub-domains at this line.

Maximum principal curvature. Color Plate A.5 shows a color map of the maximum principal curvature κ_{max}, which has a range of $-1.665 \le \kappa_{max} \le 9$. The global maximum is located at two corners $(0,0)$ and $(1,0)$, and global minimum is located at $(0.789, 0.303)$ which is a stationary point inside the domain. There is also a local maximum inside the domain at $(0.187, 0.440)$ with $\kappa_{max} = 6.607$ and four saddle points inside the domain at $(0.082, 0.802)$, $(0.114, 0.184)$, $(0.321, 0.157)$ and $(0.378, 0.851)$ with values $\kappa_{max} = 4.504, 5.127, 3.276$ and 2.470. There are two local maxima and six local minima along the boundary. The locations are $(0.478, 0)$, $(0.491, 1)$ with $\kappa_{max} = 4.569$ and $\kappa_{max} = 2.704$ for the local maxima and $(0.211, 0)$, $(0.789, 0)$, $(0.211, 1)$, $(0.789, 1)$, $(0, 0.440)$, $(1, 0.440)$ with $\kappa_{max} = 1.047, 0, 0.242, 0, 1.213, 0$ for the local minima. We subdivide the domain along the iso-parametric line $u = 0.187$ and $u = 0.789$ which contain the local maximum and minimum. The reason we choose the u iso-parametric line is that the minimum size of each domain in the u direction is larger than in the v direction.

Minimum principal curvature. Color Plate A.6 shows a color map of the minimum principal curvature κ_{min}, which has a range of $-9 \le \kappa_{min} \le 1.665$. The global maximum is located inside the domain at $(0.211, 0.303)$ with $\kappa_{min} = 1.665$. The global minimum is located at two corners $(0,0)$, $(1,0)$ with $\kappa_{min} = -9$. There are other stationary points within the domain, a minimum at $(0.813, 0.440)$ with the value $\kappa_{min} = -6.607$ and four saddle points inside the domain at $(0.622, 0.851)$, $(0.679, 0.157)$, $(0.886, 0.184)$ and $(0.918, 0.802)$ with values $\kappa_{min} = -2.470, -3.276, -5.127$ and -4.504. There are six local maxima and two local minima along the domain boundaries. The maxima are located at $(0.211, 0)$, $(0.789, 0)$, $(0.211, 1)$, $(0.789, 1)$, $(0, 0.440)$, $(1, 0.440)$ with $\kappa_{min} = 0, -1.047, 0, -0.242, 0, -1.213$. The local minima are located at $(0.478, 0)$, $(0.491, 1)$ with the value $\kappa_{min} = -4.299, -2.688$. For the same reason as in maximum principal curvature, the domain is subdivided at $u = 0.211$ and $u = 0.813$.

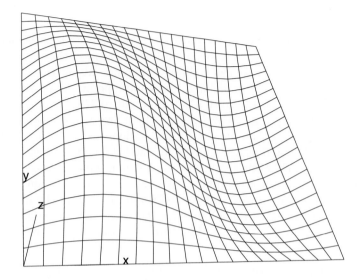

Fig. 8.11. Wave-like integral Bézier surface patch (adapted from [255])

9. Umbilics and Lines of Curvature

9.1 Introduction

An *umbilic* is a point on a surface where all normal curvatures are equal in all directions, and hence principal directions are indeterminate. Thus the orthogonal net of lines of curvature, which is described in Sect. 3.4, becomes singular at an umbilic. An obvious example of a surface consisting entirely of umbilical points is the sphere. Actually, spheres and planes are the only surfaces all of whose points are umbilics. The number of umbilics on a surface is often finite and they are isolated [1] [165, 412]. Umbilics have generic features and may act as fingerprints for shape recognition. At an umbilic, the directions of principal curvature can no longer be evaluated by second order derivatives and higher order derivatives are necessary to compute the lines of curvature near the umbilic. Monge (1746-1818), who with Gauss can be considered as the founder of differential geometry of curves and surfaces, first computed the lines of curvature of the ellipsoid (1796) which has four umbilics [320].

There exists an analogy between normal curvature and stress in elasticity theory [185]. For 2-D problems for example, it is well known that whatever the state of stress at a point, there will always be two orthogonal directions through the point in each of which the shear stress is zero. These two directions are called the axes of principal stress. The curve which lies along one of the axes of principal stress at all its points is called line of principal stress. Such lines form an orthogonal net. The point where two principal stresses are equal is called isotropic point. The state of stress at such point is that of a radial compression or tension, uniform in all directions. The lines of principal stress and isotropic points are analogous to the lines of curvature and umbilics, respectively.

There are a number of papers which deal with lines of curvature. Martin [265] introduced so called the *principal patches* whose sides are lines of curvature for use in geometric modeling. Principal patches can be created by imposing two conditions to the boundary curves known as position and frame matching [265, 290]. Among the principal patches, Dupin's cyclide patches

---

[1] Non-isolated umbilics can be found along an inflection line of a developable surface (see Sect. 9.7).

whose lines of curvature are all circular arcs are used for blending surface applications (see Pratt [332], Dutta et al. [83]).

Beck et al. [22], Farouki [96, 98], Hosaka [173], Maekawa et al. [257] provide a method to construct a net of lines of curvature on a B-spline surface. Lines of curvature are of considerable importance to plate-metal-based manufacturing [279]. When a sheet is to be shaped by rolling, then it is fed into the rolls according to a principal direction and the rolls are adjusted according to the principal curvature.

The generic features of the lines of curvature near an umbilic are fully discussed in classic work by Darboux (1896) [71], and more recently by Porteous [320, 321, 322], Maekawa et al. [257], and Gutierrez and Sotomayor [143]. Berry and Hannay [24] calculate the average density of umbilics for a surface whose deviation from a plane is specified by a Gaussian random surface, and showed the rarity of the *monstar* pattern. Maekawa and Patrikalakis [255] and Maekawa [246] describe a robust computational method to locate all isolated umbilics on a polynomial parametric surface and investigate the generic features of umbilics and the behavior of lines of curvature which pass through an umbilic. In computer vision, Brady et al. [37] compute the lines of curvature and regions of umbilics from range images. Sander and Zucker [364] extracted umbilics from an image by computing the index of the principal direction fields. Sinha and Besl [397] compute the lines of curvature from a range image and construct a quadrilateral mesh except at the umbilics.

A non-flat umbilic occurs at an elliptic point, while it never occurs at a hyperbolic point. From (3.31), where we assume convention (b) (see Fig. 3.7 (b) and Table 3.2), it is apparent that at an umbilic I and II are proportional because $\kappa = constant$, and hence we have the following relation at the umbilic

$$L = -\kappa E, \quad M = -\kappa F, \quad N = -\kappa G . \tag{9.1}$$

This result coincides with (3.45) where $k = -\kappa$. At umbilical points only, the principal directions are indeterminate and the net of lines of curvature may have singular properties. The lines of curvature depend only on the shape of the surface, and not the parametrization. Lines of curvature provide a method to describe the variation of principal curvatures across a surface. Lines of curvature can be obtained by integrating (3.41), which will be discussed further in Sect. 9.4.

In this chapter we employ sign convention (b) (see Fig. 3.7 (b) and Table 3.2) for the normal curvature.

9.2 Lines of curvature near umbilics

It is easily verified from (9.1) that $L + \kappa E$, $M + \kappa F$ and $N + \kappa G$ simultaneously vanish at the umbilics. Therefore for all du, dv, (3.41) is satisfied, and hence we cannot determine the direction of the lines of curvature which pass through the umbilic. In this section we investigate the pattern of the lines of curvature near generic umbilics. *Generic umbilics are stable with respect to small perturbations of the function representing the surface, while non-generic umbilics are unstable* [24, 364, 397]. Darboux [71] has described three generic features of lines of curvature in the vicinity of an umbilic. The three generic features are called *star, (le) monstar* and *lemon* based on the pattern of the net of lines of curvature. Color Plate A.7 illustrates these three patterns of the net of lines of curvature at the umbilic. The red solid line corresponds to the maximum principal curvature lines and the dotted blue line corresponds to minimum principal curvature lines, where convention (b) of sign of normal curvature is used (see Fig. 3.7 (b) and Table 3.2). Three lines of curvature pass through the umbilic for monstar and star, while only one passes for the lemon. The criterion distinguishing monstar from star is that all three directions of lines of curvature through an umbilic are contained in a right angle, whereas in the star case they are not contained in a right angle. There are no other patterns except for non-generic cases. An example of a non-generic umbilic can be offered by the two poles of a convex closed surface of revolution [165]. Figure 9.1 (a) shows the non-generic umbilic of a paraboloid of revolution $z = x^2 + y^2$ which has an umbilic that infinite number of lines of curvature pass through. If we perturb a coefficient in the function representing the surface slightly to $z = \frac{x^2}{8/7} + y^2$ corresponding to an elliptic paraboloid then the non-generic umbilic splits into two lemon-type generic umbilics as shown in Fig. 9.1 (b).

Consider a surface in Monge form

$$\mathbf{r} = [x, y, h(x, y)]^T , \qquad (9.2)$$

where $h(x, y)$ is a C^3 smooth function, i.e. it has continuous derivatives up to order three. We can Taylor expand the z component of the surface as in (8.73). Suppose the surface $\mathbf{r}$ has an umbilic at the origin and its tangent plane coincides with the xy plane, then it is apparent that $h(0, 0) = h_x(0, 0) = h_y(0, 0) = 0$. Evaluating the coefficients of the first and second fundamental forms of the explicit surface at the origin, which are given in (3.63) through (3.65), we obtain

$$E(0,0) = 1, \quad F(0,0) = 0, \quad G(0,0) = 1 , \qquad (9.3)$$
$$L(0,0) = h_{xx}(0,0), \quad M = h_{xy}(0,0), \quad N = h_{yy}(0,0) . \qquad (9.4)$$

Then it is apparent from (9.1) that $h_{xx}(0,0) = h_{yy}(0,0) = -\kappa(0,0)$ and $h_{xy}(0,0) = 0$. Consequently we can rewrite (8.73) into a simpler form:

$$h(x, y) = -\frac{\kappa(0, 0)}{2}(x^2 + y^2) \qquad (9.5)$$

$$+ \frac{1}{6}[x^3 h_{xxx}(0,0) + 3x^2 y h_{xxy}(0,0) + 3xy^2 h_{xyy}(0,0) + y^3 h_{yyy}(0,0)]$$

$$+ R(x, y)(|x, y|^3) .$$

From (9.5), we can observe that the equation of the surface near the umbilic is governed by the cubic form $h_c(x, y)$

$$h_c(x, y) = \frac{1}{6}(\alpha x^3 + 3\beta x^2 y + 3\gamma x y^2 + \delta y^3),\qquad(9.6)$$

where

$$\alpha = h_{xxx}(0,0),\ \beta = h_{xxy}(0,0),\ \gamma = h_{xyy}(0,0),\ \delta = h_{yyy}(0,0).\qquad(9.7)$$

Note that α, β, γ and δ vanish for a paraboloid of revolution $\mathbf{r}(u, v) = [x, y, a(x^2 + y^2)]^T$ at $(x, y) = (0, 0)$.

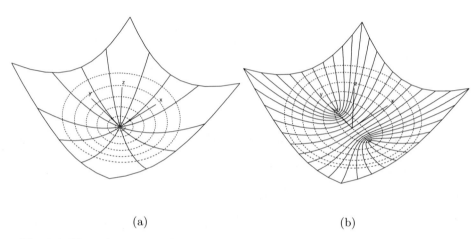

(a)   (b)

Fig. 9.1. Lines of curvature of paraboloid of revolution and elliptic paraboloid. *Solid lines* and *dotted lines* represent maximum and minimum principal curvature lines respectively: (**a**) paraboloid of revolution ($z = x^2 + y^2$) has a non-generic umbilic at $(0,0,0)$, (**b**) elliptic paraboloid ($z = \frac{x^2}{8/7} + y^2$) has two lemon-type umbilics at $(0,\pm0.1890,0.0357)$ (adapted from [257])

To study the behavior of the umbilics we can express (9.6) in polar coordinates $x = r\cos\theta$ and $y = r\sin\theta$ for a fixed radius $r = \sqrt{x^2 + y^2}$ [24]:

$$h_c(\theta) = \frac{r^3}{6}(\alpha\cos^3\theta + 3\beta\cos^2\theta\sin\theta + 3\gamma\cos\theta\sin^2\theta + \delta\sin^3\theta).\qquad(9.8)$$

It can be easily verified that $h_c(\theta+\pi) = -h_c(\theta)$. Therefore the cubic function is an antisymmetric function of θ. The roots of $dh_c/d\theta = 0$ will give the angles where local maxima and minima of $h_c(\theta)$ may occur around the umbilic, depending on the multiplicity of the roots. When there are three distinct roots, each of the roots gives the local extremum. When there are two equal roots, the double roots will provide neither maxima nor minima, however the single root gives an extremum. When there are three equal roots, the triple roots give an extremum. Since it is an antisymmetric function, maxima and minima of h_c occur on the same straight line which passes through the umbilic. Differentiating (9.8) with respect to θ and setting the equation equal to zero yields

$$\frac{dh_c(\theta)}{d\theta} = \frac{r^3}{2}(\beta\cos^3\theta - (\alpha - 2\gamma)\sin\theta\cos^2\theta + (\delta - 2\beta)\sin^2\theta\cos\theta - \gamma\sin^3\theta)$$
$$= 0. \tag{9.9}$$

When one of the roots of (9.9) is $\theta = 0$ or π then β must be zero, and when $\theta = \frac{\pi}{2}$ or $\frac{3}{2}\pi$ then γ must be zero. Conversely we can say that when $\beta = 0$ one of the roots is $\theta = 0$ or π, and when $\gamma = 0$ one of the roots is $\theta = \frac{\pi}{2}$ or $\frac{3}{2}\pi$. Consequently when $\beta \neq 0$, we can divide (9.9) by $\beta\sin^3\theta$ resulting in

$$t^3 - \frac{\alpha - 2\gamma}{\beta}t^2 + \frac{\delta - 2\beta}{\beta}t - \frac{\gamma}{\beta} = 0, \tag{9.10}$$

where $t = \cot\theta$. Similarly when $\gamma \neq 0$, we can divide (9.9) by $\gamma\cos^3\theta$ resulting in

$$\bar{t}^3 - \frac{\delta - 2\beta}{\gamma}\bar{t}^2 + \frac{\alpha - 2\gamma}{\gamma}\bar{t} - \frac{\beta}{\gamma} = 0, \tag{9.11}$$

where $\bar{t} = \tan\theta$. These cubic equations may be reduced by the substitution

$$t = s + \frac{\alpha - 2\gamma}{3\beta} \qquad \bar{t} = s + \frac{\delta - 2\beta}{3\gamma}, \tag{9.12}$$

to the normal form [403]

$$s^3 + 3ps + 2q = 0, \tag{9.13}$$

where

$$\text{when} \quad \beta \neq 0 \quad p = \frac{3\beta(\delta - 2\beta) - (\alpha - 2\gamma)^2}{9\beta^2}, \tag{9.14}$$

$$q = \frac{(2\gamma - \alpha)[2(\alpha - 2\gamma)^2 - 9(\delta - 2\beta)\beta] - 27\beta^2\gamma}{54\beta^3}, \tag{9.15}$$

$$\text{when} \quad \gamma \neq 0 \quad p = \frac{3\gamma(\alpha - 2\gamma) - (\delta - 2\beta)^2}{9\gamma^2}, \tag{9.16}$$

$$q = \frac{(2\beta - \delta)[2(\delta - 2\beta)^2 - 9(\alpha - 2\gamma)\gamma] - 27\beta\gamma^2}{54\gamma^3}. \tag{9.17}$$

The solutions to the cubic equation are given by:

- When $q^2 + p^3 > 0$; there are three distinct roots, one is real root and the other two are conjugate complex roots. The real root gives a function extremum and is given by

$$s = \sqrt[3]{-q + \sqrt{q^2 + p^3}} + \sqrt[3]{-q - \sqrt{q^2 + p^3}} \,, \qquad (9.18)$$

- When $q^2 + p^3 = 0$; there are three real roots at least two of which are equal and are given by

$$s = \mp 2\sqrt{-p}, \ \pm\sqrt{-p}, \ \pm\sqrt{-p} \,, \qquad (9.19)$$

where the upper sign is to be used if q is positive and the lower sign if q is negative. Therefore there is at most one root which will provide a function extremum. This is a non-generic case, since small perturbation will yield the case either above or below.

- When $q^2 + p^3 < 0$; there are three unequal real roots, which provide three function extrema, and are given by

$$s = 2\sqrt{-p} \, \cos\left(\frac{\tau}{3}\right), \ 2\sqrt{-p} \, \cos\left(\frac{\tau}{3} + \frac{2\pi}{3}\right), \ 2\sqrt{-p} \, \cos\left(\frac{\tau}{3} + \frac{4\pi}{3}\right) \,,$$
$$(9.20)$$

where $\cos\tau = \mp\sqrt{-\frac{q^2}{p^3}}$ and the upper sign is to be used if q is positive and the lower if q is negative.

Consequently there is either one single angle (lemon) or three different angles (star, monster) corresponding to one maximum opposite one minimum or three maxima opposite three minima for generic case. Corresponding to these angles there are lines of curvature either one or three passing through the umbilics.

Another way of classifying an umbilic is to compute the *index* around it [24, 364]. The lemon and monstar have the same index $+\frac{1}{2}$, while the star has the index $-\frac{1}{2}$. The index is defined as an amount of rotation that a straight line tangent to lines of curvature experiences when rotating in the counterclockwise direction around a small closed path around the umbilic. To compute the index of the umbilic, we can evaluate the angle ψ_i, which is the angle of principal direction, at n points along a boundary curve which surrounds the umbilic. The angle ψ_i is obtained by using the first of (3.41) (see Table 3.2 for sign convention) as

$$\tan\psi_i = \frac{dv}{du} = -\frac{L + \kappa E}{M + \kappa F} \quad or \quad \psi_i = \arctan\left(-\frac{L + \kappa E}{M + \kappa F}\right) \,, \qquad (9.21)$$

where $-\frac{\pi}{2} \le \psi_i \le \frac{\pi}{2}$. Since ψ_i can also be obtained from the second equation of (3.41) we also get

$$\tan\psi_i = \frac{dv}{du} = -\frac{M + \kappa F}{N + \kappa G} \quad or \quad \psi_i = arctan\left(-\frac{M + \kappa F}{N + \kappa G}\right) . \qquad (9.22)$$

If both $L + \kappa E$ and $M + \kappa F$ are zero or small in absolute value, we use (9.22) otherwise we use (9.21), or if $|L+\kappa E| > |M+\kappa F|$ we can invert (9.21) and solve for $\tan(\bar{\psi}_i) = du/dv$. Consequently the index can be computed by

$$Index = \frac{1}{2\pi}\sum_{i=0}^{n}\Delta\psi_i , \qquad (9.23)$$

where

$$\Delta\psi_i = \psi_{(i+1) \bmod n} - \psi_i \quad and \quad -\frac{\pi}{2} \le \Delta\psi_i \le \frac{\pi}{2} , \qquad (9.24)$$

and *mod* is the modulo operator. It is used to account for the first point which is also the last point at which the direction field is evaluated. For the examples in this book, 20 points per boundary curve were adequate for estimation of the index. Fig. 9.2 illustrates the direction field of the maximum principal curvature around the star, monstar and lemon type umbilics.

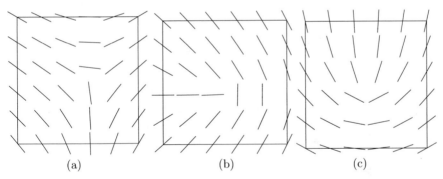

(a)   (b)   (c)

Fig. 9.2. Direction field near umbilics: (**a**) star-type, (**b**) monstar-type, (**c**) lemon-type (adapted from [257])

9.3 Conversion to Monge form

To compute the angles that the tangents to the lines of curvature at the umbilics make with the axes, the surface has to be set in Monge form for each umbilic separately. Therefore for each umbilic on the surface, a coordinate transformation is needed. But before we conduct the transformation, we need to locate all umbilical points. The principal curvature functions κ are defined in (3.49), (3.50) as $\kappa(u,v) = H(u,v) \pm \sqrt{H^2(u,v) - K(u,v)}$. If we set

$$W(u, v) = H^2(u, v) - K(u, v) , \qquad (9.25)$$

then umbilic occurs precisely at a point where the function $W(u, v)$ is zero. Since κ is a real valued function, it follows that $W(u, v) \geq 0$. Consequently, an umbilic occurs where the function $W(u, v)$ has a global minimum.

If the surface representation is non-degenerate and C^k smooth, then $H(u, v)$, $K(u, v)$ and hence $W(u, v)$ are C^{k-2} smooth. Although we are particularly interested in Bézier surfaces which are C^∞, we relax the continuity assumption for $W(u, v)$ by assuming that $W(u, v)$ is at least C^2 smooth which is guaranteed if the surface is C^4. Already the assumption of differentiability for $W(u, v)$, which is weaker than C^2, and the condition that $W(u, v)$ has a global minimum at the umbilic implies that $\nabla W = \mathbf{0}$ at an umbilic. Therefore the governing equation for locating the umbilics are given by

$$W(u, v) = 0, \quad W_u(u, v) = 0, \quad W_v(u, v) = 0 . \qquad (9.26)$$

If the surface $\mathbf{r}(u, v)$ is a polynomial parametric surface patch (e.g. a Bézier patch), then we denote $W(u, v) = \frac{P_N(u,v)}{P_D(u,v)}$ where $P_N(u, v)$ and $P_D(u, v)$ are polynomials in u, v. Hence (9.26) reduce to an overconstrained system of nonlinear polynomial equations (see also Sect. 8.2.3)

$$P_N(u, v) = 0, \quad \frac{\partial P_N(u, v)}{\partial u} = 0, \quad \frac{\partial P_N(u, v)}{\partial v} = 0 . \qquad (9.27)$$

A robust and efficient solution technique based on the interval projected polyhedron algorithm to solve a system of nonlinear polynomial equations is discussed in Chap. 4.

Consider a global frame $o\text{-}xyz$ and a surface $\mathbf{r} = [x(u, v), y(u, v), z(u, v)]^T$ with an umbilical point O as illustrated in Fig. 9.3. The umbilical point is represented by a position vector $\mathbf{r}_o$ given by:

$$\mathbf{r}_o = (x_o, y_o, z_o)^T = [x(u_o, v_o), y(u_o, v_o), z(u_o, v_o)]^T . \qquad (9.28)$$

To represent the surface in the Monge form at the umbilic O, we need to attach an orthogonal Cartesian reference frame to it, say $O\text{-}XYZ$, and we represent a surface point $\mathbf{r}(u, v)$ in the frame $O\text{-}XYZ$ as $\mathbf{R}(u, v)$. We choose unit vectors $\frac{\mathbf{r}_u}{|\mathbf{r}_u|}$, $\mathbf{N} \times \frac{\mathbf{r}_u}{|\mathbf{r}_u|}$, $\mathbf{N}$ as directions of X, Y and Z axes as shown in Fig. 9.3, where $\mathbf{r}_u$ is the tangential vector in u direction and $\mathbf{N} = (N_x, N_y, N_z)^T$ is the unit normal vector of the surface at the umbilic.

If we concatenate these three unit vectors $\frac{\mathbf{r}_u}{|\mathbf{r}_u|}$, $\mathbf{N} \times \frac{\mathbf{r}_u}{|\mathbf{r}_u|}$, $\mathbf{N}$ in a single matrix, we obtain a description of the orientation of the Monge form with respect to the frame $o\text{-}xyz$ which is called a *rotation matrix* Ω

$$\Omega = \begin{pmatrix} \frac{x_u}{|\mathbf{r}_u|} & \frac{N_y z_u - N_z y_u}{|\mathbf{r}_u|} & N_x \\ \frac{y_u}{|\mathbf{r}_u|} & \frac{N_z x_u - N_x z_u}{|\mathbf{r}_u|} & N_y \\ \frac{z_u}{|\mathbf{r}_u|} & \frac{N_x y_u - N_y x_u}{|\mathbf{r}_u|} & N_z \end{pmatrix}_{(u_o, v_o)} . \qquad (9.29)$$

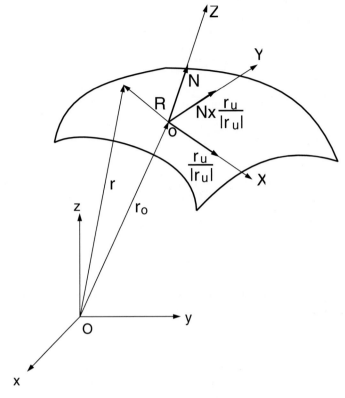

Fig. 9.3. Definition of coordinate system

Then the relation between $\mathbf{r}(u,v)$ and $\mathbf{R}(u,v)$ is:

$$\mathbf{r}(u,v) = \mathbf{r}_o + \Omega \mathbf{R}(u,v) \; . \tag{9.30}$$

Using (9.30), we can solve for $\mathbf{R}(u,v)$ as a function of $\mathbf{r}(u,v)$ that is the coordinate of P expressed in frame $O\text{-}XYZ$ as a function of the coordinate of point P expressed in $o\text{-}xyz$ frame as

$$\mathbf{R}(u,v) = \Omega^{-1}(\mathbf{r}(u,v) - \mathbf{r}_o) \; , \tag{9.31}$$

where Ω^{-1} is the inverse matrix of Ω. Since Ω is an orthonormal matrix, Ω^{-1} can be replaced by the transpose matrix Ω^T, therefore

$$\mathbf{R}(u,v) = \begin{pmatrix} X(u,v) \\ Y(u,v) \\ h(u,v) \end{pmatrix} = \Omega^T [\mathbf{r}(u,v) - \mathbf{r}_o] \; , \tag{9.32}$$

or equivalently

$$X = X(u, v) = \left(\frac{\mathbf{r}_u}{|\mathbf{r}_u|}\right)_o \cdot [\mathbf{r}(u, v) - \mathbf{r}_o] , \tag{9.33}$$

$$Y = Y(u, v) = \left(\frac{(\mathbf{r}_u \times \mathbf{r}_v) \times \mathbf{r}_u}{|\mathbf{r}_u \times \mathbf{r}_v||\mathbf{r}_u|}\right)_o \cdot [\mathbf{r}(u, v) - \mathbf{r}_o] , \tag{9.34}$$

$$Z = h(u, v) = \left(\frac{\mathbf{r}_u \times \mathbf{r}_v}{|\mathbf{r}_u \times \mathbf{r}_v|}\right)_o \cdot [\mathbf{r}(u, v) - \mathbf{r}_o] , \tag{9.35}$$

where subscript o denotes that the expressions are evaluated at (u_o, v_o). In (9.33) through (9.35), $\mathbf{r}(u, v)$ is the only term that is the function of u and v, whereas all terms involving $\mathbf{r}_u$ and $\mathbf{r}_v$ are evaluated at (u_o, v_o). Now we want to express u and v as functions of X and Y, i.e. $u = u(X, Y)$ and $v = v(X, Y)$, using (9.33) and (9.34), so that $\mathbf{R}$ can be written as the same form as (9.2):

$$\mathbf{R} = (X, Y, Z)^T = [X, Y, h(u(X, Y), v(X, Y))]^T . \tag{9.36}$$

According to the *inverse function theorem* [76, 305], this is possible if and only if

$$\begin{vmatrix} X_u & X_v \\ Y_u & Y_v \end{vmatrix} \neq 0 . \tag{9.37}$$

If we set $\mathbf{I}$ and $\mathbf{J}$ as

$$\mathbf{I} = \left(\frac{\mathbf{r}_u}{|\mathbf{r}_u|}\right)_{(u_o, v_o)} , \tag{9.38}$$

$$\mathbf{J} = \left(\frac{(\mathbf{r}_u \times \mathbf{r}_v) \times \mathbf{r}_u}{|\mathbf{r}_u \times \mathbf{r}_v||\mathbf{r}_u|}\right)_{(u_o, v_o)} , \tag{9.39}$$

then the determinant can be evaluated using the vector identity (3.15) as follows:

$$\begin{aligned} X_u Y_v - Y_u X_v &= (\mathbf{I} \cdot \mathbf{r}_u)(\mathbf{J} \cdot \mathbf{r}_v) - (\mathbf{J} \cdot \mathbf{r}_u)(\mathbf{I} \cdot \mathbf{r}_v) , \\ &= (\mathbf{I} \times \mathbf{J}) \cdot (\mathbf{r}_u \times \mathbf{r}_v) , \\ &= \mathbf{N}(u_o, v_o) \cdot (\mathbf{r}_u \times \mathbf{r}_v) . \end{aligned} \tag{9.40}$$

Since we are assuming a regular surface, (9.40) will never vanish, and hence we can apply the inverse function theorem. To evaluate α, β, γ, δ in (9.6) we need to compute h_{XXX}, h_{XXY}, h_{XYY}, h_{YYY} which can be computed using the chain rule as follows:

$$h_X = h_u u_X + h_v v_X ,$$
$$h_Y = h_u u_Y + h_v v_Y ,$$
$$h_{XX} = h_{uu} u_X^2 + 2h_{uv} u_X v_X + h_{vv} v_X^2 + h_u u_{XX} + h_v v_{XX} ,$$
$$h_{XY} = h_{uu} u_X u_Y + h_{uv}(u_X v_Y + u_Y v_X) + h_{vv} v_X v_Y + h_u u_{XY} + h_v v_{XY} ,$$
$$h_{YY} = h_{uu} u_Y^2 + 2h_{uv} u_Y v_Y + h_{vv} v_Y^2 + h_u u_{YY} + h_v v_{YY} ,$$

$$
\begin{aligned}
h_{XXX} =\ & h_{uuu}u_X^3 + 3h_{uuv}u_X^2 v_X + 3h_{uvv}u_X v_X^2 + h_{vvv}v_X^3 \\
& +3(h_{uu}u_X u_{XX} + h_{uv}u_X v_{XX} + h_{uv}u_{XX}v_X + h_{vv}v_X v_{XX}) \\
& +h_u u_{XXX} + h_v v_{XXX}\ , \\
h_{XXY} =\ & h_{uuu}u_X^2 u_Y + h_{uuv}u_X(2u_Y v_X + u_X v_Y) + h_{uvv}v_X(2u_X v_Y + u_Y v_X) \\
& +h_{vvv}v_X^2 v_Y + h_{uu}(2u_X u_{XY} + u_{XX}u_Y) \\
& +h_{uv}(2u_X v_{XY} + u_{XX}v_Y + u_Y v_{XX} + 2u_{XY}v_X) \\
& +h_{vv}(2v_X v_{XY} + v_{XX}v_Y) + h_u u_{XXY} + h_v v_{XXY}\ , \\
h_{XYY} =\ & h_{uuu}u_X u_Y^2 + h_{uuv}u_Y(2u_X v_Y + u_Y v_X) + h_{uvv}v_Y(2u_Y v_X + u_X v_Y) \\
& +h_{vvv}v_X v_Y^2 + h_{uu}(2u_{XY}u_Y + u_X u_{YY}) \\
& +h_{uv}(2u_{XY}v_Y + u_X v_{YY} + u_{YY}v_X + 2u_Y v_{XY}) \\
& +h_{vv}(2v_{XY}v_Y + v_X v_{YY}) + h_u u_{XYY} + h_v v_{XYY}\ , \\
h_{YYY} =\ & h_{uuu}u_Y^3 + 3h_{uuv}u_Y^2 v_Y + 3h_{uvv}u_Y v_Y^2 + h_{vvv}v_Y^3 \\
& +3(h_{uu}u_Y u_{YY} + h_{uv}u_Y v_{YY} + h_{uv}u_{YY}v_Y + h_{vv}v_Y v_{YY}) \\
& +h_u u_{YYY} + h_v v_{YYY}\ .
\end{aligned}
\tag{9.41}
$$

The partial derivatives of h with respect to u and v can be obtained easily from (9.35). We can determine u_X, u_Y, v_X and v_Y by using the inverse function theorem [76, 305] as follows:

$$
\begin{bmatrix} u_X & u_Y \\ v_X & v_Y \end{bmatrix} = \begin{bmatrix} X_u & X_v \\ Y_u & Y_v \end{bmatrix}^{-1} .
\tag{9.42}
$$

Hence

$$
u_X = \frac{Y_v}{X_u Y_v - Y_u X_v}, \quad u_Y = \frac{-X_v}{X_u Y_v - Y_u X_v} ,
\tag{9.43}
$$

$$
v_X = \frac{-Y_u}{X_u Y_v - Y_u X_v}, \quad v_Y = \frac{X_u}{X_u Y_v - Y_u X_v} .
\tag{9.44}
$$

We can also evaluate the higher-order derivatives such as u_{XX}, v_{XX}, u_{XY}, v_{XX}, u_{YY}, v_{YY}, u_{XXX}, v_{XXX}, u_{XXY}, v_{XXY}, u_{XYY}, v_{XYY}, u_{YYY}, v_{YYY} using the chain rule. Once h_{XXX}, h_{XXY}, h_{XYY}, h_{YYY} are obtained, we can compute the angles of tangent lines to the lines of curvature passing through the umbilic using (9.10) to (9.20). Since the angles are evaluated in the XY-plane we need to map back to the parametric uv-space for integration. Consider a point on the tangent line which passes through the origin and lies on the XY-plane, say $(r\cos\theta,\ r\sin\theta)$. Then the point can be expressed in terms of u, v using the vectors along the X and Y axes:

$$
\begin{aligned}
& r\cos\theta \frac{\mathbf{r}_u}{|\mathbf{r}_u|} + r\sin\theta \frac{\mathbf{r}_u \times \mathbf{r}_v}{|\mathbf{r}_u \times \mathbf{r}_v|} \times \frac{\mathbf{r}_u}{|\mathbf{r}_u|} \\
&= r\left[\frac{\cos\theta}{|\mathbf{r}_u|} - \frac{\sin\theta(\mathbf{r}_u \cdot \mathbf{r}_v)}{|\mathbf{r}_u \times \mathbf{r}_v||\mathbf{r}_u|}\right]\mathbf{r}_u + r\left[\frac{\sin\theta|\mathbf{r}_u|}{|\mathbf{r}_u \times \mathbf{r}_v|}\right]\mathbf{r}_v \\
&= \sigma\mathbf{r}_u + \rho\mathbf{r}_v\ .
\end{aligned}
\tag{9.45}
$$

Therefore the angle between u-axis and the tangent of the line of curvature in the uv parametric space is given by

$$\phi = arctan\left(\frac{\rho}{\sigma}\right). \tag{9.46}$$

9.4 Integration of lines of curvature

A line of curvature is a curve on a surface that has tangents which are principal directions at all of its points as we discussed in Sect. 3.4. The principal directions at a given point are those directions for which the normal curvature takes on minimum and maximum values. If the point is not an umbilic the principal directions are orthogonal. A line of curvature indicates a directional flow for the maximum or the minimum curvature across the surface. It is advantageous to express the curvature line with an arc length parametrization as $u = u(s)$ $v = v(s)$. Every principal curvature direction vector must fulfill (3.41). Hence from the first equation of (3.41) (see Table 3.2) we get

$$u' = \frac{du}{ds} = \eta(M + \kappa F) ,$$
$$v' = \frac{dv}{ds} = -\eta(L + \kappa E) , \tag{9.47}$$

where η is an arbitrary nonzero factor. At first sight, one may expect to obtain the lines of curvature by integrating (9.47). However the subsequent considerations show that a simple integration of (9.47) is generally not sufficient to compute the principal curvature lines even in situations where one does not encounter an umbilic. Namely there may occur several problems, including cases A and B below, and the criterion in (9.52) may be used to control the orientation while integrating along the curvature line. Since a principal curvature direction vector must also fulfill the second equation of (3.41) we also get

$$u' = \frac{du}{ds} = \mu(N + \kappa G) ,$$
$$v' = \frac{dv}{ds} = -\mu(M + \kappa F) . \tag{9.48}$$

The solutions u', v' of the first and the second equations of (3.41) are linearly dependent, because the system of linear equations given by (3.41) has a rank smaller than 2. It is possible:

A. That the coefficients in one of the equations can both be zero while they are not both zero in the other equation.
B. That both coefficients in one equation are small in absolute value while the other equation contains one coefficient which is large in absolute value.

Case B is encountered more often than case A. In case A, using the equation with zero coefficients yields an incorrect result, because this equation does not contain enough information to find the principal curvature direction. In case B, using the equation with the small coefficients may yield numerical inaccuracies which could be avoided by using the other equation. Alourdas [5] has developed an algorithm which makes the choice of the equation dependent on the size of the coefficients. Since $M + \kappa F$ is a common coefficient, if $|L + \kappa E| \geq |N + \kappa G|$ we solve (9.47) otherwise we solve (9.48).

We want to point out that also case A may easily occur. Therefore one needs provisions in the algorithm which takes this into account. We give now a simple example illustrating case A using a parabolic cylinder $\mathbf{r}(u, v) = (u, v, v^2)^T$. Clearly the maximum principal curvature on the parabolic cylinder is zero everywhere. Also it is apparent that $\mathbf{r}_{uu} = \mathbf{0}$ and $\mathbf{r}_{uv} = \mathbf{0}$, hence $L = M = 0$. Therefore $L + \kappa_{max} E$ and $M + \kappa_{max} F$ become zero, while $N + \kappa_{max} G \neq 0$, which can be seen by an easy computation or using the fact that a parabolic cylinder has no umbilics. Farouki [98] proved that one of the solutions (u', v') defining a principal direction (i.e. (9.47) or (9.48)) becomes indeterminate at a nonumbilic point if and only if the principal direction is tangent to a surface parameter line at that point as in this example.

It remains to determine factors η and μ. If the curvature line is arc length parametrized, the first fundamental form provides the normalization condition

$$E \left(\frac{du}{ds} \right)^2 + 2F \frac{du}{ds} \frac{dv}{ds} + G \left(\frac{dv}{ds} \right)^2 = 1 . \qquad (9.49)$$

Substituting (9.47) into (9.49), η is determined to be

$$\eta = \frac{\pm 1}{\sqrt{E(M + \kappa F)^2 - 2F(M + \kappa F)(L + \kappa E) + G(L + \kappa E)^2}} . \qquad (9.50)$$

Likewise μ is determined to be

$$\mu = \frac{\pm 1}{\sqrt{E(N + \kappa G)^2 - 2F(N + \kappa G)(M + \kappa F) + G(M + \kappa F)^2}} . \qquad (9.51)$$

The sign of η or μ determines the direction in which the solution proceeds. Choosing a fixed sign for η or μ does not guarantee that the vector (u', v') would not change direction. The need to adjust the sign of η or μ becomes even more obvious if one determines the principal curvature vector always by the numerically preferable equation in the system (3.41). The vectors obtained from (9.47) and (9.48) are linearly dependent but they do not need to have the same orientation.

The criterion which is employed in order to determine the sign of η or μ is given by the following inequality

$$| - (u'^p \mathbf{r}_u^p + v'^p \mathbf{r}_v^p) - (u' \mathbf{r}_u + v' \mathbf{r}_v)| < |(u'^p \mathbf{r}_u^p + v'^p \mathbf{r}_v^p) - (u' \mathbf{r}_u + v' \mathbf{r}_v)| ,$$

$$(9.52)$$

where $\mathbf{r}$ is a curvature line represented by the parametric form $\mathbf{r}(s) = \mathbf{r}(u(s),$ $v(s))$ and the superscript p means evaluation at the previous time step during the integration of the curvature line. It is obvious that inequality (9.52) is true if and only if the tangent vector $(u' \mathbf{r}_u + v' \mathbf{r}_v)$ reverses direction because (9.52) says that the negative tangent vector of the preceding time step is closer to the new tangent vector than the positive tangent vector of the preceding time step. When inequality (9.52) is true, the sign of η or μ should be changed to assure that the solution path does not reverse direction. Farouki [98] derives another criterion for preventing the reversal of integration direction.

We can trace the lines of curvature by integrating the initial value problem for a system of coupled nonlinear ordinary differential equations using standard numerical techniques [69, 126] such as Runge-Kutta method or a more sophisticated variable stepsize and variable order Adams method. Starting points for lines of curvature passing through the umbilics are obtained by slightly shifting outwards in the directions given by (9.46) from the umbilic. Accuracy of the lines of curvature depends on the number of integrated points used to represent the contour line by straight line segments.

9.5 Local extrema of principal curvatures at umbilics

In this section we discuss a criterion which assures the existence of local extrema of the principal curvature functions κ_{max} and κ_{min} at umbilical points of the surface [257]. The problem of detecting local extrema of principal curvature functions is motivated by engineering applications. When a ball-end mill cutter is used for NC machining, the cutter radius must be smaller than the smallest concave radius of curvature of the surface to be machined to avoid local overcut (gouging) (see Sect. 11.1.2). Gouging is the one of the most critical problems in NC machining of free-form surfaces. Therefore, we must determine the distribution of the principal curvatures of the surface, which are upper and lower bounds of the normal curvature at a given point, to select the cutter size. A natural approach to locate local extrema of the functions κ_{max} and κ_{min} would in principle be to search for zeros of the gradient vector field $\nabla \kappa_{max}$ and $\nabla \kappa_{min}$ and then use tools from differential calculus to decide if at those zeros the principal curvature functions attain extrema. The problem with this approach however is that the curvature functions κ_{max} and κ_{min} are generally not differentiable at the umbilics although those points may also be candidates for local principal curvature extrema. We will present a *necessary and sufficient* criterion, which always *detects* the existence of a local extremum of the principal curvature functions κ_{max} and κ_{min} at an umbilic, except in presence of rare well defined and easily computable conditions. Under such rare condition, the criterion will become only sufficient. This criterion is practical because it is almost always applicable and easily evaluated.

We discuss the local behavior of the functions κ_{max} and κ_{min} in the neighborhood of an umbilic. First let us consider a Taylor expansion around an umbilic (u_o, v_o) for the function defined in (9.25). We obtain

$$W(u,v) = W(u_o, v_o) + [(u - u_o)W_u(u_o, v_o) + (v - v_o)W_v(u_o, v_o)] \quad (9.53)$$
$$+ \frac{1}{2!}[(u - u_o)^2 W_{uu}(u_o, v_o) + 2(u - u_o)(v - v_o)W_{uv}(u_o, v_o)$$
$$+ (v - v_o)^2 W_{vv}(u_o, v_o)] + R(u - u_o, v - v_o)|(u - u_o, v - v_o)|^2 ,$$

with

$$\lim_{u \to u_o, v \to v_o} R(u - u_o, v - v_o) = 0 . \quad (9.54)$$

Note that (9.53) describes the remainder term in case of a second order Taylor approximation of a C^2 smooth function which is guaranteed if the surface is C^4. In the special case where all the second partial derivatives of W vanish, the condition $W(u, v) \geq 0$ implies that the third order partial derivatives must also vanish. If we consider the total number of possibilities where we have non-vanishing partial derivatives up to third order, the case where all partial derivatives vanish is statistically very rare. Therefore, we focus our attention now on the generic case where at least one of the second order partial derivatives of W does not vanish. Using (9.26), we obtain $W(u_o, v_o) = 0$ and $\nabla W(u_o, v_o) = \mathbf{0}$ at the umbilic, therefore (9.53) reduces to

$$W(u,v) = W_Q(u,v) + R(u - u_o, v - v_o)|(u - u_o, v - v_o)|^2 , \quad (9.55)$$

where

$$W_Q(u,v) = \frac{1}{2}(u - u_o, v - v_o) \begin{pmatrix} W_{uu} & W_{uv} \\ W_{uv} & W_{vv} \end{pmatrix} (u - u_o, v - v_o)^T$$

$$= \frac{1}{2} \frac{(u - u_o, v - v_o)}{|(u - u_o, v - v_o)|} \begin{pmatrix} W_{uu} & W_{uv} \\ W_{uv} & W_{vv} \end{pmatrix} \frac{(u - u_o, v - v_o)^T}{|(u - u_o, v - v_o)|} |(u - u_o, v - v_o)|^2 . \tag{9.56}$$

Now we can Taylor expand $\sqrt{W(u,v)}$ up to first order [2]

$$\sqrt{W(u,v)} = \sqrt{W_Q} + \frac{R(u - u_o, v - v_o)}{2\sqrt{W_Q}}|(u - u_o, v - v_o)|^2$$

$$= [C(u,v) + \frac{R(u - u_o, v - v_o)}{2C(u,v)}]|(u - u_o, v - v_o)| , \quad (9.57)$$

[2] Note that here the Taylor expansion of the square root first yields an approximation instead of the equal sign in (9.57). However absorbing here the error term of this square root Taylor expansion in the remainder of (9.57) justifies the equality sign.

where

$$C(u,v) = \sqrt{\frac{1}{2} \frac{(u-u_o, v-v_o)}{|(u-u_o, v-v_o)|} \begin{pmatrix} W_{uu} & W_{uv} \\ W_{uv} & W_{vv} \end{pmatrix} \frac{(u-u_o, v-v_o)^T}{|(u-u_o, v-v_o)|}} .$$

(9.58)

Next we Taylor expand the mean curvature $H(u,v)$ as follows:

$$H(u,v) = H(u_o,v_o) + (H_L(u,v) + R(u-u_o, v-v_o))|(u-u_o, v-v_o)| ,$$

(9.59)

where

$$H_L(u,v) = [H_u(u_o,v_o), H_v(u_o,v_o)] \frac{(u-u_o, v-v_o)^T}{|(u-u_o, v-v_o)|} .$$

(9.60)

Although the function $R(u-u_o, v-v_o)$ in the remainder terms are different in (9.55), (9.57), (9.59), we nonetheless use the same notation for simplicity, since we are essentially interested in the common property described in (9.54).

Consequently $\kappa(u,v)$ in equation $\kappa(u,v) = H(u,v) \pm \sqrt{H^2(u,v) - K(u,v)}$ can be expanded in the vicinity of an umbilic (u_o, v_o) as follows:

$$\kappa(u,v) = H(u_o,v_o) + (H_L(u,v) \pm C(u,v)$$
$$+ R(u-u_o, v-v_o))|(u-u_o, v-v_o)|$$
$$= H(u_o,v_o) + \bar{H}_L(u,v) \pm \bar{C}(u,v) + \bar{R}(u-u_o, v-v_o) , \quad (9.61)$$

where

$$\bar{H}_L(u,v) = H_L(u,v)|(u-u_o, v-v_o)| , \tag{9.62}$$
$$\bar{C}(u,v) = C(u,v)|(u-u_o, v-v_o)| , \tag{9.63}$$
$$\bar{R}(u-u_o, v-v_o) = R(u-u_o, v-v_o)|(u-u_o, v-v_o)| . \tag{9.64}$$

Therefore $\kappa(u,v)$ can be considered as sum of the constant term $H(u_o,v_o)$, the plane $\bar{H}_L(u,v)$, which is the tangent plane of $H(u,v)$ at (u_o,v_o), and the elliptic cone $\bar{C}(u,v)$ whose axis of symmetry is perpendicular to uv-plane, since $W(u_o,v_o) = 0$, $\nabla W(u_o,v_o) = \mathbf{0}$. First we assume that $\bar{H}_L(u,v) = 0$, in other words the tangent plane of $H(u,v)$ coincides with the uv-plane. In this case $\kappa(u,v) - H(u_o,v_o)$ reduces to $\pm\bar{C}(u,v)$. Figure 9.4 shows a positive elliptic cone $+\bar{C}(u,v)$ (maximum principal curvature) having a minimum at (u_o,v_o). When the elliptic cone is negative, minimum principal curvature has a maximum at (u_o,v_o). The condition $\bar{H}_L(u,v) = 0$ occurs when all the third order partial derivatives of the height function in the Monge form are zero. This can be proved as follows [257]:

Proof : The coefficients of first and second fundamental forms of the surface in Monge form are given in (3.63) and (3.65). Their first order partial derivatives with respect to x are readily evaluated:

$$E_x = 2h_x h_{xx}, \quad F_x = h_{xx} h_y + h_x h_{xy}, \quad G_x = 2h_y h_{xy},$$

$$L_x = \frac{h_{xxx}\sqrt{(1+h_x^2+h_y^2)} + h_{xx}(1+h_x^2+h_y^2)^{-\frac{3}{2}}(h_x h_{xx} + h_y h_{xy})}{1+h_x^2+h_y^2},$$

$$M_x = \frac{h_{xxy}\sqrt{(1+h_x^2+h_y^2)} + h_{xx}(1+h_x^2+h_y^2)^{-\frac{3}{2}}(h_x h_{xx} + h_y h_{xy})}{1+h_x^2+h_y^2},$$

$$N_x = \frac{h_{xyy}\sqrt{(1+h_x^2+h_y^2)} + h_{xx}(1+h_x^2+h_y^2)^{-\frac{3}{2}}(h_x h_{xx} + h_y h_{xy})}{1+h_x^2+h_y^2}.$$

$$(9.65)$$

Now we will differentiate (3.67) [3] with respect to x

$$H_x = \frac{(2F_x M + 2FM_x - E_x N - EN_x - G_x L - GL_x)(EG - F^2)}{2(EG - F^2)^2}$$
$$- \frac{(2FM - EN - GL)(E_x G + EG_x - 2FF_x)}{2(EG - F^2)^2}. \tag{9.66}$$

If the surface is in Monge form with an umbilic at the origin, we have $h_x = h_y = 0$, which leads to $E_x = F_x = G_x = 0$. Consequently if $h_{xxx} = h_{xxy} = h_{xyy} = 0$, then $L_x = M_x = N_x = 0$ and hence $H_x = 0$. Similarly we can say that if $h_{xxy} = h_{xyy} = h_{yyy} = 0$, then $H_y = 0$. Since H_u and H_v can be written as

$$H_u = H_x x_u + H_y y_u,$$
$$H_v = H_x x_v + H_y y_v. \tag{9.67}$$

We can conclude that if $h_{xxx} = h_{xxy} = h_{xyy} = h_{yyy} = 0$ then $H_u = H_v = 0$.
∎

Note that in case $\bar{H}_L = 0$ the term $\bar{R}(u - u_o, v - v_o)$ is negligible for local extremum properties of the function $\kappa(u, v) - H(u_o, v_o)$. Consequently when $\bar{H}_L(u, v) = 0$, or alternatively when $\nabla H(u_o, v_o) = \mathbf{0}$, the function $\kappa(u, v) - H(u_o, v_o)$, hence $\kappa(u, v)$ has a local extremum at (u_o, v_o), or more precisely, κ_{max} has a local minimum and κ_{min} has a local maximum at an umbilical point (u_o, v_o).

It is also possible that $\kappa(u, v)$ may have a local extremum at the umbilic when $\bar{H}_L(u, v) \neq 0$. This is the situation when the plane $\bar{H}_L(u, v)$ is tilted against the uv-plane.[4] There are three possible cases, the plane $\bar{H}_L(u, v)$ intersects the cone $\bar{C}(u, v)$ transversally (see Fig. 9.5 (a)), the plane $\bar{H}_L(u, v)$

---

[3] Equation (3.67) is based on convention (a) of the normal curvature, while we are currently using convention (b) (see Fig. 3.7 (b) and Table 3.2).

[4] Note that we use the following observation illustrated by Fig. 9.5 (b). The term $\bar{R}(u - u_o, v - v_o)$ is negligible for investigating the local extrema properties of

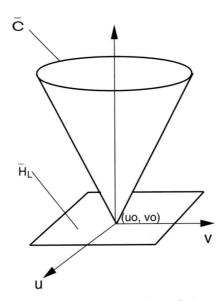

Fig. 9.4. Cone $\bar{C}(u,v)$ is perpendicular to the plane $\bar{H}_L(u,v)$ (adapted from [257])

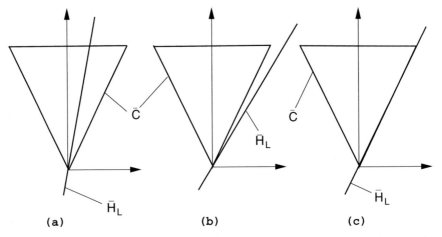

Fig. 9.5. (a) The plane intersects the cone, (b) the plane does not intersect the cone, (c) the plane and the cone are tangent to each other (adapted from [257])

does not intersect the cone $\bar{C}(u, v)$, apart from its apex, (see Fig. 9.5 (b)) and the plane $\bar{H}_L(u, v)$ and the cone $\bar{C}(u, v)$ are tangent to each other (see Fig. 9.5 (c)). Figure 9.5 (a) is the case when the plane intersects the cone in two straight lines. In this case for some directions the plane has a steeper slope than the cone, thus the sum $\bar{H}_L(u, v) \pm \bar{C}(u, v)$ does not have an extremum at (u_o, v_o), while in case (b) the plane intersects the cone only at (u_o, v_o), and the cone always has a steeper slope than the plane, thus $\bar{H}_L(u, v) \pm \bar{C}(u, v)$ has a local extremum at (u_o, v_o). Consequently we need to examine the equation $\pm \bar{C}(u, v) = -\bar{H}_L(u, v)$ which upon squaring and using (9.60) and (9.58) can be reduced to

$$
\begin{aligned}
(W_{uu} - 2H_u^2)(u - u_o)^2 &+ 2(W_{uv} - 2H_u H_v)(u - u_o)(v - v_o) \\
&+ (W_{vv} - 2H_v^2)(v - v_o)^2 = 0 \ .
\end{aligned} \tag{9.68}
$$

We can rewrite (9.68) as

$$
A(u - u_o)^2 + 2B(u - u_o)(v - v_o) + C(v - v_o)^2 = 0 \ , \tag{9.69}
$$

so that we can view (9.68) as a quadratic equation with unknown $u - u_o$ or $v - v_o$. If $B^2 - AC > 0$ there exist two distinct real roots, and thus there will be a real intersection between the plane and the cone made up of two straight lines. If $B^2 - AC = 0$ there exist two identical real roots, and thus the cone and the plane are tangent to each other, and additional evaluation of higher order terms in the Taylor expansion is necessary to decide if we have an extremum at the umbilic. If $B^2 - AC < 0$ there will be no real root, and thus there is no intersection between the cone and the plane. Consequently the criterion to have a local extremum of principal curvatures, when $\bar{H}_L(u, v) \neq 0$ or $\nabla H(u_o, v_o) \neq \mathbf{0}$, is equivalent to the condition $B^2 - AC < 0$. Hence the condition is

$$
(W_{uv} - 2H_u H_v)^2 - (W_{uu} - 2H_u^2)(W_{vv} - 2H_v^2) < 0 \ , \tag{9.70}
$$

or equivalently upon using $W(u, v) = H^2(u, v) - K(u, v)$

$$
(2HH_{uv} - K_{uv})^2 - (2HH_{uu} - K_{uu})(2HH_{vv} - K_{vv}) < 0 \ . \tag{9.71}
$$

Finally we can state the criterion as follows [257]:

Theorem 9.5.1. (Criterion for extrema of principal curvature functions at umbilics):
If we assume that $W(u, v)$ is at least C^2 smooth and at least one of the second order partial derivatives of $W(u, v)$ does not vanish then:

the function $\kappa(u, v)$ at the umbilic (u_o, v_o), provided the cone $\bar{C}(u, v)$ and the plane $\bar{H}_L = 0$ meet only at the point (u_o, v_o). Namely in that case we have a positive number α such that $|\bar{C}(u, v) - \bar{H}_L(u, v)| \geq \alpha |(u - u_o, v - v_o)|$. α is related to the smallest possible slope between plane and cone. Hence clearly $R(u - u_o, v - v_o)|(u - u_o, v - v_o)|$ is negligible to $\alpha |(u - u_o, v - v_o)|$.

1. *If $\nabla H = \mathbf{0}$ at the umbilic, then κ_{max} has a local minimum and κ_{min} has a local maximum.*
2. *If $\nabla H \neq \mathbf{0}$ at the umbilic, then κ_{max} has a local minimum and κ_{min} has a local maximum if and only if $D = (2HH_{uv} - K_{uv})^2 - (2HH_{uu} - K_{uu})(2HH_{vv} - K_{vv}) < 0$ provided $D \neq 0$. In case $D = 0$, additional evaluation of higher order terms in the Taylor expansion is necessary.*

$D = 0$ occurs when the cone and plane are tangent to each other, which is very rare. The condition $D = 0$ forces the criterion to be only sufficient and not necessary. It is quite plausible that the plane-cone tangential case (Fig. 9.5 (c)) is the rare one, while cases plane and cone are intersecting (Fig. 9.5 (a)) or plane and cone are non-intersecting (Fig. 9.5 (b)) are the generic ones.

When all the second order partial derivatives of $W(u, v)$ vanish, we need to Taylor expand up to fourth order in (9.53), since the condition $W(u, v) \geq 0$ implies that the third order partial derivatives must vanish. Also we need to Taylor expand up to second order in (9.59). Consequently $\kappa(u, v)$ can be expanded in the neighborhood of an umbilic (u_o, v_o) as sum of constant, linear and quadratic terms of mean curvature and the square root of fourth order terms of $W(u, v)$ as

$$\begin{aligned}
\kappa(u, v) &= H(u_o, v_o) + H_L(u, v)|(u - u_o, v - v_o)| \\
&\quad + (H_Q(u, v) + Q_T(u, v) + R(u - u_o, v - v_o))|(u - u_o, v - v_o)|^2 \\
&= H(u_o, v_o) + \bar{H}_L(u, v) + \bar{H}_Q(u, v) + \bar{Q}_T(u, v) + \bar{R}(u - u_o, v - v_o) ,
\end{aligned}$$
(9.72)

where $\bar{H}_Q(u, v)$ and $\bar{Q}_T(u, v)$ are the second order partial derivatives terms of the Taylor expansion of the mean curvature and the square root of fourth order partial derivatives terms of the Taylor expansion of $W(u, v)$. It is apparent that $\bar{H}_Q(u, v)$ and $\bar{Q}_T(u, v)$ have stationary point at (u_o, v_o). It follows that the plane $\bar{H}_L$ (linear term) will determine the local behavior of the function $\kappa(u, v)$. This implies now that in case $\bar{H}_L \neq 0$, $\kappa(u, v)$ cannot have a local extremum at the umbilic due to strong monotonicity behavior of the linear function. Therefore, if the second order partial derivatives of $W(u, v)$ vanish at the umbilic, then $\kappa(u, v)$ can only have an extremum in case $\bar{H}_L(u, v) = 0$. However $\bar{H}_L(u, v) = 0$ is not sufficient to guarantee a local extremum for $\kappa(u, v)$, since the point (u_o, v_o) can be a saddle point for the function $\bar{H}_Q(u, v) + \bar{Q}_T(u, v)$.

9.6 Perturbation of generic umbilics

In this section, we give a few numerical examples to demonstrate that generic umbilics are stable with respect to perturbations [257]. The example surface is a wave-like bicubic integral Bézier patch which is illustrated in Fig. 8.11. The surface is anti-symmetric with respect to $x = 0.5$. There are four spherical umbilics and one flat umbilic point on the surface. We gradually perturb the control points of the surface and observe the behavior of the lines of curvature which pass through umbilics. The control points are perturbed in the following manner. Since the example is a bicubic patch, it has 16 control points. Each control point consists of three Cartesian coordinates x, y, z, hence there are 48 components to be perturbed. A random number which varies from -1 to 1 is used to determine the 48 components. Let us denote the randomly chosen numbers for each control point as $(e_{ij}^x, e_{ij}^y, e_{ij}^z)$, $0 \leq i \leq 3$, $0 \leq j \leq 3$. We normalize the vector and add to each control point as

$$\tilde{\mathbf{P}}_{ij} = \mathbf{P}_{ij} + \zeta \frac{(e_{ij}^x, e_{ij}^y, e_{ij}^z)^T}{\sqrt{e_{ij}^{x\,2} + e_{ij}^{y\,2} + e_{ij}^{z\,2}}} , \tag{9.73}$$

where ζ is a constant. We increase the amount of perturbations gradually by increasing ζ from 0.02 by 0.02 up to 0.08. The curvature value κ, the four coefficients of the cubic terms α, β, γ, δ, angles of the tangent lines to the lines of curvature which pass through the umbilic in the tangent plane of the Monge form θ_1, θ_2, θ_3 in the 3-D space, ϕ_1, ϕ_2, ϕ_3 in the uv-space all in radians, index and the type are listed for original surface and two perturbed surfaces ($\zeta = 0.04$ and $\zeta = 0.08$) in Tables 9.1 to 9.3. The angles θ_i, ϕ_i ($1 \leq i \leq 3$) are restricted in the range $-\frac{\pi}{2} \leq \theta_i$, $\phi_i \leq \frac{\pi}{2}$. Figures 9.6 to 9.8 illustrate how the lines of curvature which only pass through the umbilic behave when the control points are perturbed. The thick solid line represents the lines of curvature for maximum principal curvature, thick dotted line represents the lines of curvature for minimum principal curvature and the thin solid lines are the iso-parametric lines of the wave-like surface.

From the figures and tables we can observe that the umbilic on the upper right jumps off from the domain but the other four umbilics remain inside the domain. *All the umbilics which stay in the domain do not change their index nor their type.* In Fig. 9.6, when the perturbation is zero, lines of curvature passing through the umbilics at lower left (0.211,0.052) and upper left (0.211,0.984) have a common line of curvature. Similarly lower right umbilic (0.789,0.052) and upper right umbilic (0.789,0.984) have a common line of curvature. As the perturbation gradually increases, they split into two lines of curvature as shown in Figs. 9.7 and 9.8. Note that in Figs. 9.7 and 9.8 the lines of curvature corresponding to the jumped off umbilic (upper right) are not shown, since we cannot compute the initial values for the integration. From these observations we can conclude that the umbilics are quite stable to the perturbation. Also the locations and the angles θ_i, ϕ_i ($1 \leq i \leq 3$) of the umbilics do not move nor rotate too much.

In computer vision, the geometric information of an object is obtained by range imaging sensors. Generally the data include noise and are processed using image processing techniques to exclude the noise, then the derivatives

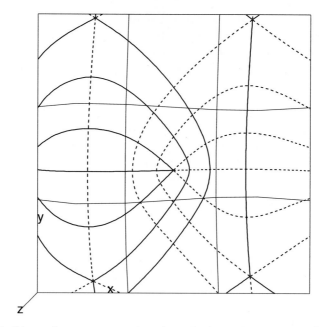

Fig. 9.6. Lines of curvature passing through the umbilics, $\zeta = 0$ (adapted from [257])

Table 9.1. Umbilics of original surface (adapted from [257])

u	0.211	0.211	0.789	0.789	0.5
v	0.052	0.984	0.052	0.984	0.440
κ	1.197	0.267	-1.197	-0.267	0.
α	4.147	0.926	4.147	0.926	6.514
β	-18.306	14.670	18.306	-14.670	0.
γ	0.	0.	0.	0.	4.2763
δ	-2.337	1.411	2.337	-1.411	0.
θ_1	0.671	0.562	0.592	0.638	0.
θ_2	-0.592	-0.638	-0.671	-0.616	-0.604
θ_3	1.571	-1.571	-1.571	1.571	0.604
ϕ_1	0.567	0.562	0.495	0.583	0.
ϕ_2	-0.495	-0.583	-0.567	-0.562	-0.752
ϕ_3	1.571	1.571	-1.571	-1.571	0.752
index	$-\frac{1}{2}$	$-\frac{1}{2}$	$-\frac{1}{2}$	$-\frac{1}{2}$	$\frac{1}{2}$
type	star	star	star	star	monstar

are directly computed from the digital data to evaluate the curvatures. What we do in the sequel is to fit a surface directly from artificial noisy data and observe the behavior of the umbilics on the fitted surface. The noisy data are produced in the following way. Evenly spaced 10×10 grid points (x, y) on $0 \le x \le 1, 0 \le y \le 1$ domain are chosen to evaluate the z-value of

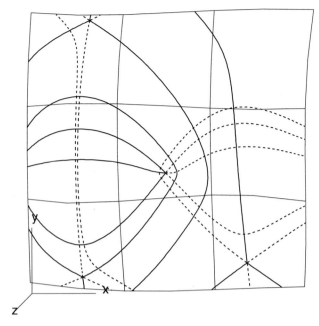

Fig. 9.7. Lines of curvature passing through the umbilics, $\zeta = 0.04$ (adapted from [257])

Table 9.2. Umbilics on perturbed surface, $\zeta = 0.04$ (adapted from [257])

u	0.190	0.214	0.794	n/a	0.492
v	0.055	0.978	0.081	n/a	0.424
κ	1.293	0.136	-1.458	n/a	0.083
α	2.390	0.551	5.014	n/a	6.351
β	-16.119	13.046	18.926	n/a	0.163
γ	0.563	-1.524	-0.360	n/a	4.666
δ	-3.182	1.711	3.234	n/a	0.510
θ_1	0.658	0.593	0.586	n/a	0.701
θ_2	-0.623	-0.667	-0.689	n/a	-0.055
θ_3	1.551	1.509	1.560	n/a	-0.644
ϕ_1	0.559	0.549	0.528	n/a	0.857
ϕ_2	-0.537	-0.589	-0.596	n/a	-0.076
ϕ_3	1.529	1.552	-1.532	n/a	-0.811
index	$-\frac{1}{2}$	$-\frac{1}{2}$	$-\frac{1}{2}$	n/a	$\frac{1}{2}$
type	star	star	star	n/a	monstar

the wave-like bicubic Bézier patch. We add randomly perturbed vectors with $\zeta = 0.05$, as introduced in (9.73), to the (x, y, z) points on the surface as noise. Then the data points (x, y, z) are fit by a bicubic Bézier patch. Figure 9.9 and Table 9.4 illustrate the results. We observe that all the umbilics stay in the domain with index and types unchanged. Also the locations and the

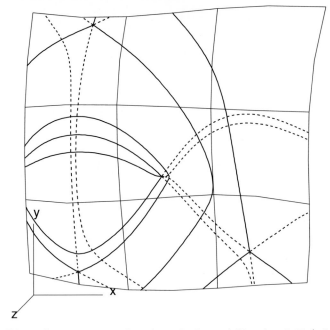

Fig. 9.8. Lines of curvature passing through the umbilics, $\zeta = 0.08$ (adapted from [257])

Table 9.3. Umbilics on perturbed surface, $\zeta = 0.08$ (adapted from [257])

u	0.167	0.217	0.795	n/a	0.474
v	0.065	0.970	0.113	n/a	0.411
κ	1.426	0.042	-1.779	n/a	0.261
α	0.701	0.374	6.355	n/a	6.356
β	-14.070	11.604	19.405	n/a	0.307
γ	1.621	-2.520	0.727	n/a	5.155
δ	-4.184	2.057	4.072	n/a	1.273
θ_1	0.632	0.573	0.594	n/a	0.773
θ_2	-0.674	-0.694	-0.692	n/a	-0.079
θ_3	1.504	1.455	-1.550	n/a	-0.655
ϕ_1	0.557	0.534	0.577	n/a	0.928
ϕ_2	-0.614	-0.594	-0.631	n/a	-0.112
ϕ_3	1.466	1.532	-1.485	n/a	-0.841
index	$-\frac{1}{2}$	$-\frac{1}{2}$	$-\frac{1}{2}$	n/a	$\frac{1}{2}$
type	star	star	star	n/a	monstar

angles θ_i, ϕ_i $(1 \leq i \leq 3)$ do not move nor rotate too much. These results provide us confidence for using the umbilics for shape recognition problems.

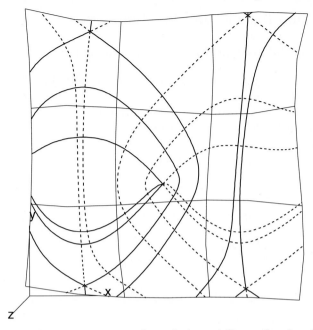

Fig. 9.9. Lines of curvature passing through the umbilics on fitted surface, $\zeta = 0.05$ (adapted from [257])

Table 9.4. Umbilics on reconstructed surface, $\zeta = 0.05$ (adapted from [257])

u	0.198	0.227	0.813	0.796	0.493
v	0.043	0.954	0.025	0.991	0.399
κ	1.278	0.147	-1.005	-0.124	0.090
α	1.748	0.999	0.678	0.357	6.572
β	-16.370	16.161	19.451	-11.981	-0.293
γ	-0.054	0.061	4.716	0.536	4.849
δ	-2.705	2.176	3.082	-1.233	-0.410
θ_1	0.656	0.622	0.686	0.622	0.094
θ_2	-0.616	-0.642	-0.575	-0.634	-0.692
θ_3	-1.569	-1.569	-1.443	1.547	0.657
ϕ_1	0.532	0.615	0.528	0.543	0.132
ϕ_2	-0.535	-0.630	-0.482	-0.554	-0.856
ϕ_3	1.497	-1.562	-1.513	1.544	0.827
index	$-\frac{1}{2}$	$-\frac{1}{2}$	$-\frac{1}{2}$	$-\frac{1}{2}$	$\frac{1}{2}$
type	star	star	star	star	monstar

9.7 Inflection lines of developable surfaces

9.7.1 Differential geometry of developable surfaces

A *ruled surface* is a curved surface which can be generated by the continuous motion of a straight line in space along a space curve called a *directrix*. This straight line is called a *generator, or ruling*, of the surface. A book by Pottmann and Wallner [330] studies line geometry from the viewpoint of scientific computation and shows the interplay between theory and applications. Any point on a parametric ruled surface can be expressed as

$$\mathbf{r}(u, v) = \boldsymbol{\alpha}(u) + v\boldsymbol{\beta}(u) , \qquad (9.74)$$

where $\boldsymbol{\alpha}(u)$ is a directrix or base curve of the ruled surface and $\boldsymbol{\beta}(u)$ is a unit vector which gives the direction of the ruling at each point on the directrix. Alternatively, the surface can be represented as a ruling joining corresponding points on two space curves. This is represented by

$$\mathbf{r}(u, v) = (1 - v)\mathbf{r}_A(u) + v\mathbf{r}_B(u), \quad 0 \le u, v \le 1 , \qquad (9.75)$$

where $\mathbf{r}_A(u)$ and $\mathbf{r}_B(u)$ are directrices, as shown in Fig. 9.10. The two representations are identical if

$$\boldsymbol{\alpha}(u) = \mathbf{r}_A(u) \quad \text{and} \quad \boldsymbol{\beta}(u) = \mathbf{r}_B(u) - \mathbf{r}_A(u) . \qquad (9.76)$$

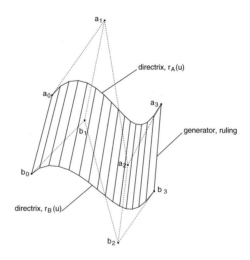

Fig. 9.10. A ruled surface

A *developable surface* is a special ruled surface which has the same tangent plane at all points along a generator [13, 222, 120, 32, 326, 252, 329]. Since surface normals are orthogonal to the tangent plane and the tangent plane along a generator is constant, all normal vectors along a generator are parallel. This is shown in Fig. 9.11.

A developable surface has following differential geometry properties [412]:

1. A developable surface can be mapped isometrically onto a plane.
2. Isometric surfaces have the same Gaussian curvature at corresponding points.
3. Corresponding curves on isometric surfaces have the same geodesic curvature at corresponding points.
4. Every isometric mapping is conformal; i.e. the angle of intersection of every arbitrary pair of intersecting arcs on a developable surface is the same as that of the corresponding inverse image in the plane at the corresponding points.
5. A geodesic on a developable surface maps to a straight line in the plane.

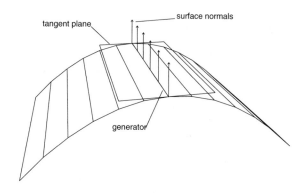

Fig. 9.11. A developable surface with its tangent plane along a ruling

A developable surface can be formed by bending or rolling a planar surface without stretching or tearing; in other words, it can be developed or unrolled isometrically onto a plane. Developable surfaces are also known as *singly curved surfaces*, since one of their principal curvatures is zero. Developable surfaces are widely used with materials that are not amenable to stretching. Applications include the formation of ship hulls, ducts, shoes, clothing and automobile parts such as upholstery, body panels and windshields [120].

As indicated by Munchmeyer and Haw [281], a developable surface can be shaped purely by rolling and should be fed to the roller so that the direction of the zero principal curvature is parallel to the rolls. However, when the sheet reaches a line of inflection, it can no longer be fed into the roller in the same direction because the direction of bending changes. Therefore, it is beneficial for planning the fabrication process to determine the lines of inflection prior to such a process.

Surface inflection of a developable surface was studied by Hoitsma [172] who showed that a surface has an inflection at a point p if and only if its mean curvature changes sign in the neighborhood of p. Maekawa and Chalfant [251] further extended this result and derived two theorems (Theorem 9.7.1 and 9.7.2).

Since the Gaussian curvature of a developable surface is zero everywhere [412, 76], the maximum and minimum principal curvatures (3.49) and (3.50) of a developable surface can be written as

$$\kappa_{max} = H + |H|, \quad \kappa_{min} = H - |H| . \tag{9.77}$$

The principal curvatures reduce to

$$\kappa_{max} = 2H, \quad \kappa_{min} = 0 \quad when \ H > 0 , \tag{9.78}$$
$$\kappa_{max} = 0, \quad \kappa_{min} = 0 \quad when \ H = 0 , \tag{9.79}$$
$$\kappa_{max} = 0, \quad \kappa_{min} = 2H \quad when \ H < 0 . \tag{9.80}$$

It is clear from (9.78) through (9.80) that at least one of the principal curvatures is zero at each point on a developable surface, which agrees with the fact that the Gaussian curvature is zero everywhere (see (3.61)). κ_{max} in (9.78) and κ_{min} in (9.80) are termed the *nonzero* principal curvature, κ^*, where $\kappa^* = 2H$.

In the following we establish some elementary differential geometry properties of developable surfaces. We assume that the developable surface is regular and the $u = const$ iso-parametric line corresponds to the generator of the developable surface or, in other words, the straight line ruling is in the v direction. With this assumption, $\mathbf{r}_{vv} = \mathbf{0}$, and hence the second fundamental form coefficient (see (3.28)) $N = \mathbf{r}_{vv} \cdot \mathbf{N}$ vanishes. From (3.46), since Gaussian curvature of a developable surface is zero,

$$K = \frac{-M^2}{EG - F^2} = 0 , \tag{9.81}$$

and hence we have $M = 0$. Therefore, the mean curvature (3.47) reduces to

$$H = \frac{-GL}{2(EG - F^2)} . \tag{9.82}$$

Recall that the nonzero principal curvature is given by $\kappa^* = 2H$ and since $G > 0$, H and hence κ^* become zero if and only if $L = 0$, otherwise $\kappa^* = 2H \neq 0$.

Next, we will show that the $u = const$ parametric straight lines become the lines of zero curvature. This can be seen from the fact that i) the $u = const$

iso-parametric straight lines have zero normal curvature, and ii) no other direction has zero normal curvature. The second fact comes from Euler's theorem introduced in Sect. 3.6. When $\kappa_2 = 0$, (3.87) reduces to $\kappa = \kappa_1 \cos^2(\alpha)$, which becomes zero only when $\alpha = \frac{\pi}{2}$ or $\frac{3\pi}{2}$, corresponding to the direction of κ_2. Similarly, when $\kappa_1 = 0$, $\kappa = 0$ only when $\alpha = 0$ or π, corresponding to the direction of κ_1.

Theorem 9.7.1. *A developable surface does not possess generic isolated flat points [5] but rather may contain a line of non-generic flat points along a generator [251].*

Proof: From (9.79) and (9.82), L vanishes at a flat point $\mathbf{r}(u_f, v_f)$ where both principal curvatures are zero. Therefore from the first equation of (3.27) we have

$$L(u_f, v_f) = -\mathbf{r}_u(u_f, v_f) \cdot \mathbf{N}_u(u_f, v_f) = 0 . \qquad (9.83)$$

From (9.81), $M = 0$ on a developable surface. Hence from the second equation of (3.27) we have

$$M(u_f, v_f) = -\mathbf{r}_v(u_f, v_f) \cdot \mathbf{N}_u(u_f, v_f) = 0 . \qquad (9.84)$$

Since $\mathbf{N}$ is a unit vector, we also have

$$\mathbf{N}(u_f, v_f) \cdot \mathbf{N}_u(u_f, v_f) = 0 . \qquad (9.85)$$

If $\mathbf{N}_u$ is not zero, then from (9.83), (9.84) and (9.85) $\mathbf{N}_u$ must be perpendicular to $\mathbf{r}_u$, $\mathbf{r}_v$ and $\mathbf{N}$. This is impossible because $\mathbf{N}$ is perpendicular to both $\mathbf{r}_u$ and $\mathbf{r}_v$, and $\mathbf{r}_u$ is not parallel to $\mathbf{r}_v$. Thus, $\mathbf{N}_u(u_f, v_f)$ must equal zero. For developable surfaces, the unit normal vector $\mathbf{N}$ is constant along a generator. Therefore, the rate of change of the unit normal vector in the u direction must also be constant along a generator. This leads us to the fact that $\mathbf{N}_u$ is not only zero at $\mathbf{r}(u_f, v_f)$ but also zero along the $u = u_f$ iso-parametric line. Therefore, for a given $u = u_f$, (9.83) becomes

$$L(u_f, v) = -\mathbf{r}_{uu}(u_f, v) \cdot \mathbf{N}(u_f, v) = 0 , \qquad (9.86)$$

for $0 \le v \le 1$. Consequently, the entire generator consists of a line of flat points. ∎

For a developable surface, the *inflection line* is a generator which consists of a line of flat points and the nonzero principal curvature changes sign. The inflection line can be detected by finding $u = u_f$ such that $L(u_f, v_n) = 0$ where v_n is an arbitrary constant between 0 and 1. $L(u, v_n) = 0$ can be written as

---

[5] A developable surface cannot possess spherical umbilics since one of the principal curvatures is always zero.

$$\mathbf{r}_{uu}(u, v_n) \cdot \frac{\mathbf{r}_u(u, v_n) \times \mathbf{r}_v(u, v_n)}{|\mathbf{r}_u(u, v_n) \times \mathbf{r}_v(u, v_n)|} = 0 \ . \tag{9.87}$$

Since we are assuming a regular surface such that $|\mathbf{r}_u \times \mathbf{r}_v| \neq 0$, we only need to set the numerator of (9.87) equal to zero. Thus, $|\mathbf{r}_{uu}(u, v_n) \quad \mathbf{r}_u(u, v_n) \quad \mathbf{r}_v(u, v_n)| = 0$. For a polynomial surface with degree n in the u direction this results in a univariate polynomial equation of degree $(3n - 4)$ in u

$$(y_u z_v - z_u y_v)x_{uu} - (x_v z_u - x_u z_v)y_{uu} + (x_u y_v - x_v y_u)z_{uu} = 0. \tag{9.88}$$

If the surface is expressed in a piecewise polynomial form such as a B-spline representation, (9.88) must be applied to each polynomial segment separately. The univariate polynomial equation can be robustly and efficiently solved by the Interval Projected Polyhedron algorithm described in Chap. 4.

The local approximation (8.73) will now be applied to developable surfaces.

Lemma 9.7.1. *A developable surface is, in general, locally a parabolic cylinder except at an inflection line, where it becomes a cubic cylinder, provided that $h_{xxx} \neq 0$ [251].*

Proof: Let us consider an orthogonal Cartesian reference frame O-XYZ attached to the surface $\mathbf{r} = \mathbf{r}(u, v)$ at an arbitrary point P with $\mathbf{r}(u_0, v_0)$ being P. We choose unit vectors $\frac{\mathbf{r}_v}{|\mathbf{r}_v|} \times \mathbf{N}$, $\frac{\mathbf{r}_v}{|\mathbf{r}_v|}$ and $\mathbf{N}$ at P as the directions of X, Y and Z axes such that the Y axis coincides with the generator $u = u_0$, the Z axis coincides with the surface normal vector and the X axis is orthogonal to both axes. Therefore the local coordinates X, Y and Z are given by

$$X = \left(\frac{\mathbf{r}_v}{|\mathbf{r}_v|} \times \mathbf{N}\right)_o \cdot [\mathbf{r}(u, v) - \mathbf{r}(u_o, v_o)] \ , \tag{9.89}$$

$$Y = \left(\frac{\mathbf{r}_v}{|\mathbf{r}_v|}\right)_o \cdot [\mathbf{r}(u, v) - \mathbf{r}(u_o, v_o)] \ , \tag{9.90}$$

$$Z = h(u, v) = \mathbf{N}(u_o, v_o) \cdot [\mathbf{r}(u, v) - \mathbf{r}(u_o, v_o)] \ , \tag{9.91}$$

where subscript o denotes that the expressions are evaluated at (u_o, v_o). In (9.89) through (9.91) all terms involving $\mathbf{r}_u$, $\mathbf{r}_v$ and $\mathbf{N}$ are evaluated at (u_o, v_o), so $\mathbf{r}(u, v)$ is the only term that is a function of u and v. If we consider u and v as functions of X and Y, i.e. $u = u(X, Y)$ and $v = v(X, Y)$, then the height function h can be represented as a function of X and Y through intermediate variables u and v, i.e. $h(u(X, Y), v(X, Y))$.

The second fundamental form coefficient M in terms of the height function $h(X, Y)$ is given in (3.65) as $M = \frac{h_{XY}}{(1 + h_X^2 + h_Y^2)^{1/2}}$. Since $M = 0$ on a developable surface (see (9.81)), we have $h_{XY} = 0$. Furthermore, all the second and higher order partial derivatives with respect to Y vanish, since the Y axis corresponds to the generator, which is linear in v. Thus (8.73) reduces to

$$h(X,Y) = \frac{1}{2}X^2 h_{XX}(0,0) + \frac{1}{6}[X^3 h_{XXX}(0,0) + 3X^2 Y h_{XXY}(0,0)]$$
$$+ R(X,Y)(|X,Y|^3) \,. \tag{9.92}$$

The second fundamental form coefficient L in terms of the height function is given in (3.65) as $L = \frac{h_{XX}}{(1+h_X^2+h_Y^2)^{1/2}}$. Since L is zero along the entire generator line, h_{XX} and its variation in the Y direction h_{XXY} become zero at a line of inflection. Thus (9.92) further reduces to

$$h(X,Y) = \frac{1}{6}(\mathbf{r}_{uuu} \cdot \mathbf{N}) \left(\sqrt{\frac{G}{EG - F^2}} \right)^3 X^3 + R(X,Y)(|X,Y|^3) \,, \tag{9.93}$$

provided that $h_{XXX}(0,0) \neq 0$. Here $h_{XXX}(0,0)$ is obtained by using the inverse function theorem [76] (see Sect. 9.3) and it can be shown that in general

$$\frac{\partial^k h}{\partial X^k} = \frac{\partial^k \mathbf{r}}{\partial u^k} \cdot \mathbf{N} \left(\sqrt{\frac{G}{EG - F^2}} \right)^k . \tag{9.94}$$

From (9.92) and (9.93) it is apparent that for small x, the quadratic term dominates except at an inflection line where the surface will become locally a cubic cylinder. ∎

When $h_{xxx}(0,0)$ becomes zero, the higher order partial derivatives must be considered, and this is studied in the following.

A developable surface is said to have *contact of order* k with the tangent plane along the generator if the Taylor expansion for $h(X,Y)$ starts with terms of degree $k + 1$. The *ordinary inflection line* (see (9.93)) thus has a contact of order $k = 2$. If the tangent plane has contact of order $k \geq 3$ with the surface along a generator, a developable surface may not look like a cubic cylinder at an inflection line.

If the developable surface has a contact of order k with the tangent plane, $\frac{\partial^i h}{\partial X^i}$ is zero or equivalently $\frac{\partial^i \mathbf{r}}{\partial u^i} \cdot \mathbf{N}$ is zero for $1 \leq i \leq k$ along the entire generator. Accordingly its variation in Y also vanishes; hence $\frac{\partial^i}{\partial X^i} \frac{\partial}{\partial Y} h = 0$ for $1 \leq i \leq k$ along the entire generator. Since all the second and higher order partial derivatives with respect to Y vanish, the Taylor expansion of the height function along the higher order contact line reduces to

$$h(X,Y) = \frac{1}{(k+1)!} \left(\frac{\partial^{k+1} \mathbf{r}}{\partial u^{k+1}} \cdot \mathbf{N} \right) \left(\sqrt{\frac{G}{EG - F^2}} \right)^{k+1} X^{k+1}$$
$$+ R(X,Y)(|X,Y|^{k+1}) \,. \tag{9.95}$$

We can observe that for an even k the height function $h(X,Y)$ changes sign when X moves across the inflection line, while for an odd k it maintains the same sign. In other words, for an even k the height function passes through the tangent plane along the inflection line, whereas for an odd k, it lies entirely on one side of the tangent plane. Therefore inflection lines exist only for even k. The order of contact can be detected by first solving (9.88), and substituting the solution into

$$(y_u z_v - z_u y_v)\frac{\partial^k x}{\partial u^k} - (x_v z_u - x_u z_v)\frac{\partial^k y}{\partial u^k} + (x_u y_v - x_v y_u)\frac{\partial^k z}{\partial u^k} = 0 , \quad (9.96)$$

to find $k \geq 3$ such that (9.96) is zero for k but nonzero for $k+1$. The integer k found by this process gives the order of contact (see (9.88), (9.94)). A simple example for the higher order odd case is given by $\mathbf{r}(u, v) = (u, v, u^4)^T$. Since $\mathbf{r}_{uu} = (0, 0, 12u^2)^T$ and $\mathbf{r}_u \times \mathbf{r}_v = (-4u^3, 0, 1)^T$, we can easily see that the line of flat points is located at $u = 0$ from (9.88). And the order of contact is found to be $k = 3$, since $\mathbf{r}_{uuu} = (0, 0, 24u)^T$ and $\mathbf{r}_{uuuu} = (0, 0, 24)^T$. In this case the line of flat points is not an inflection line, since k is odd.

9.7.2 Lines of curvature near inflection lines

At flat points the principal directions are indeterminate and the orthogonal net of lines of curvature may have singular properties. In the following we investigate the pattern of the lines of curvature near a line of non-generic flat points.

Theorem 9.7.2. *There is only one line of curvature that passes through each flat point on a line of flat points, and that line of curvature is orthogonal to the direction of the generator [251].*

Proof: By Lemma 9.7.1 the developable surface is expressed locally as a cubic cylinder at an ordinary inflection line and more generally in the form of (9.95) at a higher order contact line. If we rewrite (9.95) in terms of polar coordinates by substituting $X = r \cos \theta$ for a fixed radius $r = \sqrt{X^2 + Y^2}$ we obtain

$$h(\theta) = c \cos^{k+1} \theta , \quad (9.97)$$

where c is a constant evaluated at a point on the line of flat points given by

$$c = \frac{r^{k+1}}{(k+1)!}\left(\frac{\partial^{k+1}\mathbf{r}}{\partial u^{k+1}} \cdot \mathbf{N}\right)\left(\sqrt{\frac{G}{EG - F^2}}\right)^{k+1} . \quad (9.98)$$

If k is even, $h(\theta + \pi) = -h(\theta)$, and $h(\theta)$ is an antisymmetric function of θ, whereas if k is odd $h(\theta)$ is a symmetric function of θ. The roots of $\frac{dh}{d\theta} = 0$ will give the angles where local maxima and minima of $h(\theta)$ may occur around the flat point. The equation can be restricted to the range $0 \leq \theta < 2\pi$ without loss of generality. The roots are easily computed as $\theta = 0$, $\frac{\pi}{2}$, π and $\frac{3\pi}{2}$. Only $\theta = 0$ and $\theta = \pi$ (which coincide with the local x axis) give extrema, since

$\frac{d^2 h\left(\frac{\pi}{2}\right)}{d\theta^2} = \frac{d^2 h\left(\frac{3\pi}{2}\right)}{d\theta^2} = 0$. Thus $\theta = \frac{\pi}{2}$ and $\theta = \frac{3\pi}{2}$ (which coincide with the local y axis) provide neither a maximum nor a minimum. In other words, a line of flat points is not a line of curvature. Consequently, there is only one line of curvature passing through each flat point and it is orthogonal to the direction of the generator. For an even k the lines of maximum/minimum principal curvature switch to lines of minimum/maximum principal curvature at the inflection line since $h(\theta)$ is antisymmetric, while for odd k they remain the same, since $h(\theta)$ is symmetric. ∎

If we denote ϕ as the angle between the u axis and the direction of the nonzero principal curvature in uv parametric space, ϕ can be evaluated as follows. Since the direction of the nonzero principal curvature is orthogonal to the generator (parallel to the local x axis), its direction is given by $\mathbf{r}_v \times (\mathbf{r}_u \times \mathbf{r}_v) = (\mathbf{r}_v \cdot \mathbf{r}_v)\mathbf{r}_u - (\mathbf{r}_v \cdot \mathbf{r}_u)\mathbf{r}_v = G\mathbf{r}_u - F\mathbf{r}_v$ and hence $\phi = -\tan^{-1}\frac{F}{G}$.

We can trace the lines of curvature which pass through the flat points of an inflection line by integrating the initial value problem following the procedure described in Sect. 9.4. The starting points are obtained by slightly shifting outwards in the directions 0 and π from the flat points or, equivalently, along the positive and negative local x axis.

In generic cases, umbilics are isolated [257]; thus an inflection line, which consists of a line of flat points, is non-generic and therefore unstable. In the following we give a couple of numerical examples that demonstrate the instability of the line of flat points along the inflection line with respect to perturbations.

The example surface is a degree (3-1) integral Bézier patch which is constructed by the method developed in Chalfant [50]. The control points are given by

$$
\begin{array}{ll}
\mathbf{b}_{00} = (0, 0, 0)^T, & \mathbf{b}_{01} = (0.5, 0, 2)^T, \\
\mathbf{b}_{10} = (1.8, 3, 0)^T, & \mathbf{b}_{11} = (1.895, 2.325, 2)^T, \\
\mathbf{b}_{20} = (3.3, -2, 1.5)^T, & \mathbf{b}_{21} = (3.0575, -1.55, 3.1625)^T, \\
\mathbf{b}_{30} = (4, 0, 0)^T, & \mathbf{b}_{31} = (3.6, 0, 2)^T.
\end{array}
$$

The surface has an ordinary inflection line at $u = 0.5754$, which has been computed by solving the degree 5 univariate polynomial equation (9.88). This surface has a net of lines of curvature which is shown in Fig. 9.12(a). Solid lines represent the lines of maximum principal curvature, while dotted lines represent the lines of minimum principal curvature. The inflection line is depicted with a dash dotted line. Figure 9.12(b) shows a magnification near the inflection line. We can observe that there is only one line of curvature that passes through a flat point orthogonal to the inflection line.

We gradually perturb the control points of the surface and observe the behavior of the lines of curvature which pass through the inflection line as we did in Sect. 9.6. Since the example is a degree (3-1) patch, it has 8 control points. Each control point consists of three Cartesian coordinates x, y, z, so there are 24 components to be perturbed. We gradually increase the perturbation by increasing ζ in (9.73) from 0.02 to 0.08 in steps of 0.02.

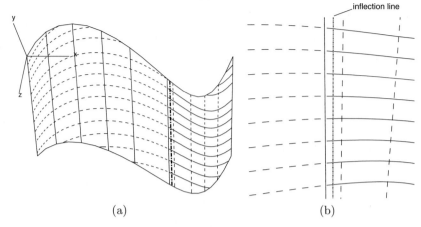

Fig. 9.12. (a) Lines of curvature of developable surface with inflection, (b) magnification near inflection line (adapted from [257])

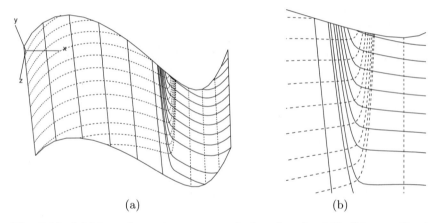

Fig. 9.13. (a) Lines of curvature on perturbed surface $\zeta = 0.08$, (b) magnification near $u=0.57$ (adapted from [257])

Figure 9.13 illustrates the behavior of the lines of curvature when the control points are perturbed ($\zeta = 0.08$). We can see from the figure that the entire inflection line, which consists of a line of flat points, disappears. Hence there is no singularity in the net of lines of curvature when a perturbation is induced. The nonzero principal curvatures on both sides of the former inflection line [6] meet at right angles near the former inflection line and make a very sharp change in direction (almost a right angle).

---

[6] Once the control points are perturbed both principal curvatures may not be nonzero, but here we are referring to the nonzero principal curvature before perturbation.

10. Geodesics

10.1 Introduction

Computation of shortest paths on free-form surfaces is an important problem in geometric design of ship hulls, robot motion planning, computation of medial axis transforms of trimmed surface patches, terrain navigation, NC machining, and cable installation on the sea floor. The history of geodesic lines begins with a study by Johann Bernoulli, who solved the problem of the shortest distance between two points on a convex surface in 1697, according to Struik [412]. He showed that the osculating plane of the geodesic line must always be perpendicular to the tangent plane. The equation of geodesics for implicit surfaces was first obtained by Euler (1732). His attention to the problem was due to Johann Bernoulli, probably through the aid of his nephew Daniel, who was at St. Petersburg with Euler [411]. Bliss [28] obtained the geodesic lines on the anchor ring, which has a torus shape, analytically. Munchmeyer and Haw [281] applied geodesic curves to geometric design of ship hulls, namely to find out the precise layout of the seams and butts on a ship hull. Beck et al. [22] performed both initial-value integration and boundary-value integration (based on shooting method) of geodesic paths, using the fourth order Runge-Kutta method on a bicubic spline surface. Patrikalakis and Bardis [296] computed geodesic offsets of curves on rational B-spline surfaces using the initial-value integration of geodesics normal to an initial progenitor curve on the surface. One application of such offsets is automated construction of linkage curves for free-form procedural blending surfaces. Sneyd and Peskin [398] investigated the computation of geodesic paths on a generalized cylinder based on an initial value problem using a second order Runge-Kutta method. Their work was motivated by constructing the great vessels of the heart out of geodesic fibers. Kimmel et al. [200] presented a numerical method for finding the shortest path on surfaces by calculating the propagation of an equal geodesic-distance contour from a point or a source region on the surface. The algorithm works on a rectangular grid using finite difference approximations. Maekawa [247] and Robinson and Armstrong [347] computed the geodesics by discretizing the governing differential equations using a finite difference approximation on a mesh of points, which reduces the problem to a set of nonlinear equations. The set of nonlinear equations can be solved by quadratically convergent Newton ite-

ration method, which starts with an initial guess and improves the solution
iteratively. This technique is referred as, *direct method, relaxation method* or
finite difference method. The shortest path problem is also very active among
the robot motion planning and terrain navigation communities, however they
usually represent the surface as a polyhedral surface and solve the problem
using techniques from the field of computational geometry [269].

A geodesic path is sometimes defined as the shortest path between two
points on a surface; however this is not always a satisfactory definition. In
this book we follow Struik [412] and define geodesics as below:

Definition 10.1.1. *Geodesics are curves of zero geodesic curvature.*

In other words, the osculating planes of a geodesic curve on a surface contain
the surface normal. From this definition we can easily see that the geodesic
between two points on a sphere is a great circle. But there are two arcs
of a great circle between two such points, and only one of them provides
the shortest distance, except when the two points are the end points of a
diameter of the sphere. This example indicates that there may exist more
than one geodesics between two points on a surface.

Let $Q(t)$ describe a moving point on a surface where t may be viewed as
a time parameter belonging to an interval beginning with t_0, and $Q[t_0, t_1]$
describe the path between points $Q(t_0)$ and $Q(t_1)$. If the point $Q(t)$ moves
away from the starting point $Q(t_0)$ along a geodesic path C, i.e. curve with
zero geodesic curvature, then it may occur that $Q(t)$ will reach a point $Q(t_R)$
such that for every $\varepsilon > 0$ the path $Q[t_0, t_R + \varepsilon]$ is no longer the shortest
surface path joining the points $Q(t_0)$ and $Q(t_R + \varepsilon)$. In other words, $Q(t_R)$
was the last time point such that the geodesic path $Q[t_0, t_R]$ is the shortest
surface path joining the points $Q(t_0)$ and $Q(t_R)$. This point $Q(t_R)$ is called
conjugate to $Q(t_0)$ on the geodesic C (see also [449]). The location, where each
extension of a shortest geodesic fails to define a shortest path from $Q(t_0)$,
belongs to the cut locus of the point $Q(t_0)$ on the surface. This geometric
locus and its generalizations (being important for considerations on shortest
paths and geodesic distance) have been studied in [447, 449].

10.2 Geodesic equation

10.2.1 Parametric surfaces

We assume that the given parametric surface $\mathbf{r}=\mathbf{r}(u, v)$ is a regular and non-
periodic NURBS surface patch. Wolter [448] shows that on a regular NURBS
surface patch there always exists a shortest path joining any two patch points.
If the surface patch is defined on a rectangular or even more generally on a
locally convex planar domain, then any shortest path in the patch joining
any two patch points must have a continuous tangent with the path being
arc length parametrized. If the shortest path (without its end points) does not

meet the patch boundary then this shortest path is a geodesic in the sense of
Definition 10.1.1 where this is proven in [448] under very weak assumptions.
We shall henceforth assume throughout this chapter that the shortest path
to be computed will not meet the patch boundary except possibly at its end
points.

Let C be an arc length parametrized regular curve on this surface which
passes through point P as shown in Fig. 3.6 and denoted by

$$\mathbf{r}(s) = \mathbf{r}(u(s), v(s)) \ . \tag{10.1}$$

Let $\mathbf{t}$ be a unit tangent vector of C at P, $\mathbf{n}$ be a unit normal vector of C
at P, $\mathbf{N}$ be a unit surface normal vector of S at P and $\mathbf{u}$ be a unit vector
perpendicular to $\mathbf{t}$ in the tangent plane of the surface, defined by $\mathbf{u} = \mathbf{N} \times \mathbf{t}$.
The $\mathbf{u}$ component of the curvature vector $\mathbf{k}$ of $\mathbf{r}(s)$ is the geodesic curvature
vector $\mathbf{k}_g$ and is given by

$$\mathbf{k}_g = (\mathbf{k} \cdot \mathbf{u})\mathbf{u} \ . \tag{10.2}$$

The scalar function

$$\kappa_g = \mathbf{k} \cdot \mathbf{u} \ , \tag{10.3}$$

is called the *geodesic curvature* of C at P, or equivalently

$$\kappa_g = \frac{d\mathbf{t}}{ds} \cdot (\mathbf{N} \times \mathbf{t}) \ . \tag{10.4}$$

The unit tangent vector of the curve C can be obtained by differentiating
(10.1) with respect to the arc length using the chain rule

$$\mathbf{t} = \frac{d\mathbf{r}(u(s), v(s))}{ds} = \mathbf{r}_u \frac{du}{ds} + \mathbf{r}_v \frac{dv}{ds} \ . \tag{10.5}$$

Thus we have

$$\frac{d\mathbf{t}}{ds} = \mathbf{r}_{uu} \left(\frac{du}{ds} \right)^2 + 2\mathbf{r}_{uv} \frac{du}{ds}\frac{dv}{ds} + \mathbf{r}_{vv} \left(\frac{dv}{ds} \right)^2 + \mathbf{r}_u \frac{d^2u}{ds^2} + \mathbf{r}_v \frac{d^2v}{ds^2} \ , \tag{10.6}$$

and hence substituting (10.5) and (10.6) into (10.4) yields

$$
\begin{aligned}
\kappa_g = &\left[(\mathbf{r}_u \times \mathbf{r}_{uu}) \left(\frac{du}{ds} \right)^3 + (2\mathbf{r}_u \times \mathbf{r}_{uv} + \mathbf{r}_v \times \mathbf{r}_{uu}) \left(\frac{du}{ds} \right)^2 \frac{dv}{ds} \right. \\
&\left. + (\mathbf{r}_u \times \mathbf{r}_{vv} + 2\mathbf{r}_v \times \mathbf{r}_{uv}) \frac{du}{ds} \left(\frac{dv}{ds} \right)^2 + (\mathbf{r}_v \times \mathbf{r}_{vv}) \left(\frac{dv}{ds} \right)^3 \right] \cdot \mathbf{N} \\
&+ (\mathbf{r}_u \times \mathbf{r}_v) \cdot \mathbf{N} \left(\frac{du}{ds}\frac{d^2v}{ds^2} - \frac{d^2u}{ds^2}\frac{dv}{ds} \right) \ .
\end{aligned}
\tag{10.7}
$$

We can easily observe that the coefficients of $\left(\frac{du}{ds}\right)^3$, $\left(\frac{du}{ds}\right)^2\frac{dv}{ds}$, $\frac{du}{ds}\left(\frac{dv}{ds}\right)^2$, $\left(\frac{dv}{ds}\right)^3$, $\left(\frac{du}{ds}\frac{d^2v}{ds^2} - \frac{d^2u}{ds^2}\frac{dv}{ds}\right)$ are all functions of the coefficients of the first fundamental form E, F and G and their derivatives, E_u, F_u, G_u, E_v, F_v, G_v. It is interesting to note that the normal curvature κ_n depends on both the first and second fundamental forms, while the geodesic curvature depends only on the first fundamental form. Using the Christoffel symbols Γ^i_{jk} $(i, j, k = 1, 2)$ defined as follows [412]

$$
\begin{aligned}
\Gamma^1_{11} &= \frac{GE_u - 2FF_u + FE_v}{2(EG - F^2)}, & \Gamma^2_{11} &= \frac{2EF_u - EE_v + FE_u}{2(EG - F^2)}, \\
\Gamma^1_{12} &= \frac{GE_v - FG_u}{2(EG - F^2)}, & \Gamma^2_{12} &= \frac{EG_u - FE_v}{2(EG - F^2)}, \\
\Gamma^1_{22} &= \frac{2GF_v - GG_u + FG_v}{2(EG - F^2)}, & \Gamma^2_{22} &= \frac{EG_v - 2FF_v + FG_u}{2(EG - F^2)},
\end{aligned}
\tag{10.8}
$$

geodesic curvature reduces to

$$
\begin{aligned}
\kappa_g = \Bigg[& \Gamma^2_{11}\left(\frac{du}{ds}\right)^3 + (2\Gamma^2_{12} - \Gamma^1_{11})\left(\frac{du}{ds}\right)^2\frac{dv}{ds} + (\Gamma^2_{22} - 2\Gamma^1_{12})\frac{du}{ds}\left(\frac{dv}{ds}\right)^2 \\
& -\Gamma^1_{22}\left(\frac{dv}{ds}\right)^3 + \frac{du}{ds}\frac{d^2v}{ds^2} - \frac{d^2u}{ds^2}\frac{dv}{ds}\Bigg]\sqrt{EG - F^2}\, .
\end{aligned}
\tag{10.9}
$$

According to the definition, we can determine the differential equation that any geodesic on a surface must satisfy by simply setting $\kappa_g = 0$ in (10.9) and obtain

$$
\begin{aligned}
\frac{du}{ds}\frac{d^2v}{ds^2} - \frac{d^2u}{ds^2}\frac{dv}{ds} = & -\Gamma^2_{11}\left(\frac{du}{ds}\right)^3 - (2\Gamma^2_{12} - \Gamma^1_{11})\left(\frac{du}{ds}\right)^2\frac{dv}{ds} \\
& + (2\Gamma^1_{12} - \Gamma^2_{22})\frac{du}{ds}\left(\frac{dv}{ds}\right)^2 + \Gamma^1_{22}\left(\frac{dv}{ds}\right)^3\, .
\end{aligned}
\tag{10.10}
$$

Alternatively, we can derive the differential equation for geodesics by considering that the surface normal $\mathbf{N}$ has the direction of a normal to the geodesic curve $\pm\mathbf{n}$

$$
\mathbf{n}\cdot\mathbf{r}_u = 0, \qquad \mathbf{n}\cdot\mathbf{r}_v = 0\, .
\tag{10.11}
$$

Since $k\mathbf{n} = \frac{d\mathbf{t}}{ds}$, (10.11) can be rewritten as

$$
\frac{d\mathbf{t}}{ds}\cdot\mathbf{r}_u = 0, \qquad \frac{d\mathbf{t}}{ds}\cdot\mathbf{r}_v = 0\, .
\tag{10.12}
$$

By substituting (10.6) into equations (10.12) we have

$$(\mathbf{r}_{uu} \cdot \mathbf{r}_u) \left(\frac{du}{ds}\right)^2 + 2(\mathbf{r}_{uv} \cdot \mathbf{r}_u)\frac{du}{ds}\frac{dv}{ds} + (\mathbf{r}_{vv} \cdot \mathbf{r}_u)\left(\frac{dv}{ds}\right)^2$$

$$+ E\frac{d^2u}{ds^2} + F\frac{d^2v}{ds^2} = 0 , \tag{10.13}$$

$$(\mathbf{r}_{uu} \cdot \mathbf{r}_v) \left(\frac{du}{ds}\right)^2 + 2(\mathbf{r}_{uv} \cdot \mathbf{r}_v)\frac{du}{ds}\frac{dv}{ds} + (\mathbf{r}_{vv} \cdot \mathbf{r}_v)\left(\frac{dv}{ds}\right)^2$$

$$+ F\frac{d^2u}{ds^2} + G\frac{d^2v}{ds^2} = 0 . \tag{10.14}$$

By eliminating $\frac{d^2v}{ds^2}$ from (10.13) using (10.14), and eliminating $\frac{d^2u}{ds^2}$ from (10.14) using (10.13) and employing the Christoffel symbols, we obtain [412]

$$\frac{d^2u}{ds^2} + \Gamma^1_{11}\left(\frac{du}{ds}\right)^2 + 2\Gamma^1_{12}\frac{du}{ds}\frac{dv}{ds} + \Gamma^1_{22}\left(\frac{dv}{ds}\right)^2 = 0 , \tag{10.15}$$

$$\frac{d^2v}{ds^2} + \Gamma^2_{11}\left(\frac{du}{ds}\right)^2 + 2\Gamma^2_{12}\frac{du}{ds}\frac{dv}{ds} + \Gamma^2_{22}\left(\frac{dv}{ds}\right)^2 = 0 . \tag{10.16}$$

Equations (10.15) and (10.16) are related by the first fundamental form $ds^2 = Edu^2 + 2Fdudv + Gdv^2$ and if we eliminate ds from both equations, the equations reduce to (10.10) with u taken as parameter. These two second order differential equations can be rewritten as a system of four first order differential equations [235]

$$\frac{du}{ds} = p , \tag{10.17}$$

$$\frac{dv}{ds} = q , \tag{10.18}$$

$$\frac{dp}{ds} = -\Gamma^1_{11}p^2 - 2\Gamma^1_{12}pq - \Gamma^1_{22}q^2 , \tag{10.19}$$

$$\frac{dq}{ds} = -\Gamma^2_{11}p^2 - 2\Gamma^2_{12}pq - \Gamma^2_{22}q^2 . \tag{10.20}$$

We can also find this result by means of the general rules of the calculus of variations [166]. We want to minimize

$$I = \int_A^B ds = \int_A^B \sqrt{E + 2F\frac{dv}{du} + G\left(\frac{dv}{du}\right)^2}\,du = \int_A^B f(u, v, \dot{v})du ,$$

$$\tag{10.21}$$

subject to the conditions

$$v(A) = v_A, \qquad v(B) = v_B , \tag{10.22}$$

where

$$f(u, v, \dot{v}) = \sqrt{E + 2F\dot{v} + G\dot{v}^2}, \quad \dot{v} = \frac{dv}{du}, \tag{10.23}$$

and v_A and v_B are given constants. It is well known from calculus of variations that the solution of Euler's equation [166]

$$\frac{\partial f}{\partial v} - \frac{d}{du}\frac{\partial f}{\partial \dot{v}} = 0, \tag{10.24}$$

gives an extreme value to the integral (10.21). When (10.23) is substituted in Euler's equation (10.24) we can derive the differential equation for geodesics.

Example 10.2.1. Let us obtain the geodesic equations for a parametric bilinear surface (hyperbolic paraboloid) $\mathbf{r}(u, v) = (u, v, uv)$ (see Fig. 3.4). We have

$$\begin{aligned} E &= 1 + v^2, \; F = uv, \; G = 1 + u^2, \\ E_u &= 0, \qquad F_u = v, \; G_u = 2u, \\ E_v &= 2v, \qquad F_v = u, \; G_v = 0, \end{aligned}$$

thus, the Christoffel symbols become

$$\Gamma_{11}^1 = \Gamma_{11}^2 = \Gamma_{22}^1 = \Gamma_{22}^2 = 0,$$
$$\Gamma_{12}^1 = \frac{v}{u^2 + v^2 + 1},$$
$$\Gamma_{12}^2 = \frac{u}{u^2 + v^2 + 1}.$$

Finally the geodesic equations for the bilinear surface are given by

$$\frac{du}{ds} = p,$$
$$\frac{dv}{ds} = q,$$
$$\frac{dp}{ds} = \frac{-2v}{u^2 + v^2 + 1}pq,$$
$$\frac{dq}{ds} = \frac{-2u}{u^2 + v^2 + 1}pq.$$

10.2.2 Implicit surfaces

We can also derive the geodesic equation for an implicit surface by finding an expression of the geodesic curvature for an implicit surface. Let us consider an arc length parametrized curve $\mathbf{r} = \mathbf{r}(s)$ or $x = x(s)$, $y = y(s)$, $z = z(s)$ on an implicit surface $f(x, y, z) = 0$. By substituting $\mathbf{t} = (x', y', z')^T$, $\frac{d\mathbf{t}}{ds} = (x'', y'', z'')^T$, $\mathbf{N} = \frac{\nabla f}{|\nabla f|}$, into (10.4), we obtain the expression for the geodesic curvature of a curve on the implicit surface

$$\kappa_g = \frac{(y'z'' - z'y'')f_x + (z'x'' - x'z'')f_y + (x'y'' - y'x'')f_z}{\sqrt{f_x^2 + f_y^2 + f_z^2}} . \quad (10.25)$$

For the sake of completeness, the geodesic curvature for a non-arc-length parametrized curve is given by

$$\kappa_g = \frac{(\dot{y}\ddot{z} - \dot{z}\ddot{y})f_x + (\dot{z}\ddot{x} - \dot{x}\ddot{z})f_y + (\dot{x}\ddot{y} - \dot{y}\ddot{x})f_z}{(\dot{x}^2 + \dot{y}^2 + \dot{z}^2)^{\frac{3}{2}}\sqrt{f_x^2 + f_y^2 + f_z^2}} . \quad (10.26)$$

Now if we set $\kappa_g = 0$, we deduce

$$(y'z'' - z'y'')f_x + (z'x'' - x'z'')f_y + (x'y'' - y'x'')f_z = 0 . \quad (10.27)$$

Since the unit tangent vector (x', y', z') and the curvature vector (x'', y'', z'') of the geodesic curve are orthogonal to each other, we have

$$x'x'' + y'y'' + z'z'' = 0 . \quad (10.28)$$

The third equation can be derived from (6.21)

$$f_{xx}(x')^2 + f_{yy}(y')^2 + f_{zz}(z')^2 + 2(f_{xy}x'y' + f_{yz}y'z' + f_{xz}x'z') \\ + f_x x'' + f_y y'' + f_z z'' = 0 . \quad (10.29)$$

Now we solve the linear system of three equations (10.27) to (10.29) in (x'', y'', z''), assuming that $(z'f_y - y'f_z)^2 + (x'f_z - z'f_x)^2 + (y'f_x - x'f_y)^2$ does not vanish, yielding

$$x'' = \frac{(x'f_z - z'f_x)z' + (x'f_y - y'f_x)y'}{(z'f_y - y'f_z)^2 + (x'f_z - z'f_x)^2 + (y'f_x - x'f_y)^2}\Lambda , \quad (10.30)$$

$$y'' = \frac{(y'f_z - z'f_y)z' + (y'f_x - x'f_y)x'}{(z'f_y - y'f_z)^2 + (x'f_z - z'f_x)^2 + (y'f_x - x'f_y)^2}\Lambda , \quad (10.31)$$

$$z'' = \frac{(z'f_y - y'f_z)y' + (z'f_x - x'f_z)x'}{(z'f_y - y'f_z)^2 + (x'f_z - z'f_x)^2 + (y'f_x - x'f_y)^2}\Lambda , \quad (10.32)$$

where $\Lambda = f_{xx}(x')^2 + f_{yy}(y')^2 + f_{zz}(z')^2 + 2(f_{xy}x'y' + f_{yz}y'z' + f_{xz}x'z')$. These three second order differential equations can be rewritten as a system of six first order differential equations:

$$x' = p , \quad (10.33)$$

$$y' = q , \quad (10.34)$$

$$z' = r , \quad (10.35)$$

$$p' = \frac{(pf_z - rf_x)r + (pf_y - qf_x)q}{(rf_y - qf_z)^2 + (pf_z - rf_x)^2 + (qf_x - pf_y)^2}\Lambda , \quad (10.36)$$

$$q' = \frac{(qf_z - rf_y)r + (qf_x - pf_y)p}{(rf_y - qf_z)^2 + (pf_z - rf_x)^2 + (qf_x - pf_y)^2}\Lambda , \quad (10.37)$$

$$r' = \frac{(rf_y - qf_z)q + (rf_x - pf_z)p}{(rf_y - qf_z)^2 + (pf_z - rf_x)^2 + (qf_x - pf_y)^2}\Lambda . \quad (10.38)$$

Figure 10.1 shows a geodesic on an ellipsoid ($\frac{x^2}{9} + \frac{y^2}{4} + z^2 = 1$) computed by integrating the above system of six first order differential equations as an initial value problem. The initial values are given by $(x, y, z) = (0, 2, 0)$, $(p, q, r) = \left(\frac{\sqrt{2}}{2}, 0, \frac{\sqrt{2}}{2}\right)$ and the integration is terminated at $(x, y, z) = (2.439, -0.726, 0.456)$.

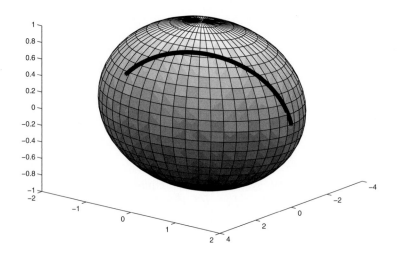

Fig. 10.1. Geodesics on an ellipsoid

10.3 Two point boundary value problem

10.3.1 Introduction

We can solve the system of four first order ordinary differential equations (10.17) to (10.20) as an initial value problem (IVP), where all four boundary conditions are given at one point, or as a boundary value problem (BVP), where four boundary conditions are specified at two distinct points. Most of the problems that arise in applications of geodesics are not IVP but BVP, which are much more difficult to solve. It is well known that the solution of an IVP is unique, however for a BVP it is possible that the differential equations have many solutions or even no solution [194]. General methods for the solutions of two-point BVPs can be found in [194, 117].

It is convenient to write the system of differential equations in vector form, since we can describe the equations for systems in terms of a single vector equation. Let us set

$$\begin{aligned}
\mathbf{y} &= (y_1, y_2, \ldots, y_n)^T, \quad \mathbf{g} = (g_1, g_2, \ldots, g_n)^T, \\
\boldsymbol{\alpha} &= (\alpha_1, \alpha_2, \ldots, \alpha_n)^T, \quad \boldsymbol{\beta} = (\beta_1, \beta_2, \ldots, \beta_n)^T, \\
s &\in [A, B],
\end{aligned} \tag{10.39}$$

where y_i, g_i are functions and α_i, β_i are constants. Then the general first order vector differential equation for a boundary value problem can be written as:

$$\frac{d\mathbf{y}}{ds} = \mathbf{g}(s, \mathbf{y}), \quad \mathbf{y}(A) = \boldsymbol{\alpha}, \quad \mathbf{y}(B) = \boldsymbol{\beta}. \tag{10.40}$$

There are two commonly used approaches to the numerical solution of BVPs. The idea of the first technique is that if all values of $\mathbf{y}(s)$ are known at $s = A$, then the problem can be reduced to an IVP. However, $\mathbf{y}(A)$ can be found only by solving the problem. Therefore an iterative procedure must be used. We assume values at $s = A$, which are not given as boundary conditions at $s = A$ and compute the solution of the resulting IVP to $s = B$. The computed values of $\mathbf{y}(B)$ will not, in general, agree with the corresponding boundary condition at $s = B$. Consequently, we need to adjust the initial values and try again. The process is repeated until the computed values at the final point agree with the boundary conditions and is referred to as shooting method. The second method is based on a finite difference approximation to $\frac{d\mathbf{y}}{ds}$ on a mesh of points in the interval $[A, B]$. This method starts with an initial guess and improves the solution iteratively and is referred to as, direct method, relaxation method or finite difference method. We have implemented both methods and found that the finite difference method is much more reliable than the shooting method. By contrast to the finite difference method, the shooting method is often very sensitive to the unknown initial values at point A. First, we briefly discuss the shooting method.

10.3.2 Shooting method

We assume a value for p_A and solve the differential equation as an IVP using the fourth order Runge-Kutta method. Using the first fundamental form (3.13), given p_A we can obtain q_A from

$$q_A = \frac{-F p_A \pm \sqrt{F^2 p_A^2 - G(E p_A^2 - 1)}}{G}. \tag{10.41}$$

Here we also have to assume the entire arc length of the geodesic path s to stop the integration. Thus the unknowns can be considered as p_A and s. If we denote the computed value of (u_B, v_B) as (u_B^*, v_B^*), the difference can be given as $(u_B^* - u_B, v_B^* - v_B)$. We need to adjust p_A and s to make the difference zero. This can be done by employing Newton's method

$$\begin{pmatrix} p_A \\ s \end{pmatrix}_{i+1} = \begin{pmatrix} p_A \\ s \end{pmatrix}_i - \begin{bmatrix} \dfrac{\partial u_B^*}{\partial p_A} & \dfrac{\partial u_B^*}{\partial s} \\ \dfrac{\partial v_B^*}{\partial p_A} & \dfrac{\partial v_B^*}{\partial s} \end{bmatrix}_i^{-1} \begin{pmatrix} u_B^* - u_B \\ v_B^* - v_B \end{pmatrix}, \tag{10.42}$$

where the Jacobian matrix is evaluated numerically. We first change p_A slightly to $p_A + \Delta p_A$ and integrate the ordinary differential equations as an IVP to evaluate the end point $(u_B^*(p_A + \Delta p_A, s), v_B^*(p_A + \Delta p_A, s))$, from which we can compute the partial derivatives $\frac{\partial u_B^*}{\partial p_A}$ and $\frac{\partial v_B^*}{\partial p_A}$ as

$$\frac{\partial u_B^*}{\partial p_A} = \frac{u_B^*(p_A + \Delta p_A, s) - u_B^*(p_A, s)}{\Delta p_A} , \qquad (10.43)$$

$$\frac{\partial v_B^*}{\partial p_A} = \frac{v_B^*(p_A + \Delta p_A, s) - v_B^*(p_A, s)}{\Delta p_A} . \qquad (10.44)$$

Similarly we change s slightly to $s + \Delta s$ and integrate the ordinary differential equations as IVP to evaluate the end point $(u_B^*(p_A, s + \Delta s), v_B^*(p_A, s + \Delta s))$, from which we can compute the partial derivatives $\frac{\partial u_B^*}{\partial s}$ and $\frac{\partial v_B^*}{\partial s}$

$$\frac{\partial u_B^*}{\partial s} = \frac{u_B^*(p_A, s + \Delta s) - u_B^*(p_A, s)}{\Delta s} , \qquad (10.45)$$

$$\frac{\partial v_B^*}{\partial s} = \frac{v_B^*(p_A, s + \Delta s) - v_B^*(p_A, s)}{\Delta s} . \qquad (10.46)$$

10.3.3 Relaxation method

The relaxation method [336, 247] starts by first discretizing the governing equations by finite differences on a mesh with m points. The computation begins with an initial guess and improves the solution iteratively or in other words relaxes to the true solution. Let us consider an arc length parametrized curve connecting A and B on the surface with a mesh of points satisfying $A = s_1 < s_2 < \ldots < s_m = B$. We approximate the n first order differential equations by the trapezoidal rule [117]

$$\frac{\mathbf{Y}_k - \mathbf{Y}_{k-1}}{s_k - s_{k-1}} = \frac{1}{2}[\mathbf{G}_k + \mathbf{G}_{k-1}], \qquad k = 2, 3, \ldots, m , \qquad (10.47)$$

with boundary conditions

$$\mathbf{Y}_1 = \boldsymbol{\alpha}, \qquad \mathbf{Y}_m = \boldsymbol{\beta} . \qquad (10.48)$$

Here the n-vectors $\mathbf{Y}_k$, $\mathbf{G}_k$ are meant to approximate $\mathbf{y}(s_k)$ and $\mathbf{g}(s_k)$. $\mathbf{Y}_1$ has n_1 known components, while $\mathbf{Y}_m$ has $n_2 = n - n_1$ known components. This discrete approximation will be accurate to the order of h^2 ($h = \max_k\{s_k - s_{k-1}\}$). Equation (10.47) forms a system of $(m-1)n$ nonlinear equations with mn unknowns $\mathbf{Y}_k = (Y_1, Y_2, \ldots, Y_n)_k^T$ ($k = 1, \ldots, m$). The remaining n equations come from boundary conditions (10.48). Let us refer to (10.47) as

$$\mathbf{F}_k = (F_{1,k}, F_{2,k}, \ldots, F_{n,k})^T = \frac{\mathbf{Y}_k - \mathbf{Y}_{k-1}}{s_k - s_{k-1}} - \frac{1}{2}[\mathbf{G}_k + \mathbf{G}_{k-1}] = \mathbf{0} ,$$

$$(10.49)$$

where $k = 2, 3, \ldots, m$, and refer to (10.48) as

$$\mathbf{F}_1 = (F_{1,1}, F_{2,1}, \ldots, F_{n_1,1})^T = \mathbf{Y}_1 - \boldsymbol{\alpha} = \mathbf{0} \ ,$$
$$\mathbf{F}_{m+1} = (F_{1,m+1}, F_{2,m+1}, \ldots, F_{n_2,m+1})^T = \mathbf{Y}_m - \boldsymbol{\beta} = \mathbf{0} \ , \quad (10.50)$$

then we have mn nonlinear equations

$$\mathbf{F} = (\mathbf{F}_1^T, \mathbf{F}_2^T, \ldots, \mathbf{F}_{m+1}^T)^T = \mathbf{0} \ . \quad (10.51)$$

This system of nonlinear equations can be solved by quadratically convergent Newton iteration, if a sufficiently accurate starting vector $\mathbf{Y}^{(0)} = (\mathbf{Y}_1^T, \mathbf{Y}_2^T, \ldots, \mathbf{Y}_m^T)^T$ is provided. The Newton iteration scheme is given by

$$\mathbf{Y}^{(i+1)} = \mathbf{Y}^{(i)} + \Delta\mathbf{Y}^{(i)} \ , \quad (10.52)$$
$$[\mathbf{J}^{(i)}]\Delta\mathbf{Y}^{(i)} = -\mathbf{F}^{(i)} \ , \quad (10.53)$$

where superscripts (i) denote i-th iteration and $[\mathbf{J}^{(i)}]$ is the mn by mn Jacobian matrix of $\mathbf{F}^{(i)}$ with respect to $\mathbf{Y}^{(i)}$.

Since the corrections are based on a first order Taylor approximation, the usual Newton method may not be sufficient for a complex nonlinear problem unless a good initial approximation is provided. If the vector norm of the correction vector is large, then it is an indication that the problem is highly nonlinear and may produce a divergent iteration. To achieve more stability we can employ a step correction procedure

$$\mathbf{Y}^{(i+1)} = \mathbf{Y}^{(i)} + \mu\Delta\mathbf{Y}^{(i)} \ , \quad (10.54)$$

where $0 < \mu \leq 1$ chosen so that $\| \Delta\mathbf{Y}^{(i+1)} \|_1 < \| \Delta\mathbf{Y}^{(i)} \|_1$, where $\| \Delta\mathbf{Y} \|_1$ is a scaled vector norm and defined as

$$\| \Delta\mathbf{Y} \|_1 = \sum_{k=1}^{m} \left(\frac{|\Delta u_k|}{M_u} + \frac{|\Delta v_k|}{M_v} + \frac{|\Delta p_k|}{M_p} + \frac{|\Delta q_k|}{M_q} \right) \ , \quad (10.55)$$

where M_u, M_v, M_p and M_q are the scale factors for each variable. Maekawa [247] used $M_u = M_v = 1$ and $M_p = M_q = 10$, since the magnitude of Δp_k and Δq_k are roughly ten times larger that of Δu_k and Δv_k as numerical experiments have shown. If $\mu = 1$ the equation reduces to the usual Newton's method, while if $\mu < 1$ the rate of convergence will be less than quadratic. Newton's method terminates when the norm of the solution vector is smaller than the pre-specified tolerance ε_N. The order of ε_N should be proportional to h^2, since we are using the trapezoidal rule (see (10.47)).

10.4 Initial approximation

10.4.1 Linear approximation

Linear approximation is the simplest and most often provides a good initial approximation, since it is a solution to the system of geodesic equations

(10.17) to (10.20) when we neglect all the nonlinear terms in the right hand side. We connect the two end points in the parameter space by a straight line and define a uniform mesh or grid by a set of $k = 1, 2, \ldots, m$ points as shown in Fig. 10.2 (a). Therefore we have

$$u_k = u_A + \frac{u_B - u_A}{m - 1}(k - 1), \tag{10.56}$$

$$v_k = v_A + \frac{v_B - v_A}{m - 1}(k - 1). \tag{10.57}$$

When the uniform mesh in the parameter space is mapped onto the surface, the corresponding arc length mesh will not be in general uniform.

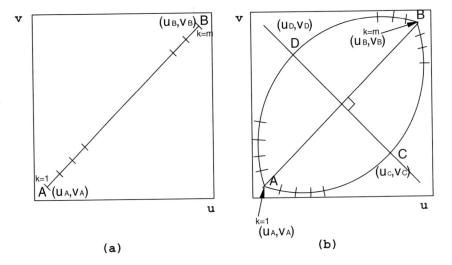

Fig. 10.2. Initial approximations (adapted from [247]): (**a**) linear approximation, (**b**) circular arc approximation

If we assume that $u_A \neq u_B$ then

$$\frac{dv}{du} = \frac{v_B - v_A}{u_B - u_A} \equiv \phi, \tag{10.58}$$

hence

$$\frac{dv}{ds} = \phi \frac{du}{ds}. \tag{10.59}$$

By substituting this relation into the first fundamental form we obtain

$$E\left(\frac{du}{ds}\right)^2 + 2F\phi\left(\frac{du}{ds}\right)^2 + G\phi^2\left(\frac{du}{ds}\right)^2 = 1. \tag{10.60}$$

Thus

$$p_k = \frac{du}{ds} = \pm \frac{1}{\sqrt{E_k + 2F_k\phi + G_k\phi^2}} \ , \tag{10.61}$$

$$q_k = \frac{dv}{ds} = \pm \frac{\phi}{\sqrt{E_k + 2F_k\phi + G_k\phi^2}} \ . \tag{10.62}$$

When $u_A = u_B$, it is easy to find that $p_k = 0$ and $q_k = \frac{1}{\sqrt{G_k}}$. It is well known that conjugate points do not exist on regions of a surface where the Gaussian curvature is negative [412]. Therefore, the linear approximation will typically provide a good initial approximation to the geodesic path in those regions.

10.4.2 Circular arc approximation

The problem of the straight line approximation is that when there are more than one path, it cannot capture the other paths. To make the method more reliable, the following algorithm has been developed [247]. First we pick two points C and D in the parameter domain, which are on the bisector of the two end points A and B, such that $\overline{AC} = \overline{AD}$ or $\overline{BC} = \overline{BD}$ as illustrated in Fig. 10.2 (b). Then we determine two circular arcs which pass through the three points A, C, B and A, D, B. If C and D are taken at a large enough distance from AB, all the geodesic paths in the parameter domain between points A and B are likely to lie within or close to the region surrounded by the two circular arcs. Notice that the algorithm fails once the circular arcs go outside the domain so these arcs are chosen such that they are entirely within the domain. The uv coordinates in the parameter domain i.e. (u_k, v_k), $k = 1, \ldots, m$ can be obtained by equally distributing the points along the circular arc in the parameter domain. Once we have a set of points in the parameter domain, we can easily evaluate p_k, q_k by using the central difference formula for a non-uniform mesh points [117], for $k = 2, \ldots, m - 1$

$$f'(s_k) = \frac{\frac{h_k}{h_{k+1}}(f_{k+1} - f_k) - \frac{h_{k+1}}{h_k}(f_{k-1} - f_k)}{h_k + h_{k+1}} \ , \tag{10.63}$$

the forward difference formula for $k = 1$

$$f'(s_1) = \frac{-\frac{h_2}{h_3}f_3 + \left(\frac{h_2}{h_3} + \frac{h_3}{h_2} + 2\right)f_2 - \left(2 + \frac{h_3}{h_2}\right)f_1}{h_2 + h_3} \ , \tag{10.64}$$

and the backward difference formula for $k = m$

$$f'(s_m) = \frac{\frac{h_m}{h_{m-1}}f_{m-2} - \left(\frac{h_{m-1}}{h_m} + \frac{h_m}{h_{m-1}} + 2\right)f_{m-1} + \left(2 + \frac{h_{m-1}}{h_m}\right)f_m}{h_{m-1} + h_m} \ , \tag{10.65}$$

where f is replaced by u or v, and the step length $h_k = s_k - s_{k-1}$ is evaluated by computing the chord length between the successive points on the surface. Even if the mesh points are equally distributed along the circular arc in the parameter domain as shown in Fig. 10.2 (b), h_k is not in general constant.

We give the flow chart of the algorithm based on circular arc approximation for computing the geodesic path between two given points in Fig. 10.3.

10.5 Shortest path between a point and a curve

In this section we solve a problem of finding a shortest path between a point and a curve on a free-form non-periodic parametric surface [247], which utilizes the method we have developed in Sects. 10.3 and 10.4. This concept is important in robot motion planning and constructing a medial axis on a free-form surface. Suppose we have a point A and a curve C on a parametric surface $\mathbf{r}(u, v)$ defined as $\mathbf{r}^c(t) = \mathbf{r}(u^c(t), v^c(t))$, we want to compute the shortest path between point A and curve C as shown in Fig. 10.4. The existence of such a shortest path follows from results shown in [448]. Let us denote the intersection point of the curve C and the shortest path by B. Wolter [449] developed a necessary condition to have a shortest path from point A to curve C, provided that the point B is not an end point of the curve C. The condition is given by the orthogonality of the tangent vector of the geodesic curve, connecting A and C, at B and the tangent vector of C at B. If point B were known, the problem could be reduced to an IVP, since at point B, u, v, p and q are all known. However, point B can be found only by solving the problem. We guess a parameter value $t = t_B$ for point B and solve the BVP. In general the unit tangent vector of the curve $\mathbf{r}^c(t)$ and the unit tangent vector of the arc length parametrized geodesic curve $\mathbf{r}^g(s)$ at the guessing point will not be orthogonal to each other. Consequently, we need to adjust the parameter value t_B and iterate until those two unit tangent vectors become orthogonal. Therefore for each iteration, we need to solve a two point boundary value problem, which also requires iterations (i.e., nested iteration). If we denote these two unit tangent vectors as $\mathbf{t}^c$ and $\mathbf{t}^g$, then they can be expressed as follows:

$$\mathbf{t}^c = \frac{\frac{d\mathbf{r}^c(t)}{dt}}{\left|\frac{d\mathbf{r}^c(t)}{dt}\right|} = \frac{\mathbf{r}_u \frac{du^c}{dt} + \mathbf{r}_v \frac{dv^c}{dt}}{\sqrt{\left(\frac{du^c}{dt}x_u + \frac{dv^c}{dt}x_v\right)^2 + \left(\frac{du^c}{dt}y_u + \frac{dv^c}{dt}y_v\right)^2}}, \quad (10.66)$$

$$\mathbf{t}^g = \frac{d\mathbf{r}^g(s)}{ds} = \mathbf{r}_u \frac{du^g}{ds} + \mathbf{r}_v \frac{dv^g}{ds}. \quad (10.67)$$

Therefore the orthogonality condition can be written as

$$\omega(t) = \mathbf{t}^c \cdot \mathbf{t}^g$$

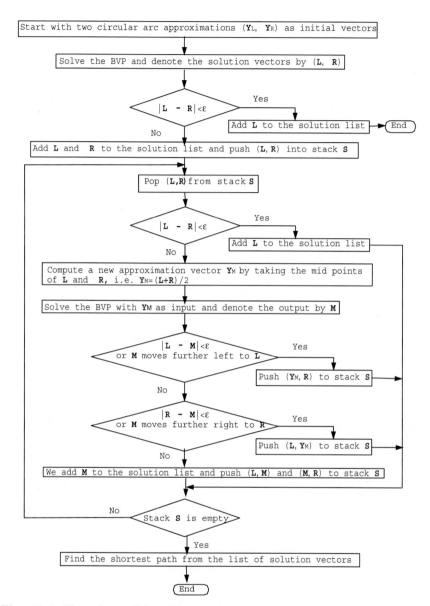

Fig. 10.3. Flow chart of the algorithm based on circular arc approximation for computing the geodesic path between two given points (adapted from [247])

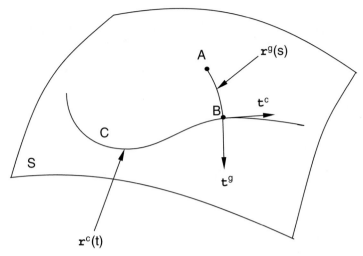

Fig. 10.4. Shortest path between point A and curve C (adapted from [247])

$$= \frac{\frac{du^g}{ds}\frac{du^c}{dt}E + \left(\frac{dv^g}{ds}\frac{du^c}{dt} + \frac{dv^c}{dt}\frac{du^g}{ds}\right)F + \frac{dv^g}{ds}\frac{dv^c}{dt}G}{\sqrt{\left(\frac{du^c}{dt}x_u + \frac{dv^c}{dt}x_v\right)^2 + \left(\frac{du^c}{dt}y_u + \frac{dv^c}{dt}y_v\right)^2}}$$

$$= 0 \, . \tag{10.68}$$

Consequently we need to find a parameter value t_B such that $\omega(t_B) = 0$. Since the relationship described above is implicit, we use the *secant method* [69] instead of Newton's method to obtain t_B. The secant method can be derived from Newton's method by replacing the derivative $\frac{d\omega^{(i)}}{dt}$ by the quotient $(\omega^{(i)} - \omega^{(i-1)})/(t^{(i)} - t^{(i-1)})$ where superscripts (i) denote i-th iteration. This leads to the following scheme

$$t^{(i+1)} = t^{(i)} + \Delta t^{(i)}, \quad \Delta t^{(i)} = -\frac{t^{(i)} - t^{(i-1)}}{\omega^{(i)} - \omega^{(i-1)}}\omega^{(i)}, \quad \omega^{(i)} \neq \omega^{(i-1)} \, .$$

$$\tag{10.69}$$

Notice that the secant method requires two initial approximations $t^{(1)}$ and $t^{(2)}$. Since the secant method has an order of convergence of $\frac{1}{2}(1+\sqrt{5}) \simeq 1.618$ [69], it converges within a reasonable number of iterations. If the correction $\Delta t^{(i)}$ is large, it is again an indication that the problem is highly nonlinear. In such case we also employ a step correction procedure

$$t^{(i+1)} = t^{(i)} + \nu \Delta t^{(i)} \, , \tag{10.70}$$

where ν is a correction factor $0 < \nu \leq 1$, determined as for the modified Newton's method. Since we do not know how many solutions exist and the corresponding parameter values of the curve beforehand, we can use a similar algorithm to the circular arc approximation. If the range of the parameter value of the curve is $0 \leq t \leq 1$, we start from both ends of the curve (i.e. $t^{(1)}=0$, $t^{(2)}=0.02$ and $t^{(1)}=1$, $t^{(2)}=0.98$). Then we recursively find the solutions. Although we may have several footpoints B in different locations of the curve, for each footpoint there is only one unique solution between A and B, since this can be viewed in the context of an initial value problem.

10.6 Numerical applications

10.6.1 Geodesic path between two points

The first example is a wave-like bicubic B-spline surface, whose control polyhedron is a lattice of 7×7 vertices with uniform knot vectors in both directions and spans $0 \leq x \leq 1$, $0 \leq y \leq 1$. Let us compute the geodesic path between two corner points, $(u_A, v_A)=(0,0)$ and $(u_B, v_B)=(1,1)$. We choose (u_C, v_C) to be $(0.7, 0.3)$ and (u_D, v_D) to be $(0.3, 0.7)$ for the circular arc approximation. The algorithm based on circular arc approximation finds three geodesic paths, as shown in Fig. 10.5 (solid thick lines). The computational conditions such as number of mesh points, tolerance and correction factor for the Newton's method ε_N, μ, as well as computational results such as number of iterations for convergence and the geodesic distances are listed in Table 10.1. Symbols Lt, Md, Rt refer to left, middle and right geodesic paths in Fig. 10.5. The middle geodesic path is not a minimal path ($s=1.865$), while the other two paths are the shortest path ($s=1.661$) due to symmetry. Figure 10.6 shows how the initial approximation path (the right-most thick solid line) converges gradually to the final solution (wavy thick solid line). The intermediate paths are illustrated by the thin solid lines.

Table 10.1. Numerical conditions and results for the computation of the geodesic path between corner points of the wave-like surface (adapted from [247])

Points	Tolerance	Correction	Iterations			Geodesic distance		
m	ε_N	factor μ	Lt	Md	Rt	Lt	Md	Rt
101	1.0E-3	0.2	22	1	22	1.661	1.865	1.661

The second example is a generalized cylinder represented by a biquadratic rational B-spline surface, whose control polyhedron is a lattice of 6×9 vertices. This surface was constructed by sweeping a circle of radius 0.5, the generatrix, along a helix ($x = \cos t$, $y = -\sin t$, $z = \frac{t}{\pi}$, $0 \leq t \leq 2\pi$), the spine, and is approximated by rational B-spline interpolating a number of generatrices. When we keep u constant, we obtain a curve on the surface which depends only on v. This curve coincides with the generatrix. Similarly $v = constant$

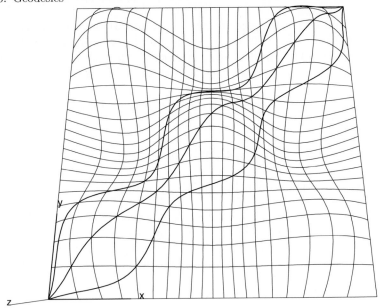

Fig. 10.5. Geodesic paths on the wave-like bicubic B-spline surface between points of two corners (adapted from [247])

represents another iso-parametric curve which is parallel to the spine. Two end points are chosen to be (u_A, v_A)=(0, 0.4) and (u_B, v_B)=(1, 0.6) as shown in Fig. 10.7. In Fig. 10.7 three initial approximations are illustrated by the thin solid lines, while the final solutions are illustrated by the thick solid lines. The two circular arcs are determined by setting $(u_C, v_C) = (0.578, 0.108)$, $(u_D, v_D) = (0.422, 0.892)$. The circular arc approximation algorithm starts with these two circular arcs and converges to the two minimal geodesic paths which are shown as thick solid lines close to the initial circular arcs. The minimal geodesic paths mapped onto the generalized cylinder are depicted in Color Plate A.8 and Fig. 10.8. Then the algorithm computes the mid-points of these solutions, which is shown as a thin straight line connecting the two end points. This initial approximation converges to a sine wave-like solution. This solution is a geodesic path but it does not provide the shortest distance (see Fig. 10.9). Table 10.2 shows the list of computational conditions and results as in Table 10.1.

10.6.2 Geodesic path between a point and a curve

Figure 10.10 shows a planar cubic Bézier curve and point A (0.3, 0.2) in the uv parameter domain, which will be mapped onto the wave-like B-spline surface

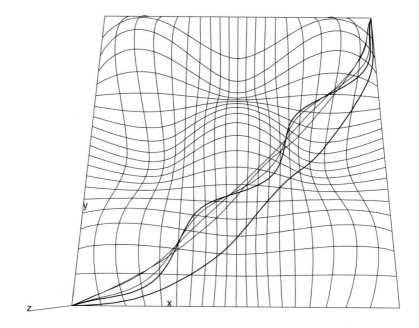

Fig. 10.6. Convergence of the right geodesic path in Fig. 10.5 (adapted from [247])

Table 10.2. Numerical conditions and results for the computation of the geodesic path between two points on generalized cylinder (adapted from [247])

Points	Tolerance	Correction	Iterations			Geodesic distance		
m	ε_N	factor μ	Lt	Md	Rt	Lt	Md	Rt
501	1.0E-2	0.2	8	10	8	5.860	6.983	5.860
1001	5.0E-3	0.2	10	10	10	5.843	6.956	5.843

as shown in Fig. 10.11. The algorithm finds three geodesic paths AB, AB' and AB'', whose tangent vectors at B, B' and B'' are orthogonal to the tangent vectors at the curve at those points. Table 10.3 shows the list of computational conditions and results. The entries t_1, t_2 and t_B are the parameter values of the curve corresponding to the first two initial approximations for the secant method and the solution value. The following entries, m, μ and ν are the number of mesh points, correction factors for the Newton and secant methods. Tolerances for the convergence of Newton and secant methods are given by ε_N, ε_S. The shortest path is given by path AB with $s=0.275$.

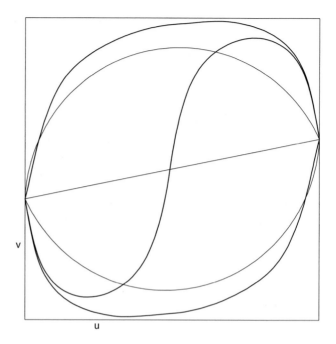

Fig. 10.7. Geodesic paths in the parameter domain of the generalized cylinder (adapted from [247])

Table 10.3. Numerical conditions and results for the computation of the geodesic path between a point and a curve on wave-like surface (adapted from [247])

t_1	t_2	t_B	m	μ	ν	ε_N	ε_S	Iter.	Geodesic distance
0	0.02	0.266	101	0.2	0.05	1.0E-3	1.0E-6	16	0.275
1	0.98	0.727	101	0.2	0.05	1.0E-3	1.0E-6	14	0.371
0.496	0.516	0.579	101	0.2	0.05	1.0E-3	1.0E-6	8	0.387

10.7 Geodesic offsets

In this section we focus on *geodesic offsets* which are different from the classical offset definition. Geodesic offsets or geodesic parallels are well known in classical differential geometry. Let us consider an arbitrary curve C on a surface. The locus of points at a constant distance measured from curve C along the geodesic curve drawn orthogonal to C is called geodesic offset (see Fig. 10.12). Patrikalakis and Bardis [296] provide an algorithm to construct such geodesic offsets on NURBS surfaces. The equations of the geodesics consist

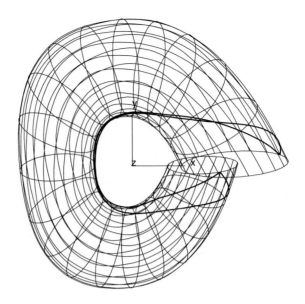

Fig. 10.8. Top view of the minimal geodesic paths on the generalized cylinder between two points of two circular edges (adapted from [247])

of four first order nonlinear ordinary differential equations (10.17) to (10.20) which are solved as an initial value problem.

Let us consider a progenitor curve lying on a parametric surface $\mathbf{r} = \mathbf{r}(u, v)$ given by $\mathbf{r}^c(t) = \mathbf{r}(u^c(t), v^c(t))$ and an arc length parametrized geodesic curve $\mathbf{r}^g(s) = \mathbf{r}(u^g(s), v^g(s))$ orthogonal to $\mathbf{r}^c$. We select n points on the progenitor curve t_i, $0 \le i \le n-1$ and compute a geodesic path for each point by a distance equal to d_g as an IVP. The initial direction $\mathbf{t}^g = (\frac{du^g}{ds}, \frac{dv^g}{ds}) = (p, q)$ can be determined by the condition that the tangent vector along the progenitor curve $\dot{\mathbf{r}}^c$ and the unit tangent vector of the geodesic curve $\mathbf{t}^g$ are orthogonal (see (10.68))

$$(\dot{u}^c E + \dot{v}^c F)p + (\dot{u}^c F + \dot{v}^c G)q = 0 , \tag{10.71}$$

and the normalization condition

$$E(p)^2 + 2Fpq + G(q)^2 = 1 , \tag{10.72}$$

leading to

$$p = \pm \frac{\omega_2}{\sqrt{E\omega_2^2 - 2F\omega_1\omega_2 + G\omega_1^2}} , \tag{10.73}$$

$$q = \mp \frac{\omega_1}{\sqrt{E\omega_2^2 - 2F\omega_1\omega_2 + G\omega_1^2}} , \tag{10.74}$$

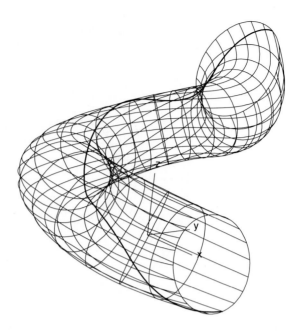

Fig. 10.9. A geodesic path on the generalized cylinder which is not the shortest path (adapted from [247])

where $\omega_1 = \dot{u}^c E + \dot{v}^c F$ and $\omega_2 = \dot{u}^c F + \dot{v}^c G$. The positive and negative signs in (10.73) and (10.74) correspond to the two possible directions of the geodesic path relative to the progenitor curve.

The terminal points of the geodesic paths, departing orthogonally from n selected points of the progenitor curve on the surface, are interpolated in the surface patch parameter space by a B-spline curve assuring that the offset curve lies entirely on the surface.

Wolter and his associates [340] compute medial curves on a surface, which is the locus of points which are equidistant from two given curves on the surface, utilizing the geodesic offset function. Their method is also applicable to the plane curve case. Also Wolter and his associates [214] applied the above method to compute a Voronoi diagram on a parametric surface instead of the Voronoi diagram in Euclidean space.

Traditionally the spacing between adjacent tool paths, which is referred to as side-step or pick-feed, has been kept constant in either the Euclidean space or in the parameter space. Recently geodesic offset curves are used to generate tool paths on a part for zig-zag finishing using 3-axis NC machining with ball-end cutter so that the scallop-height, which is the cusp height of the material removed by the cutter, will become constant [417, 366]. This

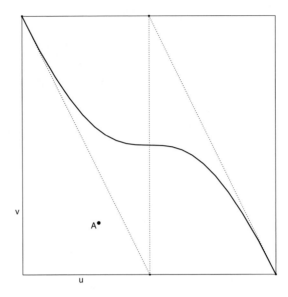

Fig. 10.10. Cubic Bézier curve in the parameter domain (adapted from [247])

leads to a significant reduction in size of the cutter location data and hence in the machining time.

10.8 Geodesics on developable surfaces

In this section it is shown that all two point BVPs for solving geodesics on developables can be reduced to IVPs using the differential geometry properties introduced in Sect. 9.7.1. Since a geodesic on a developable surface maps to a straight line on the developed plane, there is only one solution to the system (10.17)− (10.20) on a developable surface. Here we exclude periodic surfaces such as cylinders where there can be more than one solution. The basic procedure is to map the two desired points on the developable surface to a plane, draw the straight line between them and determine the angle between the generator and the geodesic line at one of the end points. The angle can be used to determine the initial direction $(u', v') = (p, q)$. Thus, all the information required for an IVP is available. Given two points A and B on the developable surface $\mathbf{r}(u, v)$ as shown in Fig. 10.13(c), the corresponding points (X_A, Y_A) and (X_B, Y_B) in the developed planar surface are required. The Frenet-Serret formulae (2.56) state that $\mathbf{t}' = -\kappa\mathbf{n}$ where $\mathbf{t}$ is the unit tangent vector to a curve, $\mathbf{n}$ is the unit normal vector to a curve and κ is the curvature. The minus sign ensures that κ is positive when $\mathbf{n}$ points away

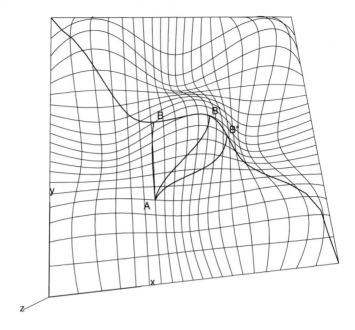

Fig. 10.11. Geodesic paths from point A to Bézier curve on the wave-like bicubic B-spline surface (adapted from [247])

from the center of curvature (see Table 3.2). For a planar curve in the (X, Y) plane, we can define the unit normal vector as $\mathbf{n} = \mathbf{t} \times \mathbf{e_z}$ where $\mathbf{e_z} = (0, 0, 1)$. Substituting this equation into the Frenet-Serret formulae yields

$$\frac{d^2 X}{ds^2} + \kappa \frac{dY}{ds} = 0, \qquad \frac{d^2 Y}{ds^2} - \kappa \frac{dX}{ds} = 0 , \qquad (10.75)$$

where (X, Y) denote the 2D coordinates on the developed plane (X,Y). If we rewrite the first equation of (10.75) in terms of the parameter u, we obtain

$$\frac{d^2 X}{ds^2} + \kappa \frac{dY}{ds} = \frac{d^2 X}{du^2} \left(\frac{du}{ds} \right)^2 + \frac{dX}{du} \left(\frac{d^2 u}{ds^2} \right) + \kappa \frac{dY}{du} \frac{du}{ds} = 0 . \quad (10.76)$$

Similarly the second equation of (10.75) can be rewritten in terms of the parameter u. Since $\frac{du}{ds} \neq 0$, (10.75) reduce to

$$\frac{d^2 X}{du^2} + \frac{\left(\frac{d^2 u}{ds^2} \right)}{\left(\frac{du}{ds} \right)^2} \frac{dX}{du} + \frac{\kappa}{\left(\frac{du}{ds} \right)} \frac{dY}{du} = 0 ,$$

$$\frac{d^2 Y}{du^2} + \frac{\left(\frac{d^2 u}{ds^2} \right)}{\left(\frac{du}{ds} \right)^2} \frac{dY}{du} - \frac{\kappa}{\left(\frac{du}{ds} \right)} \frac{dX}{du} = 0 . \qquad (10.77)$$

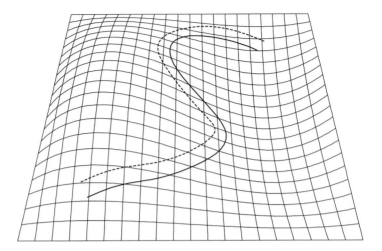

Fig. 10.12. Geodesic offset curve on a Bézier surface. *Thick solid line* represents the progenitor curve and the *thick dotted line* represents the geodesic offset curve (adapted from [250])

The *development* is based on the fact that curves on isometric surfaces have the same geodesic curvatures. Therefore, κ in (10.77) can be replaced by κ_g, the geodesic curvature of the curve on the developable surface [116]. If we choose the curve on the developable surface to be $\mathbf{r}(u, v_n)$, an iso-parametric curve in terms of u, we can replace $\frac{du}{ds}$ and $\frac{d^2 u}{ds^2}$ in (10.77) by

$$\frac{du}{ds} = \frac{1}{|\mathbf{r}_u(u, v_n)|}, \qquad \frac{d^2 u}{ds^2} = -\frac{\mathbf{r}_u(u, v_n) \cdot \mathbf{r}_{uu}(u, v_n)}{(\mathbf{r}_u(u, v_n) \cdot \mathbf{r}_u(u, v_n))^2} \ . \qquad (10.78)$$

Thus we have

$$\frac{dX}{du} = p, \qquad\qquad \frac{dY}{du} = q \ , \qquad\qquad (10.79)$$

$$\frac{dp}{du} = p\frac{(\mathbf{r}_u \cdot \mathbf{r}_{uu})}{(\mathbf{r}_u \cdot \mathbf{r}_u)} - q\kappa_g|\mathbf{r}_u|, \qquad \frac{dq}{du} = q\frac{(\mathbf{r}_u \cdot \mathbf{r}_{uu})}{(\mathbf{r}_u \cdot \mathbf{r}_u)} + p\kappa_g|\mathbf{r}_u| \ . \ (10.80)$$

To find the points A and B in the plane, we first set (X_0, Y_0) as the $(0, 0)$ point in the plane corresponding to $\mathbf{r}(0, 0)$ on the surface. We integrate the system (10.79) and (10.80) along the directrix that corresponds to $v = 0$ to determine the point $C = (u_A, 0)$, shown in Fig. 10.13(a). Since isometric maps are conformal, the angle between the directrix and the generator at $(u_A, 0)$ is the same in both representations and can be found by $\cos \theta = \frac{\mathbf{r}_u(u_A, 0)}{|\mathbf{r}_u(u_A, 0)|} \cdot \frac{\mathbf{c}}{|\mathbf{c}|}$, where $\mathbf{c} = \overrightarrow{CA}$ is a vector whose direction corresponds to the iso-parametric

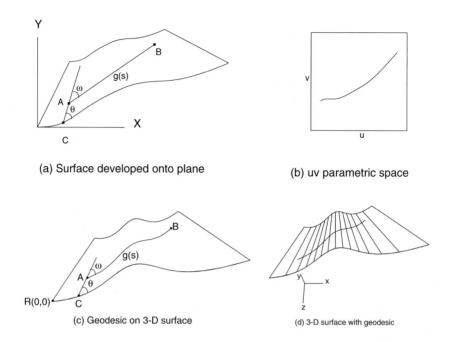

(a) Surface developed onto plane

(b) uv parametric space

(c) Geodesic on 3-D surface

(d) 3-D surface with geodesic

Fig. 10.13. Geodesic on a degree $(3,1)$ developable surface (adapted from [251])

line $\mathbf{r}(u_A, v)$ which is a straight line on the surface. Therefore it is a geodesic and will be developed into the plane as a straight line. The distance is given by $|\mathbf{c}| = \sqrt{(x_A - x_C)^2 + (y_A - y_C)^2 + (z_A - z_C)^2}$. The point A on the plane is found using C, $|\mathbf{c}|$ and θ. The point B on the plane is found by following the same procedure, and the points are connected as shown in Fig. 10.13(a). The angle ω between $\mathbf{c}$ and $\mathbf{b} = \overrightarrow{AB}$ is given by $\cos\omega = \frac{\mathbf{b}\cdot\mathbf{c}}{|\mathbf{b}||\mathbf{c}|}$. The angles ω and θ are shown in Figs. 10.13(a) and (c). This angle ω is preserved between the iso-parametric line $\mathbf{r}(u_A, v)$ and the geodesic curve $\mathbf{g}(s)$ on the developable surface at point A. Thus we have $\cos\omega = \frac{\mathbf{r}_v \cdot \mathbf{g}'(s)}{|\mathbf{r}_v||\mathbf{g}'(s)|}$, where the tangent vector to the geodesic is given by

$$\mathbf{g}'(s) = \mathbf{r}_u \frac{du}{ds} + \mathbf{r}_v \frac{dv}{ds} \ . \tag{10.81}$$

Multiplying (10.81) by $\mathbf{r}_v$ yields

$$\mathbf{r}_v \cdot \mathbf{g}'(s) = \mathbf{r}_u \cdot \mathbf{r}_v \frac{du}{ds} + \mathbf{r}_v \cdot \mathbf{r}_v \frac{dv}{ds} = \cos(\omega)|\mathbf{r}_v||\mathbf{g}'(s)| \ , \tag{10.82}$$

which (since $|\mathbf{g}'(s)| = 1$) can be reduced to

$$\frac{dv}{ds} = \frac{\cos(\omega)}{\sqrt{G}} - \frac{F}{G}\left(\frac{du}{ds}\right) , \qquad (10.83)$$

where $F = \mathbf{r}_u \cdot \mathbf{r}_v$ and $G = \mathbf{r}_v \cdot \mathbf{r}_v$ (coefficients of the first fundamental form). From the first fundamental form,

$$\mathbf{g}'(s) \cdot \mathbf{g}'(s) = E\left(\frac{du}{ds}\right)^2 + 2F\frac{du}{ds}\frac{dv}{ds} + G\left(\frac{dv}{ds}\right)^2 = 1 . \qquad (10.84)$$

Plugging (10.83) into (10.84) and solving for $\frac{du}{ds}$ yields

$$\frac{du}{ds} = \pm\sqrt{\frac{\sin^2(\omega)G}{EG - F^2}} , \qquad (10.85)$$

and thus (10.83) reduces to

$$\frac{dv}{ds} = \frac{\cos(\omega)}{\sqrt{G}} \pm \frac{1}{\sqrt{G}}\frac{F}{\sqrt{EG - F^2}}\sin(\omega) . \qquad (10.86)$$

Evaluating (10.85) and (10.86) at the initial point, we have all the initial conditions required to solve the IVP ((10.17) to (10.20)) for a geodesic. The solution to the IVP yields the uv parametric values for the geodesic that are graphed in Fig. 10.13(b). The corresponding three-dimensional coordinate values are shown in Figs. 10.13(c) and (d). The geodesic runs from (u_A, v_A) $= (0.1, 0.3)$ to $(u_B, v_B) = (0.9, 0.8)$.

11. Offset Curves and Surfaces

11.1 Introduction

11.1.1 Background and motivation

Offset *curves/surfaces*, also called parallel *curves/surfaces*, are defined as the locus of the points which are at constant distant d along the normal from the generator *curves/surfaces*. A literature survey on offset curves and surfaces was carried out by Pham [313] and more recently by Maekawa [250]. Offsets are widely used in various applications, such as tool path generation for $2\frac{1}{2}$-D pocket machining [157, 153, 349], 3-D NC machining [116, 52, 215, 366] (see Fig. 11.1), in feature recognition through construction of skeletons or medial axes of geometric models [298, 450] (see Fig. 11.2), definition of tolerance regions [93, 353, 297] (see Fig. 11.3), access space representations in robotics [237] (see Fig. 11.4), curved plate (shell) representation in solid modeling [301] (see Fig. 11.5), rapid prototyping where materials are solidified in successive two-dimensional layers [114] and brush stroke representation [198].

Because of the square root involved in the expression of the unit normal vector, offset curves and surfaces are functionally more complex than their progenitors. If the progenitor is a rational B-spline, then its offset is usually not a rational B-spline, except for special cases including cyclide surface patches [332, 83, 404], Pythagorean hodograph curves and surfaces (see Sect. 11.4) and simple solids [93]. Another difficulty arises when the progenitor has a tangent discontinuity. Then its exterior and interior offsets will become discontinuous or have self-intersections as illustrated in Fig. 11.6. Furthermore offsets may have cusps and self-intersections, even if the progenitor is regular (see Figs. 11.9, 11.25). Frequently in applications, discontinuity in offsets must be filled in and the loops arising from self-intersections must be trimmed off. In the following three sections, we will briefly review some of the literature on NC machining, medial axis transforms and tolerance regions.

11.1.2 NC machining

The purpose of milling is to remove material from a workpiece. The material is removed in the form of small chips produced by the milling cutter which

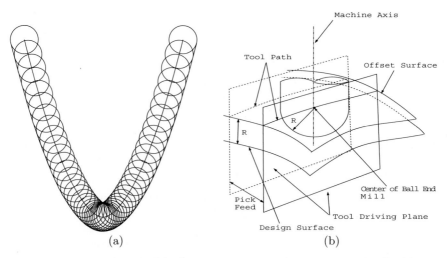

Fig. 11.1. NC machining: (**a**) $2\frac{1}{2}$-D pocket milling (adapted from [254]), (**b**) 3-D milling (adapted from [223])

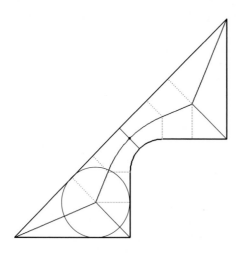

Fig. 11.2. Medial axis (adapted from [139])

rotates at a high speed. A machine tool is characterized by the motions it can perform. Such motions as changing the relative position of the tool and workpiece consist of linear translations and rotations about different axes. However, they do not include the rotation of the cutter or workpiece for maintaining cutting action. NC machines are classified as follows [291, 157]:

2-D Milling : 2-D milling refers to the contouring capability of a machine tool limited to the xy-plane. By moving along the x and y axes simul-

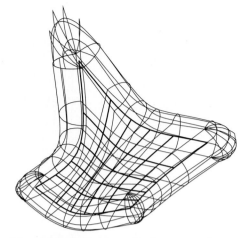

Fig. 11.3. Definition of tolerance regions.

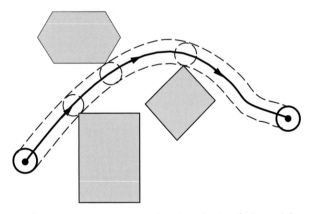

Fig. 11.4. Access space representation in robotics (adapted from [237])

taneously, while keeping z constant, a complete 360 degrees contouring capability can be achieved.

$2\frac{1}{2}$-D Milling : $2\frac{1}{2}$-D milling has a capability between 2-D and 3-D milling. In $2\frac{1}{2}$-D milling, the cutting tool can follow any arbitrary curve in the xy-plane, but can only move stepwise in the z-direction. This $2\frac{1}{2}$-D milling is also referred as *pocket machining*.

3-D Milling : 3-D milling refers to a cutting tool moving simultaneously along the x, y and z axes, but not capable of performing tool rotation with respect to the workpiece.

5-D Milling : A rotation around two of the axes x, y and z is added to x, y and z translations, hence the tool orientation can vary. The 5-D milling

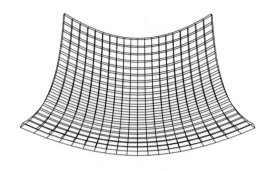

Fig. 11.5. Plate representation

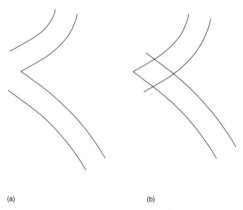

(a)   (b)

Fig. 11.6. Offsets to a tangent discontinuous curve (adapted from [254])

is suitable for large production runs, because the two additional rotations reduce the required setups significantly [287].

The success of NC milling highly depends on the availability of efficient algorithms for defining *tool paths*. The books by Marciniak [262], and Choi and Jerard [59] provide theoretical and practical information on sculptured surface NC machining. The topic of optimal tool paths for NC machining of sculptured surfaces is analyzed in [199].

The cutter motion for machining a part consists of *roughing, semi-roughing* and *finishing*, and should be considered separately, as illustrated in Fig. 11.7 [232]. For each process, an appropriate tool size and tool path needs to be determined.

Rough machining : It should be as simple as possible and preferably consist of a linear type motion only to minimize machining time. In other words, the cutter path should be as short as possible and the depth of cut and feedrate should be as large as possible.

Semi-rough machining : After rough machining, the *shoulders* left on the part should be removed.

Finishing machining : The cutter should follow the profile during these operations and the deviations of the cutter from the profile should always be maintained within a designated tolerance.

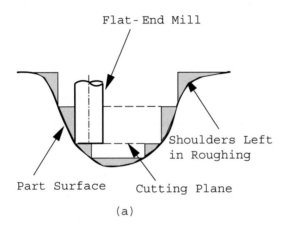

(a)

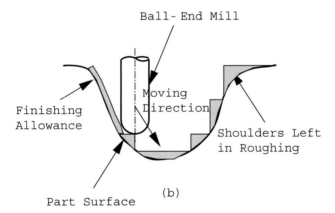

(b)

Fig. 11.7. (a) Pocket machining with flat-end mill in roughing, (b) semi-roughing with large ball-end mill (adapted from [232])

More than 80% of all mechanical parts which are manufactured by milling machines can be cut by NC pocket machining [157]. This is based on the facts that most mechanical parts consist of faces parallel or normal to a single plane, and that free-form objects are usually produced from a raw stock by $2\frac{1}{2}$-D roughing and 3-D or 5-D finishing. When a cylindrical end-mill cutter is used in $2\frac{1}{2}$-D pocket machining, tool paths are generated by offsetting at a distance equal to the radius of the cutter from the boundary curve. When the cutter is located on the side of the curve where the center of curvature lies, the cutter radius must be smaller than the smallest radius of curvature of the boundary curve of the part to be machined to avoid local overcut (gouging). Gouging is one of the most critical problems in NC pocket machining. To avoid gouging, we need to determine the distribution of the curvatures along the boundary curve to select an appropriate cutter size.

Figure 11.1 (a) shows the tool path of a cylindrical cutter pocket machining a region where the center of curvature of the parabolic boundary curve lies. The parabola $\mathbf{r}(t) = (t, t^2)^T$ has the maximum curvature at $(0,0)$ with curvature value $\kappa = 2$. Thus if the radius of the cylindrical cutter exceeds 0.5, there will be a region of gouging as depicted in Fig. 11.10 (a), where the cutter has a radius 0.8. Also the offset with $d = -0.8$ has one self-intersection and two cusps. Points on the segment of the offset bounded by the self-intersecting points on the offset have distance less than the nominal offset distance 0.8 from the generator and this fact causes gouging. Therefore, if we trim off the region of the offset bounded by the two parameter values associated with the self-intersection, the cutter will not overcut the part but will leave an undercut region as shown in Fig. 11.10 (b). The undercut region must be revisited with the smaller size cutter. Each point on the trimmed offset curve is at least distance $|d|$ from every point on the progenitor [102]. Therefore computing the self-intersection points of the offset of a progenitor curve is important.

Most of the tool path generation algorithms for $2\frac{1}{2}$-D pocket machining based on offset contouring, first approximate the input curve with a combination of straight lines and circular arc segments, since traditional CNC interpolators accommodate only such elements and also the offsets of those elements are also straight lines and circular arc segments. Then the approximated boundary curves are offset. The difficult part is to identify and remove all the loops arising from self-intersections. There are two different approaches to remove such loops, namely the Voronoi diagram method [157] and pairwise intersection method [153]. Persson's early work [307] is one of the first to study spiral pocket machining using Voronoi diagrams. A book by Held [157] reviews all the related work until roughly 1991 and introduces an algorithm for the determination of tool paths for spiral and zig-zag milling, and the optimization of tool paths. Held's spiral algorithm, based on an extension of Persson's method [307] provides a general approach for fully automated pocket machining.

In the pairwise intersection method, computing the self-intersections of the offsets reduces to computing the intersections of a straight line to a straight line, a circle to a circle or a straight line to a circle. A brute force approach takes $O(n^2)$ time for computation where n is the number of segments plus the number of reflex vertices. Reflex vertices have an interior angle larger than π. Hansen and Arbab [153] showed that careful elimination of non-intersecting segments reduces the computation time complexity to $K_P O(n \log n)$ for a given shape P.

Rohmfeld [349] developed an algorithm to generate tool paths for arbitrary simple piecewise smooth G^0 generator curves. The redundant global loops are removed by interval operations on the parameter space of the generator curves using the invariance of Gauss-Bonnet values between the generator and the so-called IGB (Invariant Gauss-Bonnet)-offset, which is equivalent to the rolling ball offset.

When a ball-end mill cutter is used in 3-D machining, the cutter will not gouge the design surface as long as the center of the ball-end mill moves on the trimmed offset surface, where loops arising from self-intersections are removed, and with the offset distance equal to the radius of the cutter (see Fig. 11.1 (b)). A detailed literature review on this topic is given in [183, 78]. Among many methods, Sakuta et al. [361], Kuragano et al. [216], Kuragano [215], Kim and Kim [197], Lartigue et al. [223] employ the *offset surface-plane intersection* method. Sakuta et al. [361] approximate an offset surface by offsetting a quadrilateral mesh of points ignoring small gaps, while Kuragano et al. [216] and Kuragano [215] generate a polygonal offset surface by connecting the offset points, where points along the normal of the free-form surface are offset by the radius of the ball-end mill, to the desired accuracy. When there is a self-intersection in the polygonal offset surface, the portion bounded by the self-intersection lines is trimmed off. Then the approximated (trimmed) offset surface is intersected with parallel planes, which are called *tool driving planes*, at a regular interval resulting in a series of intersection lines (see Fig. 5.2). The interval between two successive parallel planes is called *pick feed*. The intersection curves of the approximated polygonal offset surface with these parallel planes generate the required tool paths.

11.1.3 Medial axis

The *Medial Axis* (MA) or *skeleton* of a solid is the locus of centers of balls which are maximal within the solid, together with the limit points of this locus (a ball is maximal within a solid if it is contained in the solid but is not a proper subset of any other ball contained in the solid). The *Medial Axis Transform* (MAT) is composed of the medial axis together with the associated *radius function* which is the radius of the maximal ball with center any given point on the MA. The MA of a 2-D bracket-like region (2-D solid) is shown in Fig. 11.2. The maximal disc associated with a point at the intersection of three MA branches is also shown. Originally proposed by Blum [30, 29], the MAT has been developed extensively since then. It has several properties which neither the Boundary Representation nor the Constructive Solid Geometry directly provide. First, because it elicits important *symmetries* of an object, it facilitates the design and interrogation of symmetrical objects [29]. Second, the MAT exhibits *dimensional reduction* [54, 450]; for example, it transforms a 3-D solid into a connected set of points, curves, and surfaces, along with an associated *radius function* described in more detail below. Third, once a solid is represented with the MAT, the skeleton and radius function themselves may be manipulated, and the boundary will deform in a natural way, suggesting applications in computer animation [451]. Fourth, the skeleton may be used to facilitate the creation of coarse and fine finite element meshes of the region [139, 140, 141, 142, 298, 406, 12, 418, 338]. Fifth, the MAT determines constrictions and other global shape characteristics that are important in mesh generation, performance analysis, manufacturing simulation, and path planning [294, 298]. Sixth, the MAT can be used in document encoding [43, 42] and other image processing applications [38]. Finally, the MAT may be useful in tolerance specification [171].

The MA is closely related with equidistantial point sets, especially the well known Voronoi Diagram [335]. For a 2-D polygonal region or a 3-D polyhedral region, the Voronoi Diagram is a superset of MA, while for objects with nonlinear boundary, the Voronoi Diagram may not be a superset of MA. The major difference is in that the MA is intrinsic to a solid, while the Voronoi Diagram depends on the specific decomposition of the boundary of the solid.

Algorithms for determining the MAT or related sets. The MAT was introduced and explored by Blum [30] and further explored by Blum [29] and Blum and Nagel [31] to describe biological shape. Soon after it was introduced, various algorithms for computing the MAT were developed for special planar regions. Montanari [272] developed an algorithm to compute the MAT of a multiply-connected polygonal region. His algorithm proceeds by identifying significant *branch points* and propagating the boundary contour inward, while connecting the branch points with appropriate linear or parabolic segments. A more efficient algorithm for computing the MAT of a convex polygonal region in $O(n \log n)$ was presented by Preparata [334] along with an $O(n^2)$ algorithm for a non-convex polygonal region. Lee [229] developed an $O(n \log n)$ algorithm for polygonal regions with non-convex corners. Srinivasan and Nackman [405] presented an $O(nh + n \log n)$ algorithm for multiply connected polygonal regions with h holes. Gursoy and Patrikalakis [140, 141, 142] developed an algorithm to compute the MAT of a multiply connected planar region bounded by line segments and circular arcs, and used this algorithm to generate finite element meshes automatically and determine global shape characteristics. Guibas and Stolfi [137] investigate the relationship between the Voronoi Diagram and the Delaunay Triangulation, and develop the quad-edge data structure to represent them. Sugihara [416] investigates the use of Voronoi Diagrams to approximate various types of generalized Voronoi Diagrams. Rosenfeld [350] considers different representations of shapes based on an axis and a generation rule. Held's book [157] contains a comprehensive review of Voronoi Diagram algorithms, which he uses in the context of pocket machining. Another comprehensive review of the state of the art in Voronoi Diagram algorithms has been compiled by Aurenhammer [14].

Other work has concentrated on discrete and approximate approaches to determine the MAT or its related sets. Nackman [282] proposes a 3-D algorithm to use a polyhedral approximation of a smooth boundary and produce a polyhedral approximation to the skeleton. The algorithm is an extension of Bookstein's line skeleton approach [36] to 3-D. It takes as input a polyhedral surface made up of convex polygons and generates a connected graph of convex polygons approximating the MA of the original object; since the input polyhedron is assumed to be an approximation to a smooth curved object, the output is not the skeleton of the polyhedron itself but rather a collection of polygons approximately tangent to the skeleton of the true object. Lavender et al. [226] use an octree-based approach to determine the Voronoi Diagram. Their algorithm works on set-theoretic solid models, composed of unions, intersections, and differences of primitive regions represented by a collection of polynomial inequalities, and produces an octree (or quadtree in two dimensions) which divides space into Voronoi regions at some specified resolution. Scott et al. [370] discuss a method for determining the Symmetric Axis (a superset of the MA) which is based on a combined wave/diffusion process in the plane. Their algorithm proceeds by assigning each boundary pixel a unit displacement above the plane and every other pixel a zero displacement, and then numerically propagating a wave from the boundary. The wave is attenuated by a diffusion process to reduce numerical error, and local maxima in the wave are declared to lie on the Symmetric Axis. Although useful for binary images at low resolutions, the error may be large for higher resolutions. Memory and processing requirements for this method tend to be quite high as well.

Brandt, and Brandt and Algazi [41, 40, 39] find a continuous approximation to the skeleton in both the planar and the 3-D case by first discretizing the boundary. The boundary is sampled at a given sampling density, yielding a set of discrete points which form a pixelized or voxelized approximation to the boundary. The next step is to run an efficient discrete-point Voronoi diagram on the set of points. Finally, portions of the skeleton which result from the effects of quantization are pruned away [39]. This approach attempts to classify each of the vertices in the interior Voronoi diagram according to how many foot points the vertex has. The number of footpoints is determined by taking the associated maximal sphere at the vertex, increasing the radius slightly, and intersecting the dilated sphere with the boundary. This intersection partitions the surface of the sphere into areas which lie either inside or outside the region. Each area lying outside the boundary is assumed to correspond to a footpoint; since the most commonly occurring type of skeleton point has two footpoints, only these points are kept, and the rest are pruned away.

Chiang [54] takes a planar region bounded by piecewise C^2 curves and performs a cellular decomposition of the plane in a neighborhood of the region. Each cell is assigned an approximate distance to the nearest point on the boundary of the region using an algorithm due to Danielsson [70] which computes the Euclidean distance transform. In order for us to explain the Euclidean distance transform, we consider a given binary image S (where each pixel (i, j) is assigned 0 or 1); we call $\bar{S}$ the set of pixels with value 0. The distance map is defined as a scalar function on S

$$L(i, j) = min(d[(i, j), \bar{S}]) , \qquad (11.1)$$

where d is a distance function. If d is the Euclidean distance between two pixels

$$d_e((i, j), (h, k)) = \sqrt{(j - i)^2 + (k - h)^2} , \qquad (11.2)$$

then the distance map is called the Euclidean distance map. This information is later used to find a starting point for tracing axis branches in two dimensions and for recognizing when the tracing has passed the end of a branch. The tracing itself uses the distance information to determine on which boundary elements the footpoints of the current Medial Axis point lie. Once these elements are known, a set of simultaneous equations describing the local structure of the MA near the given point is formed. Using these equations to determine the tangent to the MA at the given point, a short distance along the tangent is traversed, the point is refined with Newton iteration, and another tracing step is taken. At each step, the distance information is used to determine whether or not the current branch has become inactive. If so, a branch point or an end point has been hit, and the tracing either proceeds along another branch or stops. Although the tracing is not extended to three dimensions, Chiang [54] notes that the same Euclidean distance transform in 3-D may be used to determine an approximation to the skeleton. One simple way of using the distance transform in this way is to identify those points which have locally maximal distance values after the distance transform is carried out. Such points are clearly close to centers of maximal disks, and so can be considered to provide an approximation to the skeleton.

Sudhalkar et al. [415] introduce a set called the *box skeleton* which they argue has the properties which make the MAT desirable as an alternate representation of shape. In particular, the box skeleton exhibits dimensional reduction, homotopic equivalence and invertibility. However, their skeleton is defined using the L_∞ norm (the box norm) instead of the Euclidean norm and thus may be quite different from the Medial Axis. Their algorithm for determining the box skeleton operates on discrete objects made of unit squares (or cubes, in 3-D) and proceeds by *thinning* the object while maintaining homotopic equivalence to the original object. In order to perform the thinning, the object is transformed into a graph; in the planar case, the boundaries between adjacent pixels are considered to be edges of the graph, and the intersections of these edges are the vertices of the graph. The first thinning step proceeds by replacing the graph by that portion of the *dual* graph which is interior to the original (primal) graph. Since this procedure alone may result in a disconnected skeleton, the boundary "shrink wraps" around the skeleton as it thins. Procedures based on these concepts are developed for both 2-D and 3-D discrete objects.

Most of the 3-D algorithms in existence (such as the ones above) are fundamentally *discrete* algorithms. To our knowledge, few *continuous* approaches have been proposed, due largely to the computational complexity involved. One of the few such techniques is developed by Hoffmann [170], who proposed a method for assembling the skeleton of a CSG object. His method proceeds by determining points of closest approach between pairs of boundary elements and checking these points to make sure they are in fact on the Medial Axis. (Each of these points are on the MA if and only if the distance to the pair of elements is less than or equal to the distances to the other boundary elements.) The points are then sorted in order of increasing distance from the boundary, and then a local analysis around each point is performed, in an attempt to identify whether the point lies on a face, edge, or vertex of the Medial Axis. This determination is made by identifying all boundary elements which lie at the same minimum distance from the point and forming a set of simultaneous equations in n variables which describe the equidistantial set that the point belongs to. Based on the rank of the Jacobian of this set of equations, the point is predicted to lie on a face, edge, or vertex of the skeleton. Then neighboring faces and edges are traced out in order of increasing distance from the boundary. The method requires intersection of equidistantial sets with one another in order to trim away portions which do not belong to the Medial Axis. Related papers by Dutta and Hoffmann [81, 82] consider the exact representation of the bisectors which appear as skeleton branches in the skeleton of CSG objects bounded by planes, natural quadrics, and torii.

Reddy and Turkiyyah [341] propose an algorithm for determining the skeleton of a 3-D polyhedron based on a generalization of the Voronoi Diagram. They compute an abstract Delaunay Triangulation of the polyhedron and use the result to obtain the dual, the generalized Voronoi Diagram. The Delaunay triangulation computed is a generalization of the usual Delaunay triangulation, which connects isolated nodes together with line segments. In the generalization, the nodes represent parts of a polyhedron (specifically, either a face or a non-convex edge or a non-convex vertex of the polyhedron) and therefore the triangulation is an abstract graph. It still maintains duality with the generalized Voronoi Diagram, and is easier to compute since the Voronoi Diagram may contain trimmed quadric surfaces. The skeleton of the polyhedron is obtained by trimming away certain elements of the generalized Voronoi Diagram which are in the Voronoi Diagram but not the skeleton (for example, the equidistantial point set between a face and a non-convex vertex bounding it). The algorithm can explicitly determine certain critical points of the skeleton, but does not contain highly accurate explicit representations of the curves and surfaces making up the skeleton. Under the same approach, Turkiyyah et al. [428] developed an accelerated algorithm using Delaunay triangulation with a local optimization scheme for the generation of accurate skeletons of 3D solid models. The accelerated algorithm has linear complexity in terms of the number of points used for skeleton approximation.

Sheehy et al. [385, 384] investigate the use of a domain Delaunay triangulation on a distribution of points on the boundary (in a manner similar to Brandt [39]) to attempt to determine the topological features of the Medial Axis of a 3-D solid. The steps of an algorithm to compute the Medial Axis from these features are outlined.

Sherbrooke et al. [393] developed an algorithm for determining the MAT of a general polyhedral solid with simply connected faces and without cavities. In [394] and [391], that algorithm was refined, simplified and extended to work for polyhedral solids of arbitrary genus without cavities, with nonconvex vertices and edges. The algorithm is based on a classification scheme that relates different pieces of the medial axis to one another even in the presence of degenerate MA points. Vertices of the MA are connected to one another by tracing along adjacent edges, and finally the faces of the axis are found by traversing closed loops of vertices and edges. The completeness, complexity and stability of the algorithm were also analyzed.

Etzion and Rappoport [91] presented an algorithm for computing the Voronoi Diagram of a 3-D polyhedron based on subdivision of space. This method enables local and partial computation of the Voronoi diagram. By separating the computation of the symbolic (Voronoi graph) and geometric parts of the diagram, the algorithm tends to be more robust. In addition, a tracing algorithm for the MAT of a 3-D polyhedron is developed by Culver et al. [68] to improve accuracy using exact arithmetic and exact geometric representations. In order to improve the efficiency in searching MA elements, spatial decomposition and linear programming are utilized. Sheets in this methods are in the form of quadrics. The method also includes a new algorithm for analysis of the topology of an algebraic plane curve and a fast numerical method for implicit geometric computation.

Wolter and his associates [340, 451] compute medial curves on a surface, which is the locus of points being equidistant from two given curves on the surface, utilizing the geodesic offset function. Their method is also applicable to the plane curve case. Also Wolter and his associates [214] applied the above method to compute a Voronoi diagram on a parametric surface instead of the usual Voronoi diagram in Euclidean space.

In addition to the work described above which discusses the determination of the MAT, there has been additional work on the inverse problem of *reconstructing* the original solid from the Medial Axis and the radius function. Reconstruction of boundary surfaces from curves and surfaces of the Medial Axis in 3-D is discussed by Gelston and Dutta [125]. Vermeer [431] considers the problem of boundary reconstruction from the MAT in 2-D and 3-D. Recently, Verroust and Lazarus [432] developed a method to extract skeletal curves from a set of scattered points that are sampled from a surface using a geodesic graph and distance map with various levels of decomposition. The skeletal curves extracted can be applied in surface reconstruction.

Work on the MAT and its applications has been somewhat limited by the difficulty of developing an algorithm which is robust, accurate, and efficient to carry out the transformation especially for curved solids. Most recent work on the 3-D problem has tried to generate a discrete approximation to the actual MAT; while such algorithms may be well-suited for some meshing applications, particularly if they involve some degree of user interaction, they are less satisfactory for modeling since they do not capture the topological structure of the shape accurately. However, because the skeleton is typically path connected, and *because branch points on the skeleton may be expressed as solutions of a set of simultaneous nonlinear polynomial equations*, we believe that a *continuous* approach to the problem is promising especially if combined with interval or even exact arithmetic methods, as for example in the recent work by Culver et al. [68] for 3-D polyhedra.

Theoretical analysis of the MAT properties. There exists considerable theoretical work on the mathematical properties of the MAT and other related sets. Wolter [450] provides a thorough analysis of topological properties of the MAT in a very general context, and establishes the relationship between the MA and related symmetry sets such as the cut locus. Principal results of his paper include his proof of homotopic equivalence between an object with a C^2 boundary and its Medial Axis, the invertibility of the MAT (under conditions more general than piecewise C^2 boundaries), and the C^1 smoothness of the distance function on the complement of the cut locus. For a 2-D solid with piecewise C^2 boundaries, the medial axis is homotopically equivalent to the 2-D solid, which implies the connectedness of the medial axis. Wolter [449, 447] also earlier analyzed the differentiability of the distance function.

Farouki and Johnstone [99, 100] have studied bisector problems between a planar curve and a point on its plane, and between two co-planar curves. They introduce a natural method for tracing the bisector between two curves by using the exact representation of the bisectors of the first curve and successive points on the second curve.

Chiang [54] and Brandt [38, 39] studied many of the mathematical properties of the MAT, primarily for the domain of 2-D regions. Chiang [54] provides a proof that for two-dimensional regions with piecewise C^2 boundaries, the MA is connected, the MAT is invertible, and, provided the 2-D solid is homeomorphic to the closed unit disk, the maximal disc of a given MA point (not an end point of a MA arc) divides the MA and the solid into two disjoint trees and two disjoint 2-D solids, respectively (which provides justification for divide-and-conquer approaches). Brandt [38, 39] computes first order and second order differential properties of the planar skeleton, and explores the determination of the skeleton under different metrics (which may be useful for determining the skeleton of binary images). He also explores the notion of skeleton point classification [39], classifying skeleton points according to the number of footpoints.

The three-dimensional problem is studied in some detail by Nackman [282] and Nackman and Pizer [283], who also derive relationships between curvatures of the boundary, the skeleton, and the associated radius function. Curvature relationships in the planar case are considered by Blum [29]. Anoshkina et al. [9, 8] consider properties of the Medial Axis in the context of an investigation of singularities of the distance function to the bounding surface.

Sherbrooke et al. [395, 391] further developed the theory of the MAT of 3-D objects. They established the relationships between the curvature of the boundary and the position of the medial axis and also set up a deformation retract between each object and its medial axis for n-dimensional submanifolds of R^n with boundaries which are piecewise C^2 and completely G^1. They demonstrated that if the object is path connected, then so is the medial axis. Specifically, they proved that path connected polyhedral solids without cavities have path connected medial axes.

Stifter [407, 408] considers the Voronoi Diagram of any subset of 3-D space characterized by certain axioms and analyzes various properties of the subset with applications to robotics.

11.1.4 Tolerance region

A tolerance region of a solid is constructed according to the ball-offset operator model as in Rossignac [351]. Farouki [93] studied the problem of finding exact offsets on the exterior of simple solids. He handled the tangent discontinuities at the edge and vertices by using the rolling ball offset definition. The class of solids studied are closed convex solids like solids of revolution and extrusion.

Patrikalakis and Bardis [297] construct a tolerance region of a quadrilateral design surface patch which is bounded by ten surfaces as illustrated in Fig. 11.3. For the shapes of interest in practical applications and the strict tolerance requirements under consideration, they assumed that there will be no self-intersections in the ten bounding surfaces. These ten surfaces are: 1) four pipe or canal surfaces (see Sect. 11.6), the offsets to the edges of the design surface; 2) two offset surfaces, one along the normal and one in the opposite direction to that of the normal to the design surface; 3) four spherical surface patches, the offsets of the corners of the design surface. They approximated the pipe surfaces with rational B-splines and the normal offsets by integral B-splines and expressed the spherical segments exactly by rational B-spline surface patches. It is interesting to note that the design surface is the medial axis of the tolerance region.

11.2 Planar offset curves

11.2.1 Differential geometry

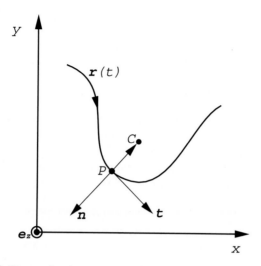

Fig. 11.8. Definitions of unit tangent and normal vectors (adapted from [254])

In this entire chapter we employ the convention (b) (see Fig. 3.7 (b) and Table 3.2) such that the curvature κ of a curve at point P is positive when the center of curvature C is on the opposite direction of the unit normal vector $\mathbf{n}$ as illustrated in Fig. 11.8. Following this convention, the Frenet-Serret formulae for a planar curve $\mathbf{r} = \mathbf{r}(t)$, $t_1 \leq t \leq t_2$ with arbitrary speed (2.57) reduce to

$$\dot{\mathbf{t}} = -v\kappa\mathbf{n}, \quad \dot{\mathbf{n}} = v\kappa\mathbf{t} , \tag{11.3}$$

where $v = |\dot{\mathbf{r}}(t)|$ is the parametric speed. The second equation of (11.3) can be rewritten as follows:

$$\dot{\mathbf{n}} = \kappa\dot{\mathbf{r}} . \tag{11.4}$$

A planar offset curve $\hat{\mathbf{r}}(t)$ with signed offset distance d to the progenitor planar curve $\mathbf{r}(t)$ is defined by

$$\hat{\mathbf{r}}(t) = \mathbf{r}(t) + d\mathbf{n}(t) . \tag{11.5}$$

The unit tangent and normal vectors and the curvature of the offset curve are given by [102]

$$\hat{\mathbf{t}} = \frac{\dot{\hat{\mathbf{r}}}}{|\dot{\hat{\mathbf{r}}}|} = \frac{1 + \kappa d}{|1 + \kappa d|} \mathbf{t} \;, \tag{11.6}$$

$$\hat{\mathbf{n}} = \hat{\mathbf{t}} \times \mathbf{e}_z = \frac{1 + \kappa d}{|1 + \kappa d|} \mathbf{n} \;, \tag{11.7}$$

$$\hat{\kappa} = \frac{\kappa}{|1 + \kappa d|} \;, \tag{11.8}$$

where (11.3) (11.4) are used for the derivation.

11.2.2 Classification of singularities

There are two types of singularities on the offset curves of a regular progenitor curve, *irregular points* and *self-intersections*. Irregular points include *isolated points* and *cusps*. A point P on a curve C is called an isolated point of C if there is no other point of C in some neighborhood of P. This point occurs when the progenitor curve with radius R is a circle and the offset is $d = -R$. A cusp is an irregular point on the offset curve where the tangent vector vanishes. Cusps at $t = t_c$ can be further subdivided into *ordinary cusps* when $\dot{\kappa}(t_c) \neq 0$ and *extraordinary points* when $\dot{\kappa}(t_c) = 0$ and $\ddot{\kappa}(t_c) \neq 0$ [102]. An isolated point and a cusp occur when $|\dot{\hat{\mathbf{r}}}(t)| = 0$, which using (11.6) reduces to

$$\kappa(t) = -\frac{1}{d} \;. \tag{11.9}$$

Note that $(1 + \kappa d)/|1 + \kappa d|$ in (11.6) and (11.7) changes abruptly from -1 to 1 when the parameter t passes through $t = t_c$ at an ordinary cusp, while at extraordinary points $(1 + \kappa d)/|1 + \kappa d|$ does not change its value (see Fig. 11.9 (b)).

Offset *curve/surface* may self-intersect *locally* when the absolute value of the offset distance exceeds the minimum radius of curvature in the concave regions (see Fig. 11.10 (a)). Also the offset *curve/surface* may self-intersect *globally* when the distance between two distinct points on the *curve/surface* reaches a local minimum (i.e. the presence of a constriction of the *curve/surface* as illustrated in Fig. 11.11). These local and global self-intersections can be visualized as machining a part using a *cylindrical/spherical* cutter whose radius is too large for $2\frac{1}{2}$-*D/3-D* milling. It is an essential task for many practical applications to detect *all* components of the self-intersection *points/curves* correctly and generate the *trimmed* offset *curve/surface*. If the cutter follows the trimmed offset, there will be no overcut or gouging, however we are left with undercut regions which must be milled with a smaller size cutter (see Fig. 11.10 (b)).

Self-intersections of offset curves include *nodes* and *tacnodes*. A node P is a point of curve C where two arcs of C pass through P and the arcs have different tangents. A tacnode is a special case of a node whose two tangents

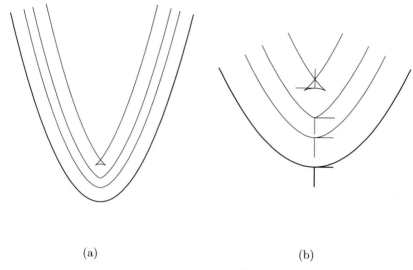

(a) (b)

Fig. 11.9. (a) Offsets to a parabola $\mathbf{r} = (t, t^2)^T$, $-2 \le t \le 2$ (*thick solid line*) with offsets d=-0.3, -0.5, -0.8 (adapted from [94]), **(b)** at d=-0.3, -0.5 the tangent and normal vectors of the offset have the same sense as the progenitor, while at $d = -0.8$ they flip directions

coincide, as illustrated in Fig. 11.11. Self-intersections of an offset curve can be obtained by seeking pairs of distinct parameter values $\sigma \ne t$ such that

$$\mathbf{r}(\sigma) + d\mathbf{n}(\sigma) = \mathbf{r}(t) + d\mathbf{n}(t) .$$
(11.10)

Example 11.2.1. (see Figs. 11.9 and 11.10)
Given a parabola $\mathbf{r} = (t, t^2)^T$, $-2 \le t \le 2$, the unit tangent and normal vectors are given by

$$\mathbf{t} = \frac{d\mathbf{r}}{ds} = \frac{d\mathbf{r}}{dt}\frac{dt}{ds} = \frac{(1, 2t)^T}{\sqrt{1+4t^2}}, \quad \mathbf{n} = \mathbf{t} \times \mathbf{e}_z = \frac{(2t, -1)^T}{\sqrt{1+4t^2}} .$$

The curvature and its first and second derivatives are given by

$$\kappa(t) = \frac{(\dot{\mathbf{r}} \times \ddot{\mathbf{r}}) \cdot \mathbf{e}_z}{|\dot{\mathbf{r}}|^3} = \frac{2}{(1+4t^2)^{\frac{3}{2}}} > 0 ,$$

$$\dot{\kappa}(t) = \frac{-24t(1+4t^2)^{\frac{1}{2}}}{(1+4t^2)^3}, \quad \ddot{\kappa}(t) = \frac{24(16t^2 - 1)}{(1+4t^2)^{\frac{7}{2}}} .$$

Thus a stationary point of curvature occurs at $t = 0$. Since $\ddot{\kappa}(0) = -24 < 0$, $\kappa(0)$ is a maximum with a curvature value $\kappa(0) = 2$. It is evident that the offset distance d has to be negative to have a cusp, since $\kappa(t)$ is always positive for any t. Now let us solve $\kappa(t) = -1/d$ for t which yields

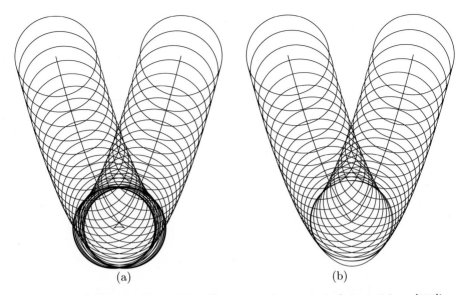

(a)   (b)

Fig. 11.10. Self-intersection of the offset curve of a parabola (adapted from [254]):
(a) offsets to the parabola $\mathbf{r}(t) = (t, t^2)^T$ with $d = -0.8$ and cutter path with
gouging, (b) trimmed offsets to the parabola $\mathbf{r}(t) = (t, t^2)^T$ with $d = -0.8$ and
cutter path with undercut

$$t = \pm \frac{\sqrt{3\sqrt{4d^2 - 1}}}{2} \, .$$

We can easily see that if $d > -\frac{1}{2}$, there is no real root. This means that there
is no singularity as long as the magnitude of the offset distance is smaller
than $\frac{1}{2}$. If $d = -\frac{1}{2}$, there exists a double root $t = 0$, while if $d < -1/2$ there
exist two symmetric values of t. When $d = -\frac{1}{2}$, at $t = 0$, we have $\dot{\kappa}(0) = 0$,
$\ddot{\kappa}(0) \neq 0$, therefore $t = 0$ is an extraordinary point, while when $d < -\frac{1}{2}$,
$\dot{\kappa}(\pm t_c) \neq 0$, so at points $t = \pm t_c$ there are ordinary cusps on the offset curve.
    The offset to the parabola $\mathbf{r} = (t, t^2)^T$ is given by

$$\hat{\mathbf{r}} = (t, t^2)^T + d \frac{(2t, -1)^T}{\sqrt{1 + 4t^2}} \, .$$

Therefore the equations for self-intersection of offset curve to the parabola in
x and y components become

$$\sigma + \frac{2d\sigma}{\sqrt{1 + 4\sigma^2}} = t + \frac{2dt}{\sqrt{1 + 4t^2}} \, ,$$

$$\sigma^2 - \frac{d}{\sqrt{1 + 4\sigma^2}} = t^2 - \frac{d}{\sqrt{1 + 4t^2}} \, .$$

It is readily observed that the offset is symmetric with respect to y-axis,
which implies that the pair of distinct parameter values forming the self-

intersection must satisfy $\sigma = -t$. The y component results in identity, while the x component yields

$$t\left(1 + \frac{2d}{\sqrt{1+4t^2}}\right) = 0 ,$$

Finally, the non-trivial solutions are $t = \pm\frac{\sqrt{4d^2-1}}{2}$.

11.2.3 Computation of singularities

Using (2.25) with $\mathbf{e}_z = (0,0,1)^T$, $\dot{\mathbf{r}}(t) = (\dot{x}(t), \dot{y}(t))^T$ and $\ddot{\mathbf{r}}(t) = (\ddot{x}(t), \ddot{y}(t))^T$, (11.9) for finding the locations of isolated points and cusps of planar offset curve reduces to [254]

$$d\left[\ddot{x}(t)\dot{y}(t) - \dot{x}(t)\ddot{y}(t)\right] - \sqrt{\dot{x}^2(t) + \dot{y}^2(t)}\left[\dot{x}^2(t) + \dot{y}^2(t)\right] = 0 . \quad (11.11)$$

Consequently, if $\mathbf{r}(t)$ is a polynomial curve, locations of irregular points can be obtained by solving the above univariate irrational function involving polynomials and square roots of polynomials, which can be transformed into two equations with two unknowns using the auxiliary variable method introduced in Sect. 4.5. The additional equation and variable result from replacing the square root involved in (11.11) leading to:

$$d\left[\ddot{x}(t)\dot{y}(t) - \dot{x}(t)\ddot{y}(t)\right] - \varrho^3 = 0 , \quad (11.12)$$
$$\varrho^2 - \left[\dot{x}^2(t) + \dot{y}^2(t)\right] = 0 . \quad (11.13)$$

By substituting the expression for the normal vector for the planar parametric curve (2.24) (see Fig. 3.7 (b) and Table 3.2) into (11.10) yields the following system for locating the self-intersections of the offset of the planar curve:

$$x(\sigma) + \frac{\dot{y}(\sigma)d}{\sqrt{\dot{x}^2(\sigma) + \dot{y}^2(\sigma)}} = x(t) + \frac{\dot{y}(t)d}{\sqrt{\dot{x}^2(t) + \dot{y}^2(t)}} ,$$
$$y(\sigma) - \frac{\dot{x}(\sigma)d}{\sqrt{\dot{x}^2(\sigma) + \dot{y}^2(\sigma)}} = y(t) - \frac{\dot{x}(t)d}{\sqrt{\dot{x}^2(t) + \dot{y}^2(t)}} . \quad (11.14)$$

If $\mathbf{r}(t)$ is a polynomial curve, (11.14) are two simultaneous bivariate irrational functions involving polynomials and square root of polynomials. Using the auxiliary variable method, system (11.14) can be transformed into four polynomial equations with four unknowns $\sigma, t, \omega, \varrho$ as follows [254]:

$$\omega\varrho\left[x(\sigma) - x(t)\right] + d\left[\varrho\dot{y}(\sigma) - \omega\dot{y}(t)\right] = 0 ,$$
$$\omega\varrho\left[y(\sigma) - y(t)\right] + d\left[-\varrho\dot{x}(\sigma) + \omega\dot{x}(t)\right] = 0 ,$$
$$\omega^2 - \left[\dot{x}^2(\sigma) + \dot{y}^2(\sigma)\right] = 0 ,$$
$$\varrho^2 - \left[\dot{x}^2(t) + \dot{y}^2(t)\right] = 0 . \quad (11.15)$$

The trivial solution $\sigma = t$ can be avoided by factoring out $\sigma - t$ from the above equations. It is apparent that the two additional equations obtained through the auxiliary variables do not contain the factor $\sigma - t$. However, the original two equations, where the square roots of polynomials have been replaced by the auxiliary variables, possess the factor $\sigma - t$ as can be seen after some algebraic manipulations [254]. Once the division operations are completed, we obtain a system of four polynomial equations with four unknowns without the trivial solution $\sigma = t$.

Similarly to (11.14), intersections of the normal offsets at distance d of two distinct planar curves $\mathbf{r}^A(\sigma)$ and $\mathbf{r}^B(t)$ can be computed by solving the following system:

$$x^A(\sigma) + \frac{\dot{y}^A(\sigma)d}{\sqrt{(\dot{x}^A)^2(\sigma) + (\dot{y}^A)^2(\sigma)}} = x^B(t) + \frac{\dot{y}^B(t)d}{\sqrt{(\dot{x}^B)^2(t) + (\dot{y}^B)^2(t)}} \ ,$$

$$y^A(\sigma) - \frac{\dot{x}^A(\sigma)d}{\sqrt{(\dot{x}^A)^2(\sigma) + (\dot{y}^A)^2(\sigma)}} = y^B(t) - \frac{\dot{x}^B(t)d}{\sqrt{(\dot{x}^B)^2(t) + (\dot{y}^B)^2(t)}} \ .$$

$$(11.16)$$

Note that for (11.14) we need to find a method to eliminate the trivial solution $\sigma = t$, while for (11.16) $\sigma = t$ can be a solution. When the input curve is a B-spline curve, we can always split it into Bézier (polynomial) segments by a knot insertion algorithm [34, 63]. In such cases not only the self-intersection in the offset of each split polynomial segment but also the intersections among the offsets of different split segments must be checked.

A system of nonlinear polynomial equations can be robustly and efficiently solved by the subdivision-based Interval Projected Polyhedron algorithm [254, 179], which was introduced in Chap. 4. A remarkable feature of this algorithm when applied to the system (11.15) is that both local and global self-intersections can be found by the same algorithm without any initial approximations (see Figs. 11.10, 11.11, 11.12).

11.2.4 Approximations

In general, an offset curve is functionally more complex than its progenitor curve because of the square root involved in the expression of the unit normal vector. If the progenitor is a NURBS curve, then its offset is usually not a NURBS curve, except for some special cases such as straight lines and circles etc. Lü [238] has shown that the offset of a parabola is a rational curve. However, this result has not been generalized to higher order curves. Farouki and Neff [101] have shown that the two-sided offsets of planar rational polynomial curves are high-degree implicit algebraic curves of potentially complex shape. These equations cannot typically be separated into two equations describing positive and negative offsets individually. The degree of this implicit offset curve is $n_o = 4n - 2 - 2m$, where n is the degree of polynomial generator curve $\mathbf{r} = (x(t), y(t))^T$ and m is the degree of $\phi(t) = GCD(\dot{x}(t), \dot{y}(t))$

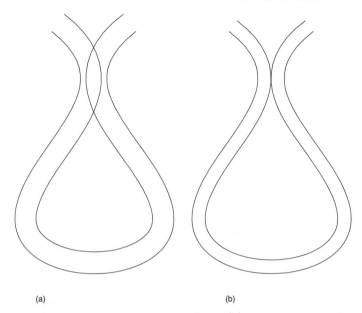

(a)   (b)

Fig. 11.11. Global self-intersections of offsets: (**a**) degree six Bézier curve and its offset with $d = -0.05$, (**b**) offset curve self-intersects forming a tacnode with $d \simeq -0.03141$ (adapted from [254])

where GCD denotes the greatest common divisor. For example the degree of the two-sided offset curve of a parabola $\mathbf{r}(t) = (t, t^2)^T$ is 6 and of a general polynomial cubic curve is 10 with $\phi(t)$ a constant.

Because of the wide application of offset curves and the difficulty in directly incorporating such entities in geometric modeling systems, due to their potential analytic and algebraic complexity, a number of researchers have developed approximation algorithms for these types of geometries in terms of piecewise polynomial or rational polynomial functions. A literature survey on approximation of planar offset curves is compiled in [313, 250]. A paper by Elber et al. [88] presented qualitative as well as quantitative comparisons for several plane offset curve approximation methods, namely the control polygon-based methods by Cobb [62], Tiller and Hanson [421], Coquillart [65], and Elber and Cohen [86], the interpolation methods by Klass [203], and Pham [312], the least square methods by Hoschek [174], the nonlinear optimization technique by Hoschek and Wissel [176], and the circle approximation method by Lee et al. [230]. They counted the number of control points of the approximated offset as a measure for efficiency, while the approximation error was within a prescribed tolerance. They found in general that the least square methods perform very well. However, when the progenitor curve is quadratic they showed that the simple method due to Tiller and Hanson [421], which translates each edge of the control polygon into the edge normal direction by an offset distance, outperforms the other methods. Therefore as

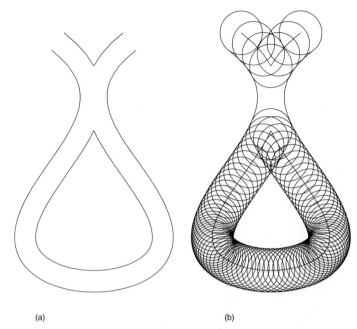

(a)   (b)

Fig. 11.12. (a) Trimmed offset of degree six Bézier curve with $d = -0.05$, (b) tool path along the trimmed offset (adapted from [254])

an example we summarize the algorithm by Tiller and Hanson [421] with an illustration in Fig. 11.13:

1. Let C be a rational B-spline progenitor curve with control vertices $\mathbf{p}_{(1)}$ in homogeneous representation and knot vector $\mathbf{T}_{(1)}$. Set $i = 1$.
2. Offset each leg of the control polygon by d.
3. Intersect consecutive legs of polygon to find new vertices $\hat{\mathbf{p}}_{(i)}$. Then the approximated offset $\hat{C}$ is defined by $\hat{\mathbf{p}}_{(i)}$ and $\mathbf{T}_{(i)}$.
4. Check deviation of the approximate offset from the true offset.
5. If the deviation is larger than the given tolerance subdivide $\mathbf{p}_{(i)}$ and $\mathbf{T}_{(i)}$ to obtain $\mathbf{p}_{(i+1)}$ and $\mathbf{T}_{(i+1)}$. Then set $i \leftarrow i + 1$ and go back to step 2. Stop if the deviation is smaller than the given tolerance.

Sederberg and Buehler [375] approximate an offset of a Bézier curve using Hermite interpolation of any even degree not less than the degree of the initial Bézier curve. The representation of the offset is a special interval Bézier curve of even degree with only the middle control point as a rectangular interval. The size of the rectangular interval indicates the tightness of the approximation.

Most of the existing offset approximation schemes generate the approximate offset curve in terms of Bézier /B-spline format in an iterative manner

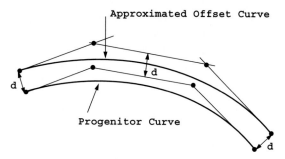

Fig. 11.13. Offset curve approximation

based on a subdivision technique to improve the accuracy of the offset curve. If local and global self-intersections exist, they must be eliminated after the approximation. Elber and Cohen [85] identify the local loops by using the fact that a pair of cusps always exist within the local self-intersection loop of the offset curve. The tangent vector of the offset $\hat{\mathbf{t}}(t)$ flips its direction at the cusps, and thus $\omega(t) = \mathbf{t}(t) \cdot \hat{\mathbf{t}}(t)$, where $\mathbf{t}(t)$ is the tangent vector of the input curve, becomes negative at the regions bounded by the cusps. Consequently, the local loops can be identified by finding the zero set of $\omega(t)$. Once a local loop is identified the curve is split into three parts, the region before the first cusp, the region between the cusps and the region after the second cusp. The offset of the curve before the first cusp and the offset of the curve after the second cusp are intersected to find the self-intersection point. The region bounded by the parameters corresponding to the self-intersection point are trimmed off.

Lee et al. [230] eliminate the local and global self-intersections by approximating the input curve by discrete points and the connecting polygon. Then the offset curve is generated by offsetting these discrete points. The local redundant polygon edges can be detected by checking the dot product between the line segment and its corresponding offset line segment. If the dot product is negative, the offset segment is eliminated. The global self-intersections are detected by a plane sweep algorithm among the offset line segments. Once a valid boundary topology on the offset polygon is computed, the valid parameter intervals of the input curve are extracted. The intersection points of polygonal approximations may not provide accurate parameter values for the intersection points of the offset curve and hence they need to be refined using, for example Newton's method.

The method introduced by Kimmel and Bruckstein [201] generates offsets via wavefront propagation methods in fluid dynamics. The algorithm works on a square grid with a resolution chosen according to the desired accuracy. Finally the contour lines (offsets) are generated based on grid values. Gurbuz and Zeid [138] employ a different approach to avoid self-intersections. Instead of offsetting analytically, the offsetting process is carried out by filling the closed balls of radius d for each point on the object. The union of the closed balls is subtracted from the object to construct the offset. An offset of a point in space is considered as a sphere. They simplified the implementation by assuming that the shape of a point is a cube and its offset is also a cube. First the object to be offset and its enclosing space are divided into small cubic cells. Therefore the accuracy is governed by the size of the cell. The cells that are on the object are referred to as boundary cells. The boundary cells separate the original set of cells into two subspaces, in and out of the object. Offset operation is carried out for each boundary cell. Chiang et al. [55] also compute offsets without self-intersections using a technique developed in image processing. First the domain is discretized into a two-dimensional grid and the progenitor curve is assigned to the nearest grid point. For each grid point, the distance to the discretized curve is evaluated. By extracting the grid points whose distances are equal to the offset distance, an offset without self-intersection can be constructed. This approach also obviously involves a trade-off between accuracy achieved and memory required.

11.3 Offset surfaces

11.3.1 Differential geometry

A parametric offset surface $\hat{\mathbf{r}}(u, v)$ is a continuum of all points at a constant distance d along normal to another parametric surface $\mathbf{r}(u, v)$ and defined as

$$\hat{\mathbf{r}}(u, v) = \mathbf{r}(u, v) + d\mathbf{N}(u, v) , \qquad (11.17)$$

where d may be a positive or negative real number and $\mathbf{N}$ is the unit normal vector of $\mathbf{r}(u, v)$ (see (3.3)). As we mentioned at the beginning of this chapter we employ the convention that the normal curvature is positive if its associated center of curvature is opposite to the direction of the surface normal (see Fig. 3.7 (b) and Table 3.2).

If $\hat{\mathbf{N}}(u, v)$ is the unit normal vector of $\hat{\mathbf{r}}(u, v)$, then the relation between $\mathbf{N}$ and $\hat{\mathbf{N}}$ is given by [444]

$$\hat{S}\hat{\mathbf{N}} = (1 + d\kappa_{max})(1 + d\kappa_{min})S\mathbf{N} , \qquad (11.18)$$

where $\hat{S} = |\hat{\mathbf{r}}_u \times \hat{\mathbf{r}}_v|$ and $S = |\mathbf{r}_u \times \mathbf{r}_v|$ or expanding the right hand side of (11.18) and using the definitions of Gaussian curvature K (3.61) and mean curvature H (3.62), (11.18) can be rewritten as follows:

$$\hat{S}\hat{\mathbf{N}} = S(1 + 2Hd + Kd^2)\mathbf{N} . \qquad (11.19)$$

If we take the norm of (11.18), we obtain

$$\hat{S} = S|(1 + d\kappa_{max})(1 + d\kappa_{min})| , \qquad (11.20)$$

and substituting $\hat{S}$ into (11.18) yields

$$\hat{\mathbf{N}} = \frac{(1 + d\kappa_{max})(1 + d\kappa_{min})}{|(1 + d\kappa_{max})(1 + d\kappa_{min})|} \mathbf{N} . \qquad (11.21)$$

If we denote by

$$e = \frac{1 + 2Hd + Kd^2}{|1 + 2Hd + Kd^2|} = \frac{(1 + d\kappa_{max})(1 + d\kappa_{min})}{|(1 + d\kappa_{max})(1 + d\kappa_{min})|} = \pm 1 , \qquad (11.22)$$

then (11.21) gives

$$\hat{\mathbf{N}} = e\mathbf{N} . \qquad (11.23)$$

From this relation we can see that $\mathbf{N}$ and $\hat{\mathbf{N}}$ are collinear but may be directed in opposite directions, if $d\kappa_{max} + 1$ and $d\kappa_{min} + 1$ have opposite signs. This occurs when the offset is taken towards the *concave* region of the progenitor surface. Offsetting towards the concave region of a surface is equivalent to taking the offset $d > 0$ where $\kappa_{min} < 0$ and $d < 0$ where $\kappa_{max} > 0$, provided the above sign convention is used. In machining, the cutter radius must not exceed the smallest concave principal radius of curvature of the surface to avoid gouging [116].

Principal curvatures of the offset surface corresponding to κ_{max} and κ_{min} of the progenitor surface can be easily obtained by adding the signed offset distance to the signed radius of principal curvature of the progenitor surface and inverting it. Taking into account the direction in which the normal vector points, we have

$$\hat{\kappa}_a = \frac{e\kappa_{max}}{1 + d\kappa_{max}} , \qquad (11.24)$$

$$\hat{\kappa}_b = \frac{e\kappa_{min}}{1 + d\kappa_{min}} . \qquad (11.25)$$

Gaussian and mean curvatures of the offset surface are readily computed by

$$\hat{K} = \hat{\kappa}_a\hat{\kappa}_b = \frac{K}{1 + 2Hd + Kd^2} , \qquad (11.26)$$

$$\hat{H} = \frac{\hat{\kappa}_a + \hat{\kappa}_b}{2} = \frac{H + Kd}{|1 + 2Hd + Kd^2|} . \qquad (11.27)$$

Given the offset distance d, the critical curvature is defined as $\kappa_{crit} = -1/d$ and three categories arise [94]:

$\kappa_{max} > \kappa_{min} > \kappa_{crit}$: The normal vector of the progenitor and its offset are directed in the same direction, since $e = 1$. Also the sign of Gaussian and principal curvatures of the offset are the same that of the progenitor.

$\kappa_{max} > \kappa_{crit} > \kappa_{min}$: The normal vector of the progenitor and its offset are directed in the opposite direction, since $e = -1$. Also the sign of Gaussian and the principal curvature of the offset corresponding to κ_{max} are opposite to that of the progenitor, while the sign of the principal curvature of the offset corresponding to κ_{min} is the same to that of the progenitor.

$\kappa_{crit} > \kappa_{max} > \kappa_{min}$: The normal vector of the progenitor and its offset are directed in the same direction ($e = 1$), while the sign of both principal curvatures of the offset are opposite to that of the progenitor and thus the sign of Gauss curvature of the offset remains the same as that of the progenitor.

11.3.2 Singularities of offset surfaces

Similar to the offset curve case, there are two types of singularities on offset surfaces, namely *irregular points and self-intersections*. It is apparent from (11.21) that offset surfaces become singular at points which satisfy

$$\kappa_{min}(u, v) = -\frac{1}{d} \quad or \quad \kappa_{max}(u, v) = -\frac{1}{d} . \tag{11.28}$$

The vector-valued mapping of a curve in the uv-parametric space, which satisfies $\kappa_{max}(u, v) = -\frac{1}{d}$ or $\kappa_{min}(u, v) = -\frac{1}{d}$, into three-dimensional coordinates using (11.17) form *cuspidal edges* of the offset surface. These curves can be viewed as contour lines of constant principal curvatures of $-\frac{1}{d}$. The detailed formulation and a robust method for tracing constant curvature lines was discussed in Chap. 8. Isolated points and cusps can be treated as a special case of cuspidal edges. When the surface is part of a sphere with radius R, then $\kappa_{min}(u, v) = \kappa_{max}(u, v) = -\frac{1}{R}$ everywhere. Therefore if the offset distance is $d = R$, offset surface degenerates to a point which is the center of the sphere. At an umbilic $\kappa_{max} = \kappa_{min} = \kappa_{umb}$, where κ_{umb} is the normal curvature at the umbilic, and if the offset is $d = -\frac{1}{\kappa_{umb}}$, the offset surface becomes singular at the point corresponding to the umbilic and forms a cusp. Therefore to detect a cusp on the offset, we need to locate all the umbilics on the progenitor surface. A robust method to locate umbilics is described in Chap. 9.

Self-intersections of an offset surface are defined by finding pairs of distinct parameter values $(\sigma, t) \neq (u, v)$ such that

$$\mathbf{r}(\sigma, t) + d\mathbf{N}(\sigma, t) = \mathbf{r}(u, v) + d\mathbf{N}(u, v) . \tag{11.29}$$

Chen and Ravani [52] present a marching algorithm to compute the self-intersection curve of an offset surface. The algorithm generates a straight-line approximation to a small portion of the intersection curve by looking at the intersection of two triangular elements representing the two tangent planes to the self-intersecting surfaces. The starting line segment is obtained by searching only the bounding curves of the patch. Aomura and Uehara [11] also developed a marching method to compute the self-intersection curves on the offset surface of a uniform bicubic B-spline surface patch. The starting points for marching are obtained by the Powell-Zangwill method based on a dense grid of points. Then the self-intersection curves are traced by integrating a system of ordinary differential equations using the Runge-Kutta-Gill method. Visualization of self-intersecting offsets of Bézier patches by means of ray tracing was studied in [430]. Self-intersection of offsets of regular Bézier surface patches due to local differential geometry and global distance function properties is investigated by Maekawa et al. [253]. The problem of computing starting points for tracing self-intersection curves of offsets is formulated in terms of a system of nonlinear polynomial equations and solved robustly by the Interval Projected Polyhedron algorithm. Trivial solutions are excluded by evaluating the normal bounding pyramids of the surface subpatches mapped from the parameter boxes computed by the polynomial solver with a coarse tolerance (see Sect. 11.3.5). Since it is an essential task for many practical applications to detect *all* components of the self-intersection curves and to trace them correctly for generating the *trimmed* offset, we will discuss this topic in greater detail in the following three sections. In Sect. 11.3.3 we discuss the self-intersection of offsets of implicit quadratic surfaces, while in Sect. 11.3.4 we discuss the self-intersection of offsets of explicit quadratic surfaces. In Sect. 11.3.5 we introduce a method to find the self-intersections of offsets of more general polynomial parametric surface patches.

11.3.3 Self-intersection of offsets of implicit quadratic surfaces

The second order algebraic surfaces (i.e. quadric surfaces) are widely used in mechanical design. Especially the natural quadrics, i.e. sphere, circular cone and circular cylinder result from machining operations such as rolling, turning, filleting, drilling and milling [149]. The offsets of the natural quadrics are also natural quadrics. *Implicit quadrics* such as ellipsoids, elliptic cones and elliptic cylinders are commonly found in die cavities and punches and are manufactured by NC machining [60]. Although Salmon [362] discussed the offsets of quadrics more than a century ago, this was not widely known in the CAGD literature until recently. Maekawa [249] showed that self-intersection curves of offsets of all the implicit quadratic surfaces are planar implicit conics and their corresponding curve on the progenitor surface can be expressed as the intersection curve between an ellipsoid, whose semi-axes are proportional to the offset distance, and the implicit quadratic surfaces themselves.

The equations of implicit quadrics including ellipsoids, hyperboloids of one and two sheets, elliptic cones, elliptic cylinders and hyperbolic cylinders can be expressed in a standard form (3.74). In the sequel we assume $a \leq b \leq c$ without loss of generality.

The components $\hat{x}$, $\hat{y}$, $\hat{z}$ of the position vector $\hat{\mathbf{r}}$ of the offset of implicit surface $f(x, y, z) = 0$ can be expressed as

$$\hat{x}(x, y, z) = x + \frac{f_x(x, y, z)}{|\nabla f(x, y, z)|} d \,, \tag{11.30}$$

$$\hat{y}(x, y, z) = y + \frac{f_y(x, y, z)}{|\nabla f(x, y, z)|} d \,, \tag{11.31}$$

$$\hat{z}(x, y, z) = z + \frac{f_z(x, y, z)}{|\nabla f(x, y, z)|} d \,, \tag{11.32}$$

where (x, y, z) satisfy $f(x, y, z) = 0$ and f_x, f_y, f_z are the x, y z components of $\nabla f(x, y, z)$. Note that $|\nabla f(x, y, z)|$ is zero at the apex of a cone and in this case the normal vector is not defined. Substituting (3.74) into (11.30), (11.31), (11.32) yields

$$\hat{x}(x, y, z) = x + \frac{\zeta x d}{a^2 \sqrt{\zeta^2 \frac{x^2}{a^4} + \eta^2 \frac{y^2}{b^4} + \xi^2 \frac{z^2}{c^4}}} \,, \tag{11.33}$$

$$\hat{y}(x, y, z) = y + \frac{\eta y d}{b^2 \sqrt{\zeta^2 \frac{x^2}{a^4} + \eta^2 \frac{y^2}{b^4} + \xi^2 \frac{z^2}{c^4}}} \,, \tag{11.34}$$

$$\hat{z}(x, y, z) = z + \frac{\xi z d}{c^2 \sqrt{\zeta^2 \frac{x^2}{a^4} + \eta^2 \frac{y^2}{b^4} + \xi^2 \frac{z^2}{c^4}}} \,. \tag{11.35}$$

It is obvious from (11.33) to (11.35) that the offsets of implicit quadratic surfaces in a standard form are symmetric with respect to xy, xz and yz-planes.

Self-intersection of offsets of implicit surfaces can be formulated by seeking pairs of distinct points on the progenitor surface $(x_1, y_1, z_1) \neq (x_2, y_2, z_2)$ such that

$$\hat{x}(x_1, y_1, z_1) = \hat{x}(x_2, y_2, z_2) \,, \tag{11.36}$$

$$\hat{y}(x_1, y_1, z_1) = \hat{y}(x_2, y_2, z_2) \,, \tag{11.37}$$

$$\hat{z}(x_1, y_1, z_1) = \hat{z}(x_2, y_2, z_2) \,. \tag{11.38}$$

Hence using (11.33) to (11.35), the equations for self-intersection reduce to

$$x_1 + \frac{\zeta x_1 d}{a^2 \sqrt{\zeta^2 \frac{x_1^2}{a^4} + \eta^2 \frac{y_1^2}{b^4} + \xi^2 \frac{z_1^2}{c^4}}} = x_2 + \frac{\zeta x_2 d}{a^2 \sqrt{\zeta^2 \frac{x_2^2}{a^4} + \eta^2 \frac{y_2^2}{b^4} + \xi^2 \frac{z_2^2}{c^4}}} \,, \tag{11.39}$$

$$y_1 + \frac{\eta y_1 d}{b^2 \sqrt{\zeta^2 \frac{x_1^2}{a^4} + \eta^2 \frac{y_1^2}{b^4} + \xi^2 \frac{z_1^2}{c^4}}} = y_2 + \frac{\eta y_2 d}{b^2 \sqrt{\zeta^2 \frac{x_2^2}{a^4} + \eta^2 \frac{y_2^2}{b^4} + \xi^2 \frac{z_2^2}{c^4}}} \,, \tag{11.40}$$

$$z_1 + \frac{\xi z_1 d}{c^2 \sqrt{\zeta^2 \frac{x_1^2}{a^4} + \eta^2 \frac{y_1^2}{b^4} + \xi^2 \frac{z_1^2}{c^4}}} = z_2 + \frac{\xi z_2 d}{c^2 \sqrt{\zeta^2 \frac{x_2^2}{a^4} + \eta^2 \frac{y_2^2}{b^4} + \xi^2 \frac{z_2^2}{c^4}}} \,. \tag{11.41}$$

The self-intersection curve of an offset can be considered as a locus of the center of a sphere, whose radius is the offset distance, rolling on the progenitor surface with two contact points. Because of the symmetry of the offsets of implicit quadratic surfaces, the center of rolling sphere must move only on the planes of symmetry and hence the self-intersection curves are on the planes of symmetry. In other words, a pair of points (x_1, y_1, z_1) and (x_2, y_2, z_2) on the progenitor surface are located symmetrically with respect to yz, xz or xy-plane and their offsets meet on the yz, xz or xy-plane.

When the offsets self-intersect in the x-direction, the self-intersection curve will lie on the yz-plane. In such case we can set $x = x_1 = -x_2$, $y = y_1 = y_2$, $z = z_1 = z_2$ and $\zeta \neq 0$, thus $\zeta^2 = 1$ and hence (11.39) reduces to

$$a^2 \sqrt{\frac{x^2}{a^4} + \eta^2 \frac{y^2}{b^4} + \xi^2 \frac{z^2}{c^4}} = -\zeta d \,, \tag{11.42}$$

while (11.40) and (11.41) reduce to identities. Similarly we obtain

$$b^2 \sqrt{\zeta^2 \frac{x^2}{a^4} + \frac{y^2}{b^4} + \xi^2 \frac{z^2}{c^4}} = -\eta d \,, \tag{11.43}$$

$$c^2 \sqrt{\zeta^2 \frac{x^2}{a^4} + \eta^2 \frac{y^2}{b^4} + \frac{z^2}{c^4}} = -\xi d \,, \tag{11.44}$$

for the y and z-directions respectively. Since all the left hand sides are positive, the right hand sides $-\zeta d$, $-\eta d$ and $-\xi d$ must be also positive. By referring to Table 3.1, we can easily find the offsetting direction (sign of d) to have self-intersection. By squaring both hand sides of (11.42) to (11.44), and summarizing the results of this section we have:

Theorem 11.3.1. *The offset of an implicit quadratic surface self-intersects in x-direction if the progenitor surface intersects the following ellipsoid provided that the offset distance is taken such that $-\zeta d$ is positive, namely*

$$EP_x : \quad \frac{x^2}{d^2} + \eta^2 \frac{y^2}{(\frac{b}{a})^4 d^2} + \xi^2 \frac{z^2}{(\frac{c}{a})^4 d^2} = 1 \,. \tag{11.45}$$

Similarly the self-intersections in y and z-directions occur if the progenitor surface intersects the following ellipsoids

$$EP_y : \quad \zeta^2 \frac{x^2}{(\frac{a}{b})^4 d^2} + \frac{y^2}{d^2} + \xi^2 \frac{z^2}{(\frac{c}{b})^4 d^2} = 1 \,, \tag{11.46}$$

$$EP_z : \quad \zeta^2 \frac{x^2}{(\frac{a}{c})^4 d^2} + \eta^2 \frac{y^2}{(\frac{b}{c})^4 d^2} + \frac{z^2}{d^2} = 1 \,, \tag{11.47}$$

provided that $-\eta d$ and $-\xi d$ are positive, respectively. The intersection curves between the progenitor surface and each ellipsoid are the foot point curves of the self-intersection curves of offsets [249].

Remark 11.3.1. When one of the coefficients ζ, η, ξ is zero, the progenitor surface reduces to either an elliptic cylinder or hyperbolic cylinder. Also the three ellipsoids (11.45), (11.46), (11.47) reduce to two elliptic cylinders.

We will not go into the details of quadric-surface intersection problems, but rather refer to many papers on this problem [233, 234, 367, 104, 443, 268]. Using (3.74), (11.33) to (11.35) and (11.42) to (11.44), it is easy to show that the self-intersection curves in the x, y and z-directions are implicit conics in the yz, xz and xy-planes given by

$$\frac{\eta b^2 - \zeta \eta^2 a^2}{(\zeta b^2 - \eta a^2)^2} y^2 + \frac{\xi c^2 - \zeta \xi^2 a^2}{(\zeta c^2 - \xi a^2)^2} z^2 = \frac{\delta a^2 - \zeta d^2}{a^2} , \tag{11.48}$$

$$\frac{\zeta a^2 - \eta \zeta^2 b^2}{(\eta a^2 - \zeta b^2)^2} x^2 + \frac{\xi c^2 - \eta \xi^2 b^2}{(\eta c^2 - \xi b^2)^2} z^2 = \frac{\delta b^2 - \eta d^2}{b^2} , \tag{11.49}$$

$$\frac{\zeta a^2 - \xi \zeta^2 c^2}{(\xi a^2 - \zeta c^2)^2} x^2 + \frac{\eta b^2 - \xi \eta^2 c^2}{(\xi b^2 - \eta c^2)^2} y^2 = \frac{\delta c^2 - \xi d^2}{c^2} . \tag{11.50}$$

Example 11.3.1. Cylindrical surfaces include elliptic cylinders and hyperbolic cylinders. Here we will only examine the hyperbolic cylinder with $\zeta = \delta = 1$, $\eta = -1$ and $\xi = 0$,

$$HC: \quad f(x,y) = \frac{x^2}{a^2} - \frac{y^2}{b^2} - 1 = 0 , \tag{11.51}$$

since the rest of the cases for cylindrical surfaces (see Table 3.1) can be derived in a similar way. The curvatures of the hyperbolic cylinder (11.51) based on the curvature sign convention (a) are given in (3.79) and (3.80). For the curvature sign convention (b), we have

$$K = 0, \quad H = \frac{-b^2 x^2 + a^2 y^2}{2 a^4 b^4 \left(\frac{x^2}{a^4} + \frac{y^2}{b^4} \right)^{\frac{3}{2}}}, \quad \kappa_{max} = 0, \quad \kappa_{min} = \frac{-b^2 x^2 + a^2 y^2}{a^4 b^4 \left(\frac{x^2}{a^4} + \frac{y^2}{b^4} \right)^{\frac{3}{2}}} ,$$

where $(x, y, z) \in HC$. The extrema of the minimum principal curvature can be computed by using the Lagrange multiplier technique described in Sect. 8.4 (see (8.89), (8.90)), which yields points $(x, y) = (\pm a, 0)$. The corresponding minimum principal curvature is $\kappa_{min}(\pm a, 0) = -\frac{a}{b^2}$.

The three ellipsoids in (11.45), (11.46) and (11.47) reduce to the following two elliptic cylinders

$$\frac{x^2}{d^2} + \frac{y^2}{(\frac{b}{a})^4 d^2} = 1 , \tag{11.52}$$

$$\frac{x^2}{(\frac{a}{b})^4 d^2} + \frac{y^2}{d^2} = 1 . \tag{11.53}$$

Since $-\zeta d$ must be positive to have self-intersection in x-direction (see (11.42)), d is forced to be negative, while d must be positive to have self-intersection in y-direction to satisfy $-\eta d > 0$ (see (11.43)). Now let us consider the self-intersection in x-direction which is illustrated in Fig. 11.14 (a).

According to Theorem 11.3.1, hyperbolic cylinder (11.51) must intersect the elliptic cylinder (11.52) to have self-intersection in the x-direction. It is apparent that these two surfaces will intersect if $d \leq -a$, since the minor axis of the elliptic cylinder is d and hyperbolic cylinder intersects the x-axis at $(\pm a, 0)$. This self-intersection is due to the global distance function property (constriction) of the hyperbola. Similarly the self-intersection in y-direction occurs if $d \geq \frac{b^2}{a}$, which corresponds to the maximum concave radius of curvature as obtained above.

Figures 11.14 (a) (b) show the cross section of self-intersections of offsets of a hyperbolic cylinder (with $a = 0.8$, $b = 1$) in x-direction with $d = -3$ and in y-direction with $d = 3$. The thick and thin solid lines represent the hyperbolic cylinder and its offset. The thick dashed dot lines represents the elliptic cylinders (11.52) and (11.53). Four thin dashed lines emanating from the intersections points and intersecting at the self-intersection points of the offset are the vector $d\mathbf{n}$. The four intersection points between the hyperbolic cylinder (11.51) and elliptic cylinder (11.52) in Fig. 11.14 (a), and the four intersection points between hyperbolic cylinder (11.51) and the elliptic cylinder (11.53) in Fig. 11.14 (b) are given by

$$\left(\pm\sqrt{\frac{b^2 d^2 + a^4}{a^2 + b^2}}, \quad \pm\frac{b^2}{a}\sqrt{\frac{d^2 - a^2}{a^2 + b^2}} \right), \quad \left(\pm\frac{a^2}{b}\sqrt{\frac{b^2 + d^2}{a^2 + b^2}}, \quad \pm\sqrt{\frac{a^2 d^2 - b^4}{a^2 + b^2}} \right) .$$

Figure 11.15 shows the self-intersecting offsets of an elliptic cylinder (with $\zeta = \eta = 1$, $\xi = 0$, $a = 0.6$ and $b = 0.8$) in the x-direction (a) with $d = -0.55$ and in the y-direction (b) with $d = -0.9$. The four intersection points for both cases are given by

$$\left(\pm\sqrt{\frac{b^2 d^2 - a^4}{b^2 - a^2}}, \quad \pm\frac{b^2}{a}\sqrt{\frac{a^2 - d^2}{b^2 - a^2}} \right), \quad \left(\pm\frac{a^2}{b}\sqrt{\frac{d^2 - b^2}{b^2 - a^2}}, \quad \pm\sqrt{\frac{b^4 - a^2 d^2}{b^2 - a^2}} \right) .$$

Example 11.3.2. Consider an ellipsoid (with $\zeta = \eta = \xi = \delta = 1$) of the form

$$EP: \quad f(x, y, z) = \frac{x^2}{a^2} + \frac{y^2}{b^2} + \frac{z^2}{c^2} - 1 = 0 . \tag{11.54}$$

The curvatures based on the curvature sign convention (a) are given in (3.82) and (3.83). For the curvature sign convention (b) we have

$$K = \frac{1}{a^2 b^2 c^2 \left(\frac{x^2}{a^4} + \frac{y^2}{b^4} + \frac{z^2}{c^4} \right)^2}, \quad H = -\frac{x^2 + y^2 + z^2 - a^2 - b^2 - c^2}{2a^2 b^2 c^2 \left(\frac{x^2}{a^4} + \frac{y^2}{b^4} + \frac{z^2}{c^4} \right)^{\frac{3}{2}}}, \tag{11.55}$$

$$\kappa = \frac{-(x^2 + y^2 + z^2 - a^2 - b^2 - c^2)}{2a^2 b^2 c^2 \left(\frac{x^2}{a^4} + \frac{y^2}{b^4} + \frac{z^2}{c^4} \right)^{\frac{3}{2}}} \tag{11.56}$$

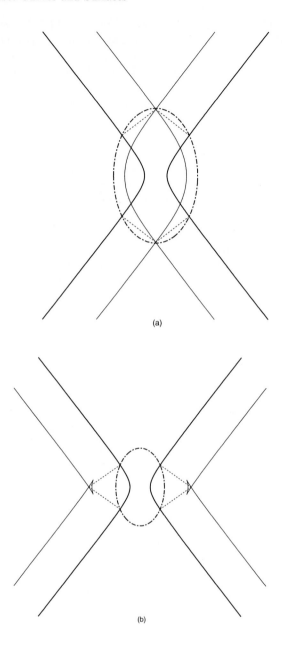

Fig. 11.14. Cross sections of self-intersecting offsets of a hyperbolic cylinder (adapted from [249]): (**a**) x-direction, (**b**) y-direction

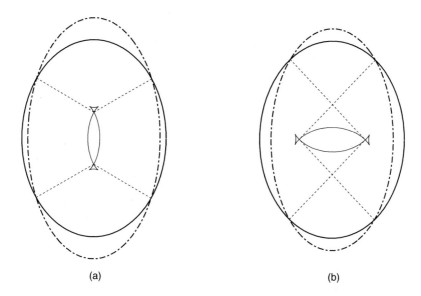

(a)   (b)

Fig. 11.15. Cross sections of self-intersecting offsets of an elliptic cylinder (adapted from [249]): (**a**) x-direction, (**b**) y-direction

$$\pm \frac{\sqrt{(x^2 + y^2 + z^2 - a^2 - b^2 - c^2)^2 - 4a^2b^2c^2 \left(\frac{x^2}{a^4} + \frac{y^2}{b^4} + \frac{z^2}{c^4}\right)}}{2a^2b^2c^2 \left(\frac{x^2}{a^4} + \frac{y^2}{b^4} + \frac{z^2}{c^4}\right)^{\frac{3}{2}}},$$

where $(x, y, z) \in EP$.

The critical points of both principal curvatures can be obtained by using the Lagrange multiplier technique described in Sect. 8.4 (see (8.89), (8.90)). Since we are assuming $a \leq b \leq c$, the maximum principal curvature has a global minimum $\frac{a}{b^2}$ at $(\pm a, 0, 0)$, a local maximum $\frac{b}{a^2}$ at $(0, \pm b, 0)$ and a global maximum $\frac{c}{a^2}$ at $(0, 0, \pm c)$, while the minimum principal curvature has a global minimum $\frac{a}{c^2}$ at $(\pm a, 0, 0)$, a local minimum $\frac{b}{c^2}$ at $(0, \pm b, 0)$ and a global maximum $\frac{c}{b^2}$ at $(0, 0, \pm c)$. Figure 11.16 shows the locations of extrema (black square), umbilics $\left(\pm a \sqrt{\frac{b^2 - a^2}{c^2 - a^2}}, \ 0, \ \pm c \sqrt{\frac{c^2 - b^2}{c^2 - a^2}}\right)$ (white circle), the maximum principal curvature lines (solid line) and the minimum principal curvature lines (dotted line).

It is apparent from (11.42) to (11.44) that d must be negative to have self-intersections in the offset. First we consider the case of self-intersection in x-direction. The two ellipsoids EP_x and EP will not intersect when EP_x is inside EP ($|d| < \frac{a^2}{c}$), or EP is inside EP_x ($|d| > a$). This leads to the conclusion that EP_x and EP intersect if

$$-a \le d \le -\frac{a^2}{c} .$$

The magnitude of the upper bound corresponds to the smallest concave radius of curvature at $(0, 0, \pm c)$, while the magnitude of the lower bound corresponds to the smallest semi-axis. With a similar discussion, we can derive the conditions for the self-intersection in the y and z-directions as

$$y\text{-direction:} \quad -\frac{b^2}{a} \le d \le -\frac{b^2}{c} ,$$
$$z\text{-direction:} \quad -\frac{c^2}{a} \le d \le -c .$$

Figure 11.17 shows two ellipsoids EP (with $a = 0.6$, $b = 0.8$, $c = 1.0$) and EP_x ((11.45) with $d = -\frac{a^2}{b} = -0.45$) which is equal to the maximum principal radius of curvature at $(0, \pm b, 0)$, intersecting each other. This is a degenerate intersection of two ellipsoids, consisting of two ellipses, which have the rational parametrization given by

$$x(t) = \pm \frac{a^2}{b} \sqrt{\frac{c^2 - b^2}{c^2 - a^2}} \frac{1 - t^2}{1 + t^2}, \quad y(t) = b\frac{2t}{1 + t^2}, \quad z(t) = \pm \frac{c^2}{b} \sqrt{\frac{b^2 - a^2}{c^2 - a^2}} \frac{1 - t^2}{1 + t^2} ,$$
$$(11.57)$$

for $-1 \le t \le 1$. The self-intersection curve of the offset in the yz-plane is an ellipse given by

$$\frac{y^2}{\frac{(b^2 - a^2)(a^2 - d^2)}{a^2}} + \frac{z^2}{\frac{(c^2 - a^2)(a^2 - d^2)}{a^2}} = 1 , \qquad (11.58)$$

which is obtained by substituting $\zeta = \eta = \xi = \delta = 1$ into (11.48). Figures 11.18 show the wireframe of the ellipsoid EP, the intersection curves of two ellipsoids (two ellipses) and pairs of vectors $d\mathbf{n}$ emanating from the intersection curves and intersecting in the yz-plane from two different view points. The locus of these intersecting points in (a) is the ellipse (11.58).

Example 11.3.3. Consider an elliptic cone ($\zeta = \eta = 1$, $\xi = -1$ and $\delta = 0$) of the form

$$EC: \quad f(x, y, z) = \frac{x^2}{a^2} + \frac{y^2}{b^2} - \frac{z^2}{c^2} = 0 . \qquad (11.59)$$

The curvatures of the elliptic cone (11.59) based on the curvature sign convention (a) are given in (3.85) and (3.86). For the sign convention (b) they are

$$K = 0, \quad H = \frac{x^2 + y^2 + z^2}{2a^2b^2c^2 \left(\frac{x^2}{a^4} + \frac{y^2}{b^4} + \frac{z^2}{c^4}\right)^{\frac{3}{2}}} , \qquad (11.60)$$

$$\kappa_{max} = \frac{x^2 + y^2 + z^2}{a^2b^2c^2 \left(\frac{x^2}{a^4} + \frac{y^2}{b^4} + \frac{z^2}{c^4}\right)^{\frac{3}{2}}} , \quad \kappa_{min} = 0 , \qquad (11.61)$$

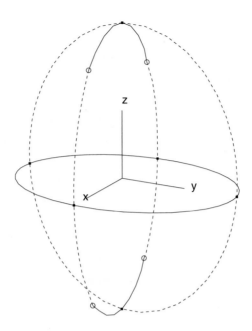

Fig. 11.16. Locations of extrema of principal curvatures (black square), umbilics (white circle) and line of curvatures of ellipsoid (a=0.6, b=0.8, c=1.0) (adapted from [249])

where $(x, y, z) \in EC$ except at the apex (0,0,0).

Since the Gaussian curvature is zero everywhere, the elliptic cone is a developable surface and hence the minimum principal curvature lines are in the ruling direction (see Sect. 9.7.1). The maximum principal curvature lines, which are orthogonal to the minimum principal curvature lines, are thus orthogonal to the ruling directions. Therefore as a point on a ruling approaches the apex, the maximum principal curvature monotonically increases and will become infinite at the apex.

It is apparent from (11.42) to (11.44) that the offset of the elliptic cone self-intersects in the x and y-directions if the offset distance d is negative, while it self-intersects in the z-direction if the offset distance d is positive. Unlike the case for ellipsoids, all the three ellipsoids intersect with the elliptic cone for all nonzero d, provided the correct sign is chosen. This observation agrees with the result that the maximum principal curvature has an infinite value at the apex.

Figure 11.19 (a) shows the elliptic cone EC (a=0.6, b=0.8 and c=1.0) intersecting the EP_x (d=-0.45). The self-intersection curve in the yz-plane is

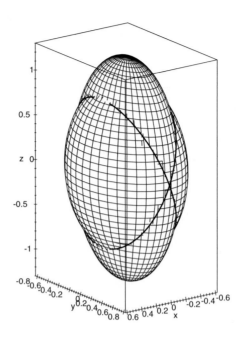

Fig. 11.17. Two intersecting ellipsoids (adapted from [249]): intersection curves, which comprise of two ellipses, represent the footpoint curve of the self-intersection curve of offset of ellipsoid

given by setting $\zeta = \eta = 1$, $\xi = -1$ and $\delta = 0$ into (11.48)

$$-\frac{y^2}{\frac{(b^2-a^2)d^2}{a^2}} + \frac{z^2}{\frac{(c^2+a^2)d^2}{a^2}} = 1 \,, \tag{11.62}$$

which is a hyperbola. Figure 11.19 (b) shows the wireframe of the elliptic cone EC, the intersection curves of EC and EP_x and pairs of vectors $d\mathbf{n}$ emanating from the intersection curves and intersecting in the yz-plane. The locus of these intersecting points is the hyperbola (11.62).

Theorem 11.3.1 provides a generalized method for obtaining the self-intersection curves of offsets and the corresponding foot point curves on the progenitor implicit quadratic surfaces. The theorem is useful for tool path generation for NC machining and other engineering applications.

11.3.4 Self-intersection of offsets of explicit quadratic surfaces

Although offset surfaces are widely used in various engineering applications, their degenerating mechanism is not well known in a quantitative manner. We

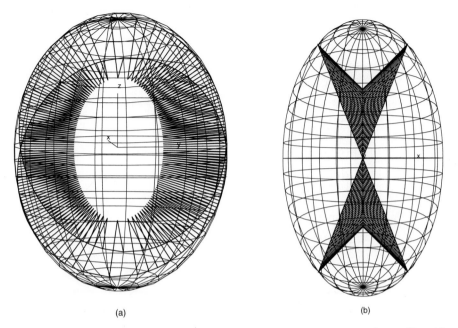

Fig. 11.18. Wireframe of the ellipsoid EP, the intersection curves of two ellipsoids (two ellipses) and pairs of vectors $d\mathbf{n}$ emanating from the intersection curves and intersecting on the self-intersection curves, adapted from [249]: (**a**) view parallel to the yz-plane, (**b**) view parallel to the xz-plane

have seen in Sect. 8.3, that any regular surface can be locally approximated in the neighborhood of a point P by an explicit quadratic surface of the form

$$\mathbf{r}(x,y) = [x, y, \frac{1}{2}(\alpha x^2 + \beta y^2)]^T , \qquad (11.63)$$

to the second order where $-\alpha$ and $-\beta$ are the principal curvatures at point P. The minus signs are consistent with curvature sign convention (b). Therefore investigations of the self-intersection mechanisms of the offsets of explicit quadratic surfaces due to differential geometry properties lead to an understanding of the self-intersecting mechanisms of offsets of regular parametric surfaces.

In the sequel we assume $d > 0$, $\beta > 0$ and $\alpha \leq \beta$ without loss of generality. According to this assumption the surface is a hyperbolic paraboloid when $\alpha < 0$, an elliptic paraboloid when $0 < \alpha < \beta$, a paraboloid of revolution when $0 < \alpha = \beta$, and a parabolic cylinder when $\alpha = 0$ as illustrated in Fig. 8.9. The paraboloid of revolution and the parabolic cylinder can be considered as degenerate cases of the elliptic paraboloid. When $\alpha = \beta$, the principal direction is not defined and the point $(0,0,0)$ will become an umbilic. If α and β vanish at the same time, the surface is part of a plane, and we do not investigate such cases.

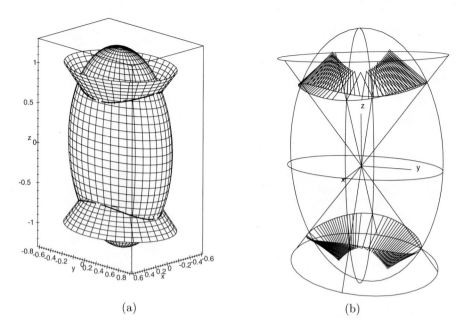

(a)   (b)

Fig. 11.19. (a) Intersection between EC ($a = 0.6$, $b = 0.8$, $c = 1.0$) and EP_x ($d = -\frac{a^2}{b}$=-0.45), (b) wireframe of the elliptic cone EC, the intersection curves of EC and EP_x and pairs of vectors $d\mathbf{n}$ emanating from the intersection curves and intersecting on the self-intersection curves (adapted from [249])

In the case for offsets of explicit quadratic surfaces, there are no self-intersections due to global distance function properties [26], thus if $d > 0$ the maximum absolute value of the negative minimum principal curvature determines the largest offset d without degeneracy. The largest magnitude of offset distance without degeneracy is called the maximum offset distance $|d_{max}|$. In Sect. 8.3 we discussed how to find the global minimum of the minimum principal curvature of explicit quadratic surfaces.

Due to Lemma 8.3.1 the global minimum of the minimum principal curvature of the explicit quadratic surface occurs at the origin, except for a parabolic cylinder ((11.63) with α=0) which has minima along the x-axis with curvature value $\kappa_{min} = -\beta$, and hence the maximum offset distance is determined to be $|d_{max}| = \frac{1}{\beta}$. If the offset distance exceeds $\frac{1}{\beta}$, the offset starts to degenerate from the point $(0, 0, \frac{1}{\beta})$ on the offset surface except for a parabolic cylinder progenitor, where the offset starts to degenerate along the line $(x, 0, \frac{1}{\beta})$.

Substitution of the expression of the offset of the explicit quadratic surface (11.63)

$$\hat{\mathbf{r}}(x,y) = [x, y, \frac{1}{2}(\alpha x^2 + \beta y^2)]^T + \frac{d}{\sqrt{1 + \alpha^2 x^2 + \beta^2 y^2}}[-\alpha x, -\beta y, 1]^T ,$$

(11.64)

into (11.29) yields a vector equation for the self-intersection curve of such offset. The x, y and z components of this vector equation are given by

$$\sigma - \frac{\alpha \sigma d}{\sqrt{1 + \alpha^2 \sigma^2 + \beta^2 t^2}} = u - \frac{\alpha u d}{\sqrt{1 + \alpha^2 u^2 + \beta^2 v^2}} ,$$

(11.65)

$$t - \frac{\beta t d}{\sqrt{1 + \alpha^2 \sigma^2 + \beta^2 t^2}} = v - \frac{\beta v d}{\sqrt{1 + \alpha^2 u^2 + \beta^2 v^2}} ,$$

$$\frac{1}{2}(\alpha \sigma^2 + \beta t^2) + \frac{d}{\sqrt{1 + \alpha^2 \sigma^2 + \beta^2 t^2}} = \frac{1}{2}(\alpha u^2 + \beta v^2) + \frac{d}{\sqrt{1 + \alpha^2 u^2 + \beta^2 v^2}} ,$$

The self-intersection curve of an offset can be considered as the locus of the center of a sphere, whose radius is the offset distance, rolling on the progenitor surface with two contact points. It is evident from (11.64) that the offsets of explicit quadratic surfaces are symmetric with respect to xz and yz planes. Because of the symmetry of the offsets of explicit quadratic surfaces, the center of the rolling sphere must move only on the planes of symmetry and hence the self-intersection curves are on the plane of symmetry. Therefore we can set $\sigma = -u$ and $t = v$ in (11.65). The y and z components will only result in identities, while the x component results in

$$\sqrt{1 + \alpha^2 \sigma^2 + \beta^2 t^2} = \alpha d ,$$

(11.66)

where the trivial solution $\sigma = u = 0$ is excluded. Similarly, we can set $\sigma = u$ and $t = -v$ in the (11.65). The x and z components will only result in identities, while the y component results in

$$\sqrt{1 + \alpha^2 \sigma^2 + \beta^2 t^2} = \beta d ,$$

(11.67)

where the trivial solution $t = v = 0$ is excluded. Therefore (11.65) have been reduced to two uncoupled equations in σ and t. Next we give a useful theorem for evaluating the self-intersection curves of offsets of explicit quadratic surfaces (11.63) and their corresponding planar curve in the xy-plane, i.e. pre-image of the self-intersection curve. In this theorem we assume that the x and y axes are taken as the directions of maximum and minimum principal curvatures.

Theorem 11.3.2. *The self-intersection curves of offsets of the explicit quadratic surfaces* $\mathbf{r}(x,y) = [x, y, \frac{1}{2}(\alpha x^2 + \beta y^2)]^T$ *and their pre-images in the xy-plane are as follows [248]:*

1. *An offset of a hyperbolic paraboloid ($\alpha < 0 < \beta$) self-intersects only in the y-direction when $\frac{1}{\beta} < d$. The resulting self-intersection curve is a parabola given by*

$$z = \frac{\alpha\beta}{2(\beta - \alpha)}x^2 + \frac{(\beta d)^2 + 1}{2\beta}, \quad y = 0, \tag{11.68}$$

$$\text{where} \quad -\frac{\beta - \alpha}{\alpha\beta}\sqrt{(\beta d)^2 - 1} \le x \le \frac{\beta - \alpha}{\alpha\beta}\sqrt{(\beta d)^2 - 1},$$

and its pre-image in the xy-plane is an ellipse when $|\alpha| \ne \beta$ or a circle when $|\alpha| = \beta$, (see Fig. 11.21 (a)) given by

$$\frac{x^2}{\left(\frac{\sqrt{(\beta d)^2 - 1}}{\alpha}\right)^2} + \frac{y^2}{\left(\frac{\sqrt{(\beta d)^2 - 1}}{\beta}\right)^2} = 1. \tag{11.69}$$

2. An offset of an elliptic paraboloid $(0 < \alpha < \beta)$ self-intersects only in the y-direction when $\frac{1}{\beta} < d < \frac{1}{\alpha}$ and self-intersects in both x and y-directions when $\frac{1}{\alpha} < d$. The self-intersection curve which self-intersects in the y-direction is a parabola (see Fig. 11.20) given by (11.68) and its pre-image in the xy-plane is an ellipse (see Figs. 11.20, 11.21 (b)) given by (11.69). The self-intersection curve which self-intersects in the x-direction is also a parabola given by

$$z = \frac{\alpha\beta}{2(\alpha - \beta)}y^2 + \frac{(\alpha d)^2 + 1}{2\alpha}, \quad x = 0, \tag{11.70}$$

$$\text{where} \quad -\frac{\alpha - \beta}{\alpha\beta}\sqrt{(\alpha d)^2 - 1} \le y \le \frac{\alpha - \beta}{\alpha\beta}\sqrt{(\alpha d)^2 - 1},$$

and its pre-image in the xy-plane is an ellipse (see Figs. 11.21 (c), (d)) given by

$$\frac{x^2}{\left(\frac{\sqrt{(\alpha d)^2 - 1}}{\alpha}\right)^2} + \frac{y^2}{\left(\frac{\sqrt{(\alpha d)^2 - 1}}{\beta}\right)^2} = 1. \tag{11.71}$$

3. An offset of a paraboloid of revolution $(0 < \alpha = \beta)$ self-intersects in all directions, when $\frac{1}{\beta} = \frac{1}{\alpha} < d$. The self-intersection curve is a point $(0, 0, \frac{(\beta d)^2 - 1}{2\beta})$, and its pre-image in the xy-plane is a circle (see Fig. 11.21 (e)) given by

$$x^2 + y^2 = \left(\frac{\sqrt{(\beta d)^2 - 1}}{\beta}\right)^2. \tag{11.72}$$

4. An offset of a parabolic cylinder $(\alpha = 0 < \beta)$ self-intersects only in the y-direction when $\frac{1}{\beta} < d$. The resulting self-intersection curve is a straight line in the xz-plane

$$z = \frac{(\beta d)^2 - 1}{2\beta}, \quad y = 0, \tag{11.73}$$

and its pre-image in the xy-plane (see Fig. 11.21 (f)) are two straight lines given by

$$y = \pm \frac{\sqrt{(\beta d)^2 - 1}}{\beta} . \tag{11.74}$$

Proof:
Case (1): Since α is negative for the hyperbolic paraboloid and we are assuming $d > 0$ and the left hand side of (11.66) is always positive, this equation cannot be used to derive the self-intersection curve. This implies that the offset of a hyperbolic paraboloid does not self-intersect in the x-direction (maximum principal direction). However, we can use (11.67) to derive the self-intersection curve in the y-direction (minimum principal direction). Upon squaring and replacing σ by x and t by y we obtain

$$\alpha^2 x^2 + \beta^2 y^2 = (\beta d)^2 - 1 . \tag{11.75}$$

This equation describes an ellipse in the xy-plane and is equivalent to (11.69). Since the left hand side of (11.75) is always positive, the equation is only valid when $\frac{1}{\beta} < d$. This indicates that there is no self-intersection unless the offset distance exceeds the maximum offset distance $|d_{max}| = \frac{1}{\beta}$. Now we can obtain the self-intersection curve in the xz-plane by mapping the ellipse in xy-plane (see (11.75)) into the 3-D coordinates using (11.64), resulting in:

$$\hat{\mathbf{r}}(x, y) = [\hat{x}(x, y),\ \hat{y}(x, y),\ \hat{z}(x, y)]^T$$
$$= [(\frac{\beta - \alpha}{\beta})x,\ 0,\ \frac{1}{2\beta}\{\alpha(\beta - \alpha)x^2 + (\beta d)^2 + 1\}]^T , \tag{11.76}$$

where $-\frac{\sqrt{(\beta d)^2 - 1}}{\alpha} \le x \le \frac{\sqrt{(\beta d)^2 - 1}}{\alpha}$. The range of parameter x comes from (11.75). If we eliminate the parameter x from (11.76) and replace $\hat{x}$ by x and $\hat{z}$ by z, we obtain the same result as (11.68).

Case (2): Since α is positive for the elliptic paraboloid, both (11.66), (11.67) can be used to obtain the self-intersection curves in the xy-plane. This implies that the offset of an elliptic paraboloid may self-intersect in both principal directions. Since we have already derived the equation from (11.67), we derive another equation from (11.66). Upon squaring and replacing σ by x and t by y we obtain

$$\alpha^2 x^2 + \beta^2 y^2 = (\alpha d)^2 - 1 , \tag{11.77}$$

which is equivalent to (11.71). Also this equation is only valid when $\frac{1}{\alpha} < d$. The self-intersection curve in 3-D coordinates can easily be obtained in a similar manner with Case (1).

Case (3): If we set $\alpha = \beta$ in (11.75) and (11.77), both equations reduce to (11.72). Also if we set $\alpha = \beta$ in (11.76), the parabola reduces to the point $(0, 0, \frac{(\beta d)^2 + 1}{2\beta})$.

Case (4): Since α is zero for the parabolic cylinder, (11.66) is not valid. Thus we set $\alpha = 0$ in (11.67) and replacing t by y we obtain $\beta^2 y^2 = (\beta d)^2 - 1$, which is equivalent to (11.74). The self-intersection curve in three dimensional coordinates can easily be obtained in a similar manner with Case (1). ∎

Note that the self-intersection curve of the offset of an elliptic paraboloid (when $\frac{1}{\beta} < d < \frac{1}{\alpha}$) has a positive quadratic term, while those of a hyperbolic paraboloid and an elliptic paraboloid (when $\frac{1}{\alpha} < d$) have negative quadratic terms.

Example 11.3.4. Consider an elliptic paraboloid $z = \frac{1}{2}(2x^2 + 4y^2)$ with offset distance $d = 0.3$. Since $\frac{1}{\beta} = \frac{1}{4} < d = 0.3 < \frac{1}{2} = \frac{1}{\alpha}$, the offset surface self-intersects only in the y-direction. The self-intersection curve is $z = 2x^2 + 0.305$ (dashed line in Fig. 11.20) and its pre-image in the xy-plane is $\frac{x^2}{0.11} + \frac{y^2}{0.0275} = 1$ (solid line in Fig. 11.20). The dot dashed line in this figure illustrates the set of footpoints of the self-intersection curve on the progenitor surface. A pair of thin solid straight lines emanating from two distinct points on the surface $\mathbf{r}(\sigma, t)$, $\mathbf{r}(u, v)$ and intersecting along the parabola are the pairs of vectors $d\mathbf{N}(\sigma, t)$ and $d\mathbf{N}(u, v)$.

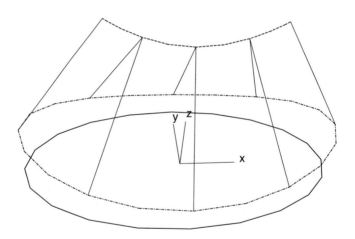

Fig. 11.20. Self-intersection curves of an offset of elliptic paraboloid ($\alpha = 2$, $\beta = 4$) with $d = 0.3$ (adapted from [248])

To illustrate Theorem 11.3.2, we plot pre-images of the self-intersection curves along with cuspidal edges in the xy-plane for the hyperbolic paraboloid ($\alpha = -2$, $\beta = 2$, $d = 0.6$), the elliptic paraboloids ($\alpha = 1.75$, $\beta = 2$, $d = 0.55$), ($\alpha = 1.75$, $\beta = 2$, $d = 0.6$) and ($\alpha = 1.75$, $\beta = 2$, $d = 0.65$), the paraboloid

($\alpha = b = 2$, $d = 0.6$) and the parabolic cylinder ($\alpha = 0$, $\beta = 2$, $d = 0.6$) as depicted in Figs. 11.21 (a) to (f).

It is interesting to note that when the progenitor surface is a hyperbolic paraboloid or an elliptic paraboloid (see Fig. 11.21 (a) to (d)), the pre-images of the self-intersection curve of its offset which self-intersects in the y-direction and the cuspidal edge $\kappa_{min}(x,y) = -\frac{1}{d}$ always intersect tangentially at $y = 0$. The pre-image of the self-intersection curve of the offset of an elliptic paraboloid (see Fig. 11.21 (c)), which self-intersects in x-direction, and the cuspidal edge $\kappa_{max}(x,y) = -\frac{1}{d}$ intersect tangentially at $x = 0$, when the two cuspidal edges intersect with the y-axis within the two umbilics. Whereas when the two cuspidal edges intersect the y-axis outside the two umbilics (see Fig. 11.21 (d)), the pre-images of the self-intersection curve and the cuspidal edge $\kappa_{min}(x,y) = -\frac{1}{d}$ intersect tangentially at $x = 0$.

It is apparent from (11.21) that the direction of the normal vector of the offset surface is opposite to that of the progenitor surface inside the loop of $\kappa_{min}(x,y) = -\frac{1}{d}$ (dashed line) in the absence of the loop of $\kappa_{max}(x,y) = -\frac{1}{d}$ (see Figs. 11.21 (a), (b), (e)), and the regions between outside the loop of $\kappa_{max}(x,y) = -\frac{1}{d}$ (dot dot dashed line) and inside the loop of $\kappa_{min}(x,y) = -\frac{1}{d}$ (see Figs. 11.21 (c), (d)), while the direction is the same within the loop of $\kappa_{max}(x,y) = -\frac{1}{d}$ (see Figs. 11.21 (c), (d)).

Figures 11.22, 11.23 and 11.24 show self-intersecting offset surfaces, self-intersection curves and cuspidal edges in 3-D space and the trimmed offset surface of a hyperbolic paraboloid ($\alpha = -2$, $\beta = 2$, $d = 0.6$), an elliptic paraboloid ($\alpha = 1.75$, $\beta = 2$, $d = 0.6$) and an elliptic paraboloid ($\alpha = 1.75$, $\beta = 2$, $d = 0.65$), respectively.

Figure 11.25 illustrates the self-intersections of the offset of a bicubic Bézier surface patch. Figure 11.25 (a) shows the pre-images of the self-intersection curve in the parameter domain. The thick line represents the numerically traced self-intersection curve, while the thin line represents the ellipses of (11.69), (11.71), which are in quite good agreement. The same bullet symbols are mapped to the same locations on the offset surface. Figure 11.25 (b) shows the mapping of the self-intersection curves in the parameter domain onto the progenitor surface. Finally, Fig. 11.25 (c) shows the offset surface with its self-intersections.

11.3.5 Self-intersection of offsets of polynomial parametric surface patches

For parametric surface patches $\mathbf{r}(u,v) = (x(u,v),\ y(u,v),\ z(u,v))^T$, the unit normal vector of a regular surface (3.3) in terms of components is given by

$$\mathbf{N}(u,v) = \frac{(y_u z_v - y_v z_u,\ x_v z_u - x_u z_v,\ x_u y_v - x_v y_u)^T}{\sqrt{(y_u z_v - y_v z_u)^2 + (x_v z_u - x_u z_v)^2 + (x_u y_v - x_v y_u)^2}}$$

$$\equiv \left(\frac{N_x(u,v)}{S(u,v)},\ \frac{N_y(u,v)}{S(u,v)},\ \frac{N_z(u,v)}{S(u,v)} \right)^T , \tag{11.78}$$

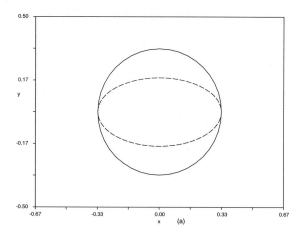

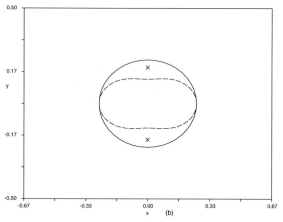

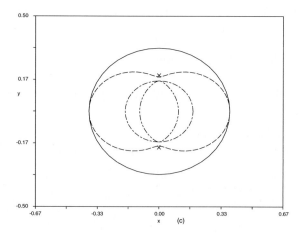

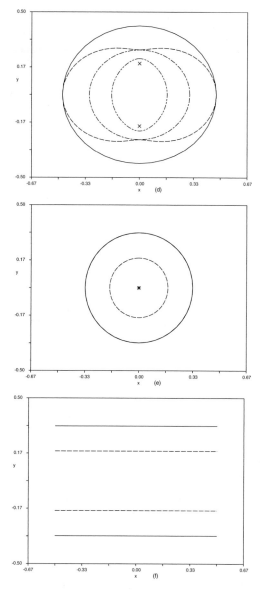

Fig. 11.21. Pre-images of self-intersection curves and cuspidal edges of offsets of explicit quadratic surfaces (adapted from [248]). The *solid lines* correspond to pre-images of the self-intersection curves for self-intersection in the y-direction. The *dashed lines* correspond to $\kappa_{min}(x, y) = -\frac{1}{d}$. The *dot dashed lines* correspond to pre-images of the self-intersection curves for self-intersection in the x-direction. The *dot dot dashed lines* correspond to $\kappa_{max}(x, y) = -\frac{1}{d}$. Symbols $\times$ and $*$ represent the locations of generic lemon type umbilics and non-generic umbilics, respectively. (**a**) Hyperbolic paraboloid ($\alpha = -2$, $\beta = 2$, $d = 0.6$), (**b**) elliptic paraboloid ($\alpha = 1.75$, $\beta = 2$, $d = 0.55$), (**c**) elliptic paraboloid ($\alpha = 1.75$, $\beta = 2$, $d = 0.6$), (**d**) elliptic paraboloid ($\alpha = 1.75$, $\beta = 2$, $d = 0.65$), (**e**) paraboloid of revolution ($\alpha = \beta = 2$, $d = 0.6$), (**f**) parabolic cylinder ($\alpha = 0$, $\beta = 2$, $d = 0.6$)

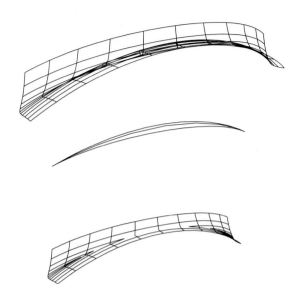

Fig. 11.22. Self-intersecting offset surface (top), region bounded by self-intersection curve (middle) and trimmed offset surface (bottom) of a hyperbolic paraboloid $z = \frac{1}{2}(-2x^2 + 2y^2)$ with $d = 0.6$ (adapted from [248])

where N_x, N_y and N_z denote the x, y and z components of $\mathbf{r}_u \times \mathbf{r}_v$ and $S = |\mathbf{r}_u \times \mathbf{r}_v|$. Consequently the vector equation for self-intersections (11.29) becomes

$$x(\sigma,t) + \frac{N_x(\sigma,t)}{S(\sigma,t)}d = x(u,v) + \frac{N_x(u,v)}{S(u,v)}d \,, \tag{11.79}$$

$$y(\sigma,t) + \frac{N_y(\sigma,t)}{S(\sigma,t)}d = y(u,v) + \frac{N_y(u,v)}{S(u,v)}d \,, \tag{11.80}$$

$$z(\sigma,t) + \frac{N_z(\sigma,t)}{S(\sigma,t)}d = z(u,v) + \frac{N_z(u,v)}{S(u,v)}d \,, \tag{11.81}$$

which is an underconstrained system with three equations with four unknowns σ, t, u, v.

We can easily trace any self-intersection curve branch if the pre-image of that branch in at least one of the parametric domains starts from the parametric domain boundary as depicted in Fig. 11.26 (a). The same symbols in σt and uv-parameter spaces are the corresponding pairs of points which give the self-intersection. It is more difficult to find starting points for tracing self-intersection curves, when the self-intersections curves are closed in both parametric domains as illustrated in Fig. 11.26 (b). This may occur due to local differential geometry properties and global distance function properties

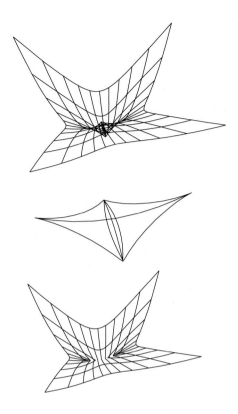

Fig. 11.23. Self-intersecting offset surface (top), region bounded by self-intersection curves (middle) and trimmed offset surface (bottom) of elliptic paraboloid $z = \frac{1}{2}(1.75x^2 + 2y^2)$ with $d = 0.6$ (adapted from [248])

of the progenitor surface. The first case occurs in the vicinity of extrema of principal curvatures in a concave region, when the offset distance exceeds the smallest radius of curvature [253]. The second case occurs in the vicinity of a pair of collinear normal points whose distance is equal or smaller than twice the offset distance. If the surface is conceptually subdivided along an iso-parametric line which contains the local extrema of principal curvature in the concave region (whose radius of curvature is smaller than the absolute offset distance $|d|$) or the collinear normal points whose distance is equal or smaller than twice the offset distance, then each sub-patch will contain simple self-intersection branches without loops.

Therefore after this subdivision process, we can find all the starting points for tracing self-intersection curves of an offset surface along iso-parametric lines made up of the boundary of all subdomains. Since we can fix one of

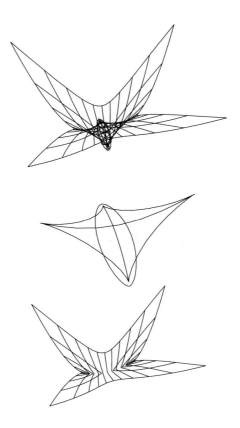

Fig. 11.24. Self-intersecting offset surface (top), region bounded by self-intersection curves (middle) and trimmed offset surface (bottom) of elliptic paraboloid $z = \frac{1}{2}(1.75x^2 + 2y^2)$ with $d = 0.65$ (adapted from [248])

the four variables (σ, t, u, v), the vector equation (11.29) reduces to three equations with three unknowns.

Let us assume that the input curve is a Bézier patch. Then the system (11.79), (11.80), (11.81) reduces to three simultaneous trivariate irrational equations involving polynomials and square root of polynomials. We can replace the square root of polynomials $S(\sigma, t)$ and $S(u, v)$ by auxiliary variables η and ζ such that $\eta^2 = S^2(\sigma, t)$ and $\zeta^2 = S^2(u, v)$. Consequently, the above system can be reduced to a nonlinear polynomial system consisting of five equations with five unknowns as follows:

$$\eta\zeta[x(\sigma, t) - x(u, v)] + d[\zeta N_x(\sigma, t) - \eta N_x(u, v)] = 0 , \qquad (11.82)$$
$$\eta\zeta[y(\sigma, t) - y(u, v)] + d[\zeta N_y(\sigma, t) - \eta N_y(u, v)] = 0 , \qquad (11.83)$$
$$\eta\zeta[z(\sigma, t) - z(u, v)] + d[\zeta N_z(\sigma, t) - \eta N_z(u, v)] = 0 , \qquad (11.84)$$

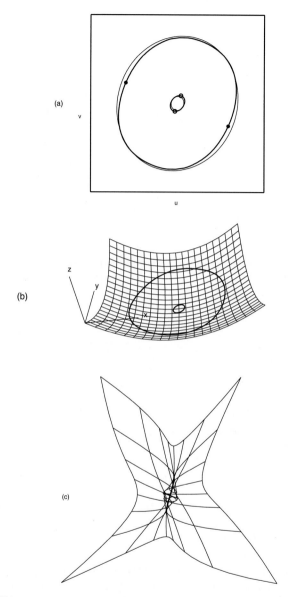

Fig. 11.25. Self-intersections curves of the offset of a bicubic surface patch when $d=0.75$ (adapted from [253]): (**a**) pre-images of the self-intersection curve in parameter domain, where the same bullet symbols are mapped to the same locations in the offset surface, (**b**) a set of footpoints of self-intersection curves on the progenitor surface, (**c**) the offset surface with self-intersections

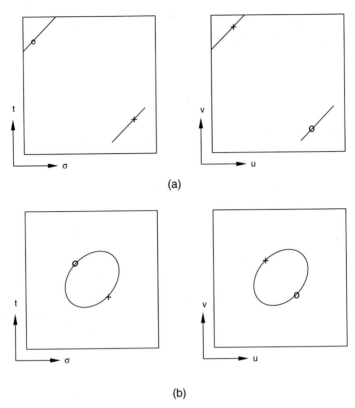

Fig. 11.26. Self-intersection curves in parameter space where the same symbols are mapped to the same locations in the offset surface (adapted from [253])

$$\eta^2 - N_x^2(\sigma, t) - N_y^2(\sigma, t) - N_z^2(\sigma, t) = 0 , \tag{11.85}$$
$$\zeta^2 - N_x^2(u, v) - N_y^2(u, v) - N_z^2(u, v) = 0 . \tag{11.86}$$

This system can be solved by the IPP method introduced in Chap. 4. Since $\sigma = u$, $t = v$ are trivial solutions, we must exclude them from the system, otherwise a Bernstein subdivision-based IPP algorithm would attempt to solve for an infinite number of roots.

A similar problem for the self-intersections of a normal offset of a planar polynomial curve has been treated in [254] by dividing out the common factor. However, for the surface case we cannot divide out these factors from the system directly, since terms $x(\sigma, t) - x(u, v)$, $y(\sigma, t) - y(u, v)$ and $z(\sigma, t) - z(u, v)$ do not necessarily exactly involve the factors $\sigma - u$ and $t - v$. Thus, the polynomial system is first solved by the Bernstein subdivision-based IPP solver with a coarse accuracy (e.g. $10^{-1} \sim 10^{-2}$). The two rectangular sub-patches on the surface for each set of roots using the de Casteljau subdivision algorithm are extracted. Then the normal rectangular pyramids [208, 209, 382], which bound normal vectors of Bézier patches, are constructed [253] and their apexes are translated to the origin. If the two pyramids intersect, the associated parameter boxes are considered as representing trivial roots and excluded from the list of roots. Figures 11.27 illustrate such non-intersecting and intersecting normal pyramids. Finally we restart the IPP solver with boxes that include the solutions but now requiring high accuracy (e.g. 10^{-8}).

When the input surface is a B-spline surface patch, we can split it into Bézier surface patches by the knot insertion algorithm [34, 63, 314]. In such cases we must check the intersections among the offsets of different split patches, where we do not need to worry about trivial solutions. Wang [440] computed intersection curves of offsets of two parametric surface patches using the orthogonal projection of the intersection curves onto the progenitor surfaces.

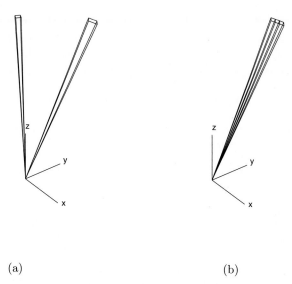

(a)   (b)

Fig. 11.27. Normal pyramids: (a) two non-intersecting normal pyramids, (b) two intersecting normal pyramids (adapted from [253])

11.3.6 Tracing of self-intersection curves

Differential equations for tracing self-intersection curves of an offset surface were first derived by Aomura and Uehara [11]. They are formulated such that the self-intersection curve is arc length parametrized in the parameter domain of the progenitor surface. Here we derive a set of ordinary differential equations following the method we introduced in Sect. 5.8.2 for tracing the surface to surface intersection curves. The tracing direction coincides with the tangential direction of the self-intersection curve $\hat{\mathbf{c}}(s)$ of the offset surface which is perpendicular to the two normal vectors at the corresponding foot points on the progenitor surfaces $\mathbf{r}(\sigma, t)$ and $\mathbf{r}(u, v)$ where $(\sigma, t) \neq (u, v)$. Therefore, the tracing direction can be obtained as follows:

$$\hat{\mathbf{c}}'(s) = \frac{\mathbf{S}(\sigma, t) \times \mathbf{S}(u, v)}{|\mathbf{S}(\sigma, t) \times \mathbf{S}(u, v)|} , \tag{11.87}$$

where $\mathbf{S}(\sigma, t)$ and $\mathbf{S}(u, v)$ are the normal vectors

$$\mathbf{S}(u, v) = \mathbf{r}_u \times \mathbf{r}_v , \tag{11.88}$$

evaluated at $\mathbf{r}(\sigma, t)$ and $\mathbf{r}(u, v)$ where $(\sigma, t) \neq (u, v)$ and $|\mathbf{S}(\sigma, t) \times \mathbf{S}(u, v)| \neq 0$. The normalization of the tangent vector forces $\hat{\mathbf{c}}(s)$ to be arc length parametrized in $\mathbf{R}^3$.

The self-intersection curve of an offset surface can be also viewed as a curve on the offset surface. If we denote the pair of the self-intersection curves in the parameter domain of the progenitor surface as $u = \sigma(s)$, $v = t(s)$ and $u = u(s)$, $v = v(s)$, where s denotes the arc length on the offset surface, then the self-intersection curve on the offset can be expressed as

$$\mathbf{r} = \hat{\mathbf{c}}(s) = \hat{\mathbf{r}}(\sigma(s), t(s)) = \hat{\mathbf{r}}(u(s), v(s)) . \tag{11.89}$$

We can derive the unit tangent vector of the self-intersection curve as a curve on the offset surface using the chain rule as:

$$\hat{\mathbf{c}}'(s) = \hat{\mathbf{r}}_u(\sigma(s), t(s))\sigma' + \hat{\mathbf{r}}_v(\sigma(s), t(s))t' , \tag{11.90}$$
$$\hat{\mathbf{c}}'(s) = \hat{\mathbf{r}}_u(u(s), v(s))u' + \hat{\mathbf{r}}_v(u(s), v(s))v' . \tag{11.91}$$

Since we know the unit tangent vector of the intersection curve from (11.87), we can find σ' and t' as well as u' and v' by taking the dot product of both sides of (11.90) with $\hat{\mathbf{r}}_u(\sigma(s), t(s))$ and $\hat{\mathbf{r}}_v(\sigma(s), t(s))$ and of (11.91) with $\hat{\mathbf{r}}_u(u(s), v(s))$ and $\hat{\mathbf{r}}_v(u(s), v(s))$, which leads to linear systems in σ', t' and u', v'. The solutions to the two linear systems have the same form except that they are evaluated at different parameter values $(\sigma(s), t(s))$ and $(u(s), v(s))$. Using the relation between $\mathbf{N}$ and $\hat{\mathbf{N}}$ (11.19), the ordinary differential equations for tracing the self-intersection curve of an offset surface are given by

$$\sigma' = \frac{det(\hat{\mathbf{c}}', \hat{\mathbf{r}}_v, \mathbf{S})}{S^2(1 + 2Hd + Kd^2)}\bigg|_{(\sigma(s), t(s))} , \tag{11.92}$$

$$t' = \frac{det(\hat{\mathbf{r}}_u, \hat{\mathbf{c}}', \mathbf{S})}{S^2(1 + 2Hd + Kd^2)}\bigg|_{(\sigma(s), t(s))} , \tag{11.93}$$

$$u' = \frac{det(\hat{\mathbf{c}}', \hat{\mathbf{r}}_v, \mathbf{S})}{S^2(1 + 2Hd + Kd^2)}\bigg|_{(u(s), v(s))}, \tag{11.94}$$

$$v' = \frac{det(\hat{\mathbf{r}}_u, \hat{\mathbf{c}}', \mathbf{S})}{S^2(1 + 2Hd + Kd^2)}\bigg|_{(u(s), v(s))}. \tag{11.95}$$

Figure 11.28 illustrates the global self-intersection of an offset without loops. As depicted in Fig. 11.28 the surface has a global constriction between two corner points and the offset surface self-intersects globally without any internal loops. The self-intersection curves can be traced by starting at the surface boundary.

The next example, in Fig. 11.29, shows global self-intersection with loops. The surface also has 4 pairs of collinear normal point with distances 0.3757, 0.3945, 0.1367, 0.3757. Therefore if the magnitude of the offset distance exceeds $\frac{0.1367}{2}$, two self-intersection loops start to grow in the parameter domain enclosing the pair of collinear normal points whose distance in $\mathbf{R}^3$ is 0.1367 [253].

11.3.7 Approximations

Farouki [94] studied the problem of computing approximate offsets of general parametric surfaces. His method involves finding the unique bicubic Hermite interpolant surface that has the exact position, slopes and cross-derivatives as the exact offset surface at the corners of some quadrilateral subdomain of the surface being offset. The accuracy of the offset is then increased by decreasing the size of the subdomain chosen. A uniform subdivision methods is implemented, although a nonuniform subdivision can be formulated to enhance the efficiency.

Patrikalakis and Prakash [301] addressed the representation of plates within the framework of the boundary representation method in a solid modeling environment (see Figure 11.5). Plates are defined as the volume bounded by a progenitor surface, its offset surface and ruled surfaces for the sides. Offset surfaces of rational B-spline/Bézier surfaces cannot in general be represented exactly within the same class of functions describing the progenitor surface. Therefore, if the offset is to be represented in the same form as the progenitor surface, approximation is required. Such approximation assists in integrating offsets in a NURBS-based modeler (at least in an approximate sense).

The steps of the approximation algorithm in [301] are summarized below.

1. Let R be a progenitor rational B-spline surface with control vertices $\mathbf{p}_{ij}$, weights $w_{ij} > 0$ and two knot vectors $\mathbf{U}$ and $\mathbf{V}$ associated with each parameter u and v.
2. Offset each vertex of the control polyhedron by a distance d along the normal vector given by (see Fig. 11.30)

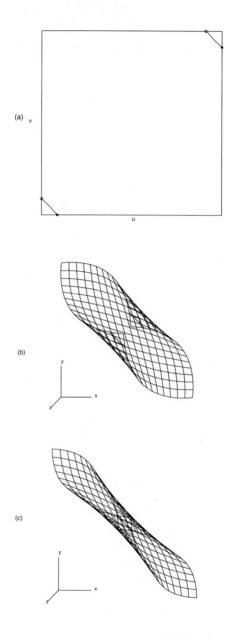

Fig. 11.28. Self-intersection curves of the offset of a bicubic surface patch when $d=0.09$ (adapted from [253]): (**a**) pre-images of the self-intersection curves in parameter domain where the same symbols are mapped to the same points in the offset surface, (**b**) the mapping of the self-intersection curves in the parameter domain onto the progenitor surface, (**c**) the offset surface and the self-intersection curve

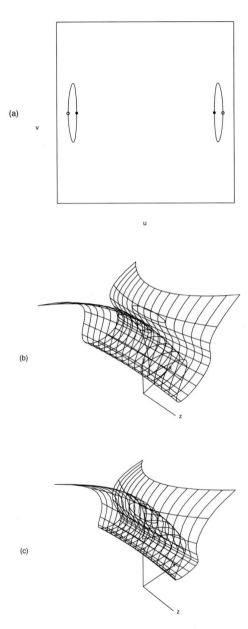

Fig. 11.29. Self-intersection curves of the offset of a bisextic surface patch when d=-0.08 (adapted from [253]): (**a**) pre-images of the self-intersection curves in parameter domain where the same symbols are mapped to the same points in the offset surface, (**b**) the mapping of the self-intersection curves in the parameter domain onto the progenitor surface, (**c**) the offset surface and the self-intersection curve

$$\mathbf{N}_{ij} = \frac{1}{8} \sum_{k=1}^{8} \mathbf{n}_k \,, \tag{11.96}$$

so that the offset control point is given by

$$\hat{\mathbf{p}}_{ij} = \mathbf{p}_{ij} + d\frac{\mathbf{N}_{ij}}{|\mathbf{N}_{ij}|} \,, \tag{11.97}$$

where $\mathbf{n}_k$ are unit normal vectors on the triangular facets of the control polyhedron around $\mathbf{P}_{ij}$ as in Fig. 11.30. Then the approximated offset $\hat{R}$ is defined by $\hat{\mathbf{p}}_{ij}$, w_{ij}, $\mathbf{U}$ and $\mathbf{V}$.

3. Check deviation of the approximate offset with the true offset for every $(u_i, v_j) \in \mathbf{U} \times \mathbf{V}$. If it is good at all points, the checking proceeds to the next stage. If it is not good at some point, then new knots are added at left and right midspans of both the u and v-directions. Knots are not added at those points where a new knot has been currently added to avoid unnecessary knots that could, possibly, lower the order of continuity. In the second stage, the surface is further checked progressively at its u and v midspans, one-third spans, etc., to some prespecified level of interior checking. A new u and v knot is added at places where the check fails. If the check passes at all points, the approximate offset is considered good enough.

4. Evaluate a new control polyhedron corresponding to the finer knot vector using the Oslo algorithm [63] and go back to step 2.

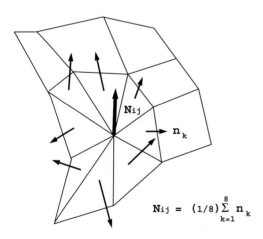

$$\mathbf{N}_{ij} = (1/8)\sum_{k=1}^{8} \mathbf{n}_k$$

Fig. 11.30. Offset surface approximation (adapted from [301])

An alternate way to compute offsets of NURBS curves and surfaces was addressed by Piegl and Tiller [316]. The approach consists of four steps: 1) recognition of special curves such as a straight line or a circle, and special surfaces such as a plane, a surface of revolution (sphere, torus, cone, cylinder) or a general surface of revolution, or a ruled surface which in special cases can be an extrusion; 2) sampling of the offset curve or surface based on bounds on second derivatives; 3) interpolation of these points by B-spline curves and surfaces; 4) removal of all extraneous knots so that the error does not exceed the tolerance.

11.4 Pythagorean hodograph

11.4.1 Curves

Farouki and Sakkalis [108] introduced a class of special planar polynomial curves called *Pythagorean hodograph* (*PH*) curves $\mathbf{r}(t) = (x(t), y(t))^T$, whose hodograph (derivative) components $\dot{x}(t)$, $\dot{y}(t)$ and a polynomial $\tau(t)$ form a Pythagorean triple $\dot{x}^2(t) + \dot{y}^2(t) = \tau^2(t)$. Thus, the *PH* curve has polynomial parametric speed $\tau(t)$; accordingly its offset is a rational curve and its arc length is a polynomial function $s(t)$ of the parameter t. The Pythagorean condition is satisfied by

$$\dot{x}(t) = (a^2(t) - b^2(t))c(t), \quad \dot{y}(t) = 2a(t)b(t)c(t), \quad \tau(t) = (a^2(t) + b^2(t))c(t) ,$$
$$(11.98)$$

where $a(t)$, $b(t)$ and $c(t)$ are polynomials satisfying $GCD(a, b) = 1$ and $max(deg(a), deg(b)) \geq 1$ [108, 114] where GCD denotes greatest common divisor. This condition is only a sufficient condition for a polynomial curve to have rational offset. For most applications $c(t)$ is chosen to be 1. The lowest degree *PH* curve occurs when the polynomials $a(t)$ and $b(t)$ are linear and thus from (11.98) its degree is cubic. However, the resulting *PH* curve cannot possess an inflection point and hence is not practical. When $a(t)$ and $b(t)$ are quadratic, the *PH* curve will be a quintic and is the lowest degree curve to have enough flexibility for practical use. The *PH* quintics can inflect and can interpolate arbitrary first-order Hermite data [103]. The degree of the offset is $2m - 1$ for a degree m *PH* curve. Therefore, the lowest degree of the offset of a *PH* curve for practical use is nine.

Farouki and Shah [112] developed a real-time *CNC* interpolator for *PH* curves using the fact that the arc length $s(t)$ of a *PH* curve is a polynomial function. As a consequence the generation of reference points along a *PH* curve is reduced to a sequence of polynomial rootfinding problems.

The planar *PH* curves can be easily generalized to space *PH* curves [110] by setting the four real polynomials $a(t)$, $b(t)$, $c(t)$, $\varepsilon(t)$ in the form

$$\dot{x}(t) = (a^2(t) - b^2(t) - c^2(t))\varepsilon(t), \quad \dot{y}(t) = 2a(t)b(t)\varepsilon(t), \quad \dot{z}(t) = 2a(t)c(t)\varepsilon(t) ,$$
$$(11.99)$$

which leads to a polynomial parametric speed $\tau(t) = (a^2(t) + b^2(t) + c^2(t))\varepsilon(t)$. A more thorough review of *PH* curves can be found in [114].

Pottmann [324] generalized the concept of PH curves to the full class of rational curves with rational offsets, by utilizing the projective dual representation. A rational planar curve is obtained as the envelope of its tangent line which is described as

$$g(t) : n_x(t)x + n_y(t)y = h(t) ,\qquad (11.100)$$

where $h(t)$ is the signed distance of the tangent line $g(t)$ from the origin and is a rational function. The vector $(n_x(t), n_y(t))^T$ is a rational unit normal of the tangent line g(t) and is given by

$$n_x(t) = \frac{2a(t)b(t)}{a^2(t) + b^2(t)}, \quad n_y(t) = \frac{a^2(t) - b^2(t)}{a^2(t) + b^2(t)} ,\qquad (11.101)$$

where (11.98) is used so that the unit normal vector becomes rational. The envelope of the one-parameter family of $g(t)$ can be obtained by solving a linear system consisting of (11.100) and its derivative $\dot{g}(t)$ for x and y as a function of t, resulting in:

$$(x(t), y(t)) = \left(\frac{X(t)}{W(t)}, \frac{Y(t)}{W(t)} \right) ,\qquad (11.102)$$

where

$$X = 2ab(\dot{a}b - a\dot{b})ef - \frac{1}{2}(a^4 - b^4)(\dot{e}f - e\dot{f}) ,$$
$$Y = (a^2 - b^2)(\dot{a}b - a\dot{b})ef + ab(a^2 + b^2)(\dot{e}f - e\dot{f}) ,$$
$$W = (a^2 + b^2)(\dot{a}b - a\dot{b})f^2 .\qquad (11.103)$$

Here the rational function $h(t)$ is replaced by $\frac{e(t)}{f(t)}$. The offset to (11.102) is obtained by simply replacing $h(t)$ by $h(t) + d$ or equivalently $e(t)$ by $e(t) + f(t)d$, and thus the degree of the offset remains the same as that of (11.102), which is an advantage over PH curves. The rational Bézier representation can be easily derived by prescribing the polynomials $a(t)$, $b(t)$, $e(t)$ and $f(t)$ and expressing the resulting polynomials X, Y and W in Bernstein form.

The form of (11.102) and (11.103) becomes simpler if the dual Bézier representation is used [324]. A plane dual Bézier curve is defined by a family of tangent lines which has the form

$$\mathbf{U}(t) = (u_o(t); u_1(t); u_2(t))^T = \sum_{i=0}^{n} \mathbf{U}_i B_{i,n}(t) ,\qquad (11.104)$$

where $\mathbf{U}_i$ are the *Bézier lines* (constant line vectors) and $B_{i,n}(t)$ is the i-th Bernstein polynomial of degree n. A line vector $\mathbf{U} = (u_o; u_1; u_2)^T$ determines a straight line $u_o + u_1 x + u_2 y = 0$. From the homogeneous representation of (11.100) in the form, $u_o W + u_1 X + u_2 Y = 0$, the dual representation in terms of projective geometry is given by

$$u_o : u_1 : u_2 = -(a^2 + b^2)e : 2abf : (a^2 - b^2)f . \qquad (11.105)$$

When f has a factor $a^2 + b^2$, there exists a common divisor in the dual representation, thus it is convenient to set $f = (a^2 + b^2)p$ which leads to

$$u_o : u_1 : u_2 = -e : 2abp : (a^2 - b^2)p . \qquad (11.106)$$

The control lines $\mathbf{U}_i$ in (11.104) are easily obtained by expressing (11.106) in Bernstein form.

Lü [238] showed that the offset to a parabola is rational; its singular point at infinity was studied by Farouki and Sederberg [111]. In [238] Lü proved that although the offset (to a parabola) is not rational in the parameter t, it may be expressed as a rational form in a new parameter, say u, via a parameter transformation. The reparametrizing function $t = t(u)$ is a rational function of the form $t = \frac{f(u)}{u}$ where $f(u)$ is a quadratic polynomial in u. The transformed curve $\bar{x}(u) = x(t(u))$, $\bar{y}(u) = y(t(u))$ is not parametrized properly, since there are two values of u, which are the roots of the quadratic equation $f(u) - tu = 0$, for each corresponding point $(x, y) = (x(t), y(t))$. While the curve $\bar{\mathbf{r}}(u)$ is traced twice in opposite directions as u increases from $-\infty$ to 0 and from 0 to $+\infty$, $\bar{\mathbf{r}}(u) + \bar{\mathbf{n}}(u)d$ defines a two-sided offset, i.e. the inward offset for $u < 0$ and the outward offset for $u > 0$. The resulting rational curve is of degree 6. Lü [239] further derives a necessary and sufficient condition for a polynomial or more generally rational planar parametric curve to have rational parametric speed.

11.4.2 Surfaces

Pottmann [324] applied the same principle of the rational curve with rational offsets to the rational surface with rational offsets. While the tangent lines are used in the curve case, a two-parametric set of tangent planes

$$g(u, v) : N_x(u, v)x + N_y(u, v)y + N_z(u, v)z = h(u, v) , \qquad (11.107)$$

is used for the surface, where $(N_x, N_y, N_z)^T$ is a rational unit normal of the tangent plane and $h(u, v)$ is a rational distance function from the origin. The rest of the discussions are analogous to the curve case.

A developable surface has a constant tangent plane along a generator. Therefore its tangent plane depends on only one parameter, say u. In other words, a developable surface can be considered as the envelope of a one parameter family of planes $g(u)$. The cross product of the normal vectors of the two planes $g(u)$ and $\dot{g}(u)$ provides a vector of the generator at parameter u. An explicit representation of rational developable surfaces with rational offsets has also been given in [324].

Lü [240] studied the rationality of offsets of quadrics. The key idea is to transform the problem of rational offsets of quadrics to a simple problem on the rationality of a cubic algebraic surface and use existing results in algebraic geometry [362]. He showed that the offsets of paraboloids, ellipsoids and hyperboloids can be rationally parametrized, while cylinders and cones except for parabolic cylinders, cylinders of revolution and cones of revolution do not possess any rational offset.

Pottmann et al. [327] proved that offsets of a nondevelopable rational ruled surface in the whole space always admit a rational parametrization. Even though the offsets to ruled surfaces are rational in the whole space where they are defined, the offset patch to a rational patch may not be expressible as a rational patch. Therefore, further research is needed for applying this technique to a finite patch.

Peternell and Pottmann [308] construct PH surfaces from arbitrary rational surfaces with the aid of a geometric transformation which describes a change between two models of Laguerre geometry. The two fundamental elements of Laguerre geometry are oriented planes and cycles. A cycle represents an oriented sphere or a point which is a degenerate sphere with zero radius. The orientation of the fundamental elements is determined by a unit normal vector field or equivalently by a signed radius for spheres. An oriented sphere and an oriented plane are said to be in oriented contact, if they are tangent to each other and their unit normals coincide at the point of contact. Laguerre geometry studies properties which are invariant under Laguerre transformations. If we consider a surface as an envelope of its oriented tangent planes, a dilatation, which is a Laguerre transformation that adds a constant $d \neq 0$ to the signed radius of each cycle without moving its center, maps the surface onto its offset at distance d.

11.5 General offsets

In 3-axis NC machining, not only ball-end cutters but also cylindrical and toroidal (fillet-end) cutters are used as shown in Fig. 11.31. While the center of a ball-end cutter moves along an offset surface, the reference point on cylindrical and toroidal cutters moves along the so-called *general offset*. General offset surfaces were first introduced by Brechner [44] and have been extended further, from the differential geometric as well as algebraic points of view, by Pottmann [325]. If we denote by $\mathbf{c}(u, v)$ the parametric representation of the cutter in the initial position, where the reference point on the axis of the cutter is chosen to be at the origin of the Cartesian coordinate system, then $-\mathbf{c}(u, v)$ represents a so-called *reflected cutter*. Then the general offset is given by

$$\hat{\mathbf{r}}_g(u, v) = \mathbf{r}(u, v) - \mathbf{c}(\mu, \nu) , \qquad (11.108)$$

and u, v, μ, ν are chosen such that there are parallel tangent plane at $\mathbf{r}(u, v)$ and $\mathbf{c}(\mu, \nu)$ [325]. As a consequence, the tangent planes at corresponding points $\mathbf{r}$ and $\hat{\mathbf{r}}_g$ of the progenitor surface and its general offset are parallel.

Thus, the general offset is the sum of the progenitor surface and the reflected cutter. If both surfaces are convex, the general offset is the Minkowski sum of the progenitor surface and the reflected cutter. The general offset surface for a cylindrical cutter is given by [325]

$$\hat{\mathbf{r}}_g(u,v) = \mathbf{r}(u,v) + d\frac{(\mathbf{e} \times \mathbf{n}(u,v)) \times \mathbf{e}}{|\mathbf{e} \times \mathbf{n}(u,v)|} \ , \tag{11.109}$$

where d is the radius of the cutter, $\mathbf{e}$ is a unit vector along the tool axis and $\frac{(\mathbf{e} \times \mathbf{n}(u,v)) \times \mathbf{e}}{|\mathbf{e} \times \mathbf{n}(u,v)|}$ is a unit vector parallel to the bottom silhouette line of the cutter. The general offset surface for a toroidal cutter it is given by [439, 355]

$$\hat{\mathbf{r}}_g(u,v) = \mathbf{r}(u,v) + c\mathbf{n}(u,v) + (d-c)\frac{(\mathbf{e} \times \mathbf{n}(u,v)) \times \mathbf{e}}{|\mathbf{e} \times \mathbf{n}(u,v)|} \ , \tag{11.110}$$

where d is the radius of the toroidal cutter, c is the corner radius of the cutter. The first two terms construct the classical offset with offset distance c at the cutter contact point and the third term is a vector parallel to the bottom silhouette line of the cutter with magnitude $d-c$, which is the radius of the spine circle of the torus.

Pottmann et al. [331] and Glaeser et al. [127] investigate collision-free 3-axis milling of free-form surfaces based on general offsets. They show that if some conditions on the curvature of the surface are fulfilled locally, and in certain cases also globally, there will be no unwanted collision of the cutting tool with the surface.

11.6 Pipe surfaces

11.6.1 Introduction

Pipe surfaces were first introduced by Monge [271] and are defined as follows: Given a space curve $\mathbf{c}(t)$ and a positive number r, the *pipe* surface with *spine* curve $\mathbf{c}(t)$ is defined to be the envelope of the set of spheres with radius r which are centered at $\mathbf{c}(t)$. Pipe surfaces can be considered as the *natural* generalization of the offset of a space curve in 3-D space. Pipe surfaces have many practical applications, such as in shape reconstruction [383], construction of blending surfaces [304, 113], transition surfaces between pipes [304], and in NC verification [438, 25]. They also have theoretical applications as well; for example, doCarmo uses them in the proof of two very important theorems in Differential Geometry concerning the total curvature of simple space curves, [76], pp. 399–402.

If we assume that the spine curve $\mathbf{c}(t)$ is regular, i.e. $\mathbf{c}(t)$ is simple and $|\dot{\mathbf{c}}(t)| \neq 0$, there exist two kinds of singularities on pipe surfaces: those that arise from local differential geometry properties of the surface and those that

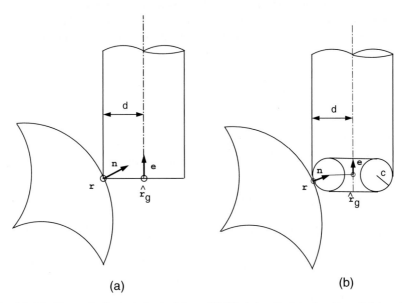

Fig. 11.31. General offsets (adapted from [250]): (**a**) cylindrical cutter, (**b**) toroidal cutter

come from global distance properties of the surface. The first type of singularity occurs when the radius of the pipe surface r exceeds the minimum radius of curvature of the spine curve; we refer to this singularity as *local self-intersection*. The second one happens, for example, when twice the radius r of the pipe surface is larger than the minimum distance between two interior points (excluding the two end points) on the spine curve; we refer to this singularity as *global self-intersection*. Kreyszig [206], doCarmo [76] and Rossignac [351] derive the condition for local self-intersection of a pipe surface and Shani and Ballard [383] describe a method to prevent local self-intersection of a generalized cylinder. So far the discussion was based on "Given a spine curve and a radius, does the pipe surface self-intersect? If so, where does it self-intersect?" However in some applications, we may encounter the case "Given a spine curve, what is the maximum radius such that the pipe surface does not self-intersect?" More precisely, given a regular space curve $\mathbf{c}(t)$ we want to find the maximum $R, R > 0$, so that the pipe surface $\mathbf{p}(r)$ is non-singular, whenever $r < R$. In [256] it is discussed how to find this maximum possible radius R.

One immediate application lies in the area of finding a topologically reliable approximation of a space curve [57]. More precisely, suppose we are given a regular space curve $\mathbf{c}(t)$, and would like to approximate $\mathbf{c}(t)$ with another curve $\mathbf{g}(t)$–within a prescribed tolerance– in a natural way; that is, there is a *space* homeomorphism $h : \mathbf{R}^3 \to \mathbf{R}^3$ that carries $\mathbf{c}(t)$ onto $\mathbf{g}(t)$ [256]. One important consequence of such a homeomorphism is that $\mathbf{c}(t)$ and $\mathbf{g}(t)$ have the same *knot* type. To do this, we first construct a nonsingular pipe surface $\mathbf{p}(r)$. Then, we construct a curve $\mathbf{g}(t)$ that lies inside $\mathbf{p}(r)$, and "looks like" $\mathbf{c}(t)$. By taking r to be the tolerance we have a reliable approximation of the given curve. Sakkalis and Charitos [359] apply the concepts of pipe surfaces and alpha shapes [84] to ambiently approximate a nonsingular space curve with a piecewise linear curve.

11.6.2 Local self-intersection of pipe surfaces

The pipe surface $\mathbf{p}(r)$ can be parametrized using the Frenet-Serret trihedron $(\mathbf{t}(t), \mathbf{n}(t), \mathbf{b}(t))$ [76, 351] as follows:

$$\mathbf{p}(t, \theta) = \mathbf{c}(t) + r[\cos\theta \mathbf{n}(t) + \sin\theta \mathbf{b}(t)] , \qquad (11.111)$$

where $t \in [0, 1]$ and $\theta \in [0, 2\pi]$. Its partial derivative with respect to t is given by

$$\mathbf{p}_t(t, \theta) = \dot{\mathbf{c}}(t) + r[\cos\theta \dot{\mathbf{n}}(t) + \sin\theta \dot{\mathbf{b}}(t)] . \qquad (11.112)$$

Equation (11.112) can be rewritten using the Frenet-Serret formulae (2.57) as

$$\mathbf{p}_t(t, \theta) = |\dot{\mathbf{c}}(t)|(1 - \kappa(t) r \cos\theta)\mathbf{t}(t) - r|\dot{\mathbf{c}}(t)|\tau(t) \sin\theta \mathbf{n}(t) + r|\dot{\mathbf{c}}(t)|\tau(t) \cos\theta \mathbf{b}(t) , \qquad (11.113)$$

where $\kappa(t)$ and $\tau(t)$ are the curvature and torsion of the spine curve given by (2.26) and (2.48), respectively. Similarly we can derive $\mathbf{p}_\theta$ as

$$\mathbf{p}_\theta(t, \theta) = r[-\sin\theta \mathbf{n}(t) + \cos\theta \mathbf{b}(t)] . \qquad (11.114)$$

The surface normal of the pipe surface can be obtained by taking the cross product of (11.113) and (11.114) yielding

$$\mathbf{p}_t \times \mathbf{p}_\theta = -|\dot{\mathbf{c}}(t)|r[1 - \kappa(t) r \cos\theta][\sin\theta \mathbf{b}(t) + \cos\theta \mathbf{n}(t)] . \qquad (11.115)$$

It is easy to observe [206, 76, 351] that the pipe surface becomes singular when $1 - \kappa(t) r \cos\theta = 0$. Since $\cos\theta$ varies between -1 and 1, there will be no local self-intersection if $\kappa(t)r < 1$. Therefore, to avoid local self-intersection we need to find the largest curvature κ_a of the spine curve and set the radius of the pipe surface such that $r < 1/\kappa_a$.

The curvature $\kappa(t)$ of a space curve $\mathbf{c}(t)$ is given in (2.26). Thus, to find the largest curvature κ_a we need to locate the critical points of $\kappa(t)$, i.e. solve the equation $\dot{\kappa}(t) = 0$ (8.20), and decide whether they are local maxima (see Sect. 7.3.1). Then we compare these local maxima with the curvature at the end points, i.e. $\kappa(0)$ and $\kappa(1)$, and obtain the largest curvature. This problem can be solved by elementary calculus. If the spine curve is given by a rational Bézier curve, equation $\dot{\kappa}(t) = 0$ reduces to a single univariate nonlinear polynomial equation (8.21) for a planar spine curve and (8.22) for a 3-D spine curve. In the case where the spine curve is a rational B-spline, we can extract the rational Bézier segments by knot insertion [175, 314].

Example 11.6.1. The parabola $y = x^2$ has its largest curvature $\kappa = 2$ at $x = 0$. Therefore in order to have no local self-intersection the radius should be $r < \frac{1}{2}$. Figure 11.32 shows the local self-intersection of the pipe surface with the above parabolic spine curve and with radius 0.8. Obviously, there is a local self-intersection on the pipe surface corresponding to the point $x = 0$ at the spine curve.

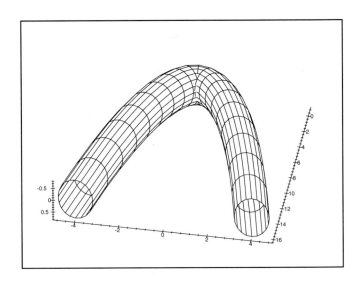

Fig. 11.32. Local self-intersection of a pipe surface ($r = 0.8$) (adapted from [256])

11.6.3 Global self-intersection of pipe surfaces

In this section we will consider how to find the maximum possible radius of a pipe surface such that it will not self-intersect in a global manner. Global self-intersection can be categorized into three types [256].

1. *End circle to end circle*: Two end circles of the pipe surface touch each other.
2. *Body to body*: Two different body portions of the pipe surface touch each other.
3. *End circle to body*: One of the end circles touches the body.

End circle to end circle global self-intersection. Let us consider the plane which contains the end point $\mathbf{c}(0)$ and is perpendicular to $\dot{\mathbf{c}}(0)$. If we denote a point on the plane as $\mathbf{x} = (x, y, z)^T$, then the equation of that plane becomes

$$[\mathbf{x} - \mathbf{c}(0)] \cdot \dot{\mathbf{c}}(0) = 0 \,. \tag{11.116}$$

Similarly the equation of the plane that contains the other end of the pipe is given by

$$[\mathbf{x} - \mathbf{c}(1)] \cdot \dot{\mathbf{c}}(1) = 0 \,. \tag{11.117}$$

The self-intersection occurs along the intersection of these two planes as shown in Fig. 11.33. It also lies on the bisecting plane of the line segment $\mathbf{c}(0)\mathbf{c}(1)$. Thus if $\mathbf{x}$ is a self-intersection point, then

$$\left(\mathbf{x} - \frac{\mathbf{c}(0) + \mathbf{c}(1)}{2} \right) \cdot (\mathbf{c}(1) - \mathbf{c}(0)) = 0 \,. \tag{11.118}$$

Equations (11.116), (11.117), (11.118) form a system of three linear equations with the three components of $\mathbf{x}$ as unknowns as follows:

$$\begin{pmatrix} \dot{c}^x(0) & \dot{c}^y(0) & \dot{c}^z(0) \\ \dot{c}^x(1) & \dot{c}^y(1) & \dot{c}^z(1) \\ c^x(1) - c^x(0) & c^y(1) - c^y(0) & c^z(1) - c^z(0) \end{pmatrix} \begin{pmatrix} x \\ y \\ z \end{pmatrix} = \begin{pmatrix} d_1 \\ d_2 \\ d_3 \end{pmatrix} \,, \tag{11.119}$$

where superscripts denote x, y, and z components, and

$$\begin{aligned}
d_1 &= c^x(0)\dot{c}^x(0) + c^y(0)\dot{c}^y(0) + c^z(0)\dot{c}^z(0) \,, \\
d_2 &= c^x(1)\dot{c}^x(1) + c^y(1)\dot{c}^y(1) + c^z(1)\dot{c}^z(1) \,, \\
d_3 &= \frac{(c^x(1))^2 + (c^y(1))^2 + (c^z(1))^2 - (c^x(0))^2 - (c^y(0))^2 - (c^z(0))^2}{2} \,.
\end{aligned} \tag{11.120}$$

The determinant of the matrix is readily computed as

$$D = \dot{\mathbf{c}}(0) \times \dot{\mathbf{c}}(1) \cdot (\mathbf{c}(1) - \mathbf{c}(0)) \,. \tag{11.121}$$

We now consider the following cases:

Case 1. $\mathbf{c}(1) \neq \mathbf{c}(0)$. In that case, if $D \neq 0$, then $r_{ee} = |\mathbf{x} - \mathbf{c}(0)|$, where $\mathbf{x}$ is the unique solution of the above system. If $D = 0$, and the system has no solution, we take $r_{ee} = \infty$. If the system has an infinte number of solutions, then we take $r_{ee} = \min |\mathbf{x} - \mathbf{c}(0)|$. This minimum is always positive since $|\mathbf{c}(1) - \mathbf{c}(0)| > 0$.

Case 2. $\mathbf{c}(1) = \mathbf{c}(0)$. In that case, if $\dot{\mathbf{c}}(0) \times \dot{\mathbf{c}}(1) \neq \mathbf{0}$, the pipe $\mathbf{p}(r)$ is always singular for every $r > 0$, and thus $r_{ee} = 0$. If $\dot{\mathbf{c}}(0) \times \dot{\mathbf{c}}(1) = \mathbf{0}$, we take $r_{ee} = \infty$.

Example 11.6.2. Figure 11.34 illustrates the case when end circles are touching each other. The control points of the spine curve, which is a cubic integral Bézier curve, are given by (2.9, 3.0, 4.1), (0.0, 1.0, 2.0), (5.0, -2.0, 1.0) and (3.0, 3.1, 4.0). The linear system (11.116), (11.117), (11.118) gives us the intersection point as (2.918, 3.055, 4.023) with radius $r = 0.0963$.

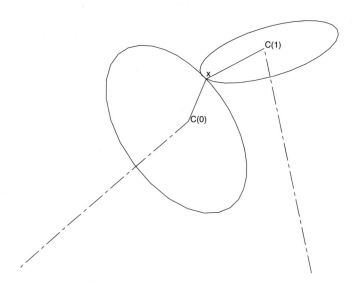

Fig. 11.33. Two end circles globally self-intersecting at point **x** (adapted from [256])

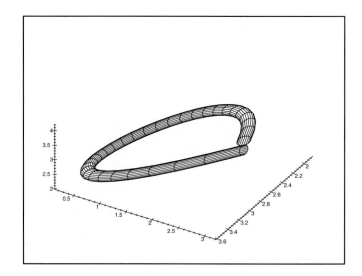

Fig. 11.34. End circle to end circle global self-intersection ($r=0.0963$) (adapted from [256])

Body to body global self-intersection. The body to body global self-intersection case can be reformulated in terms of two different points on the spine curve that have a minimum distance. This minimum distance should not be understood as the distance between two points whose parameters (σ and t) are close enough, which amounts to a distance approaching zero, i.e. the trivial solution. Rather, these two points should make the distance function stationary. Therefore, the body to body global self-intersection problem is equivalent to the minimum distance problem (see also [256]).

We assume the spine curve can be given by a rational B-spline curve, which can be split into rational Bézier curves by knot insertion [34, 314]. The minimum distance problem can be decomposed into the minimum distance between two points on different Bézier curves and the minimum distance between two points on the same Bézier curve. The first problem is discussed in Sect. 7.2 as well as in Zhou et al. [461], so we focus on the second problem here.

Let the spine curve be given by $\mathbf{c}(t) = [x(t), y(t), z(t)]^T$. Assume that the curve is regular, and $\dot{\mathbf{c}}(t)$ is continuous. The squared distance function between two points on the spine curve with parameters σ and t is given by [256]

$$D(\sigma, t) = |\mathbf{c}(\sigma) - \mathbf{c}(t)|^2 = [x(\sigma) - x(t)]^2 + [y(\sigma) - y(t)]^2 + [z(\sigma) - z(t)]^2 . \tag{11.122}$$

The stationary points of $D(\sigma, t)$ satisfy the following equations

$$D_\sigma(\sigma, t) = D_t(\sigma, t) = 0 , \tag{11.123}$$

which can be rewritten as

$$[\mathbf{c}(\sigma) - \mathbf{c}(t)] \cdot \dot{\mathbf{c}}(\sigma) = 0 , \tag{11.124}$$
$$[\mathbf{c}(\sigma) - \mathbf{c}(t)] \cdot \dot{\mathbf{c}}(t) = 0 . \tag{11.125}$$

The geometrical interpretation of (11.124) and (11.125) is that the line connecting the two points $\mathbf{c}(\sigma)$ and $\mathbf{c}(t)$ is *orthogonal* to the spine curve at both points. We assume that $\mathbf{c}(t)$ is given as a rational Bézier curve, that is

$$\mathbf{c}(t) = \frac{\sum_{i=0}^n w_i \mathbf{b}_i B_{i,n}(t)}{\sum_{i=0}^n w_i B_{i,n}(t)} \equiv \frac{\mathbf{R}(t)}{W(t)} . \tag{11.126}$$

Substituting (11.126) into (11.124) gives

$$\left[\frac{\mathbf{R}(\sigma)}{W(\sigma)} - \frac{\mathbf{R}(t)}{W(t)} \right] \cdot \left[\frac{\dot{\mathbf{R}}(\sigma)W(\sigma) - \mathbf{R}(\sigma)\dot{W}(\sigma)}{W^2(\sigma)} \right] = 0 . \tag{11.127}$$

Multiplying by its own denominator we obtain

$$[\mathbf{R}(\sigma)W(t) - \mathbf{R}(t)W(\sigma)] \cdot \left[\dot{\mathbf{R}}(\sigma)W(\sigma) - \mathbf{R}(\sigma)\dot{W}(\sigma) \right] = 0 . \tag{11.128}$$

Similarly (11.125) reduces to

$$[\mathbf{R}(\sigma)W(t) - \mathbf{R}(t)W(\sigma)] \cdot \left[\dot{\mathbf{R}}(t)W(t) - \mathbf{R}(t)\dot{W}(t)\right] = 0 . \qquad (11.129)$$

The first brackets of (11.128) and (11.129) can be rewritten as

$$\sum_{i=0}^{n}\sum_{j=0}^{n} w_i w_j \mathbf{b}_i [B_{i,n}(\sigma)B_{j,n}(t) - B_{j,n}(\sigma)B_{i,n}(t)] . \qquad (11.130)$$

Since

$$\frac{B_{i,n}(\sigma)B_{j,n}(t) - B_{j,n}(\sigma)B_{i,n}(t)}{\sigma - t} \qquad (11.131)$$

$$= B_{j,n}(t)\frac{B_{i,n}(\sigma) - B_{i,n}(t)}{\sigma - t} - B_{i,n}(t)\frac{B_{j,n}(\sigma) - B_{j,n}(t)}{\sigma - t} ,$$

we can easily factor out $(\sigma - t)$ from the first brackets of (11.128) and (11.129).
    Therefore the system of equations (11.124), (11.125) for the rational Bézier curve reduces to a system of coupled bivariate polynomial equations with degree $(3n - 2)$ in σ, $(2n - 1)$ in t and degree $(2n - 1)$ in σ, $(3n - 2)$ in t. The system can be robustly and efficiently solved by the IPP algorithm introduced in Chap. 4. If we substitute all the solutions computed by the polynomial solver into (11.122) and choose the minimum squared distance, then the maximum possible upper limit of the radius r_{bb} such that body and body of the pipe surface will not globally self-intersect is given by $r_{bb} = \sqrt{\min_I D(\sigma, t)}/2$, where $I = [0, 1] \times [0, 1]$ and $D(\sigma, t)$ is the squared distance between two points which make the distance function stationary. If there are no such points, then we set $r_{bb} = +\infty$.

Example 11.6.3. Figures 11.35 show two different views of the minimum distance between points on a rational Bézier curve of degree 4. The solid squares indicate the five control points $(-0.3, 0.8, 0.1)$, $(0.3, 0.15, -0.45)$, $(0, 0, 0.2)$, $(-0.2, 0.1, 0.8)$, $(0.3, 0.8, -0.6)$ with weights 1, 2, 0.5, 3, 1. The minimum distance between two points on the spine curve can be obtained as 0.157556, which is between the points of the parameters $(t = 0.102506)$ and $(\sigma = 0.952132)$. The spine curve has a global maximum curvature at $t = 0.70618$ with curvature value $\kappa = 48.7601$. Therefore the pipe surface starts to self-intersect locally when $r = 0.0205$ and globally when $r = 0.078778$. The situation when $r = 0.078778$ is shown in Fig. 11.36 where two different parts of the body of the pipe surface touch each other and the surface also locally self-intersects.

End circle to body global self-intersection. Finally we consider the case of end circle to body global self-intersection. This case can be considered as a special case of body to body global self-intersection. We can substitute $\sigma = 1$ into (11.125) which gives [256]

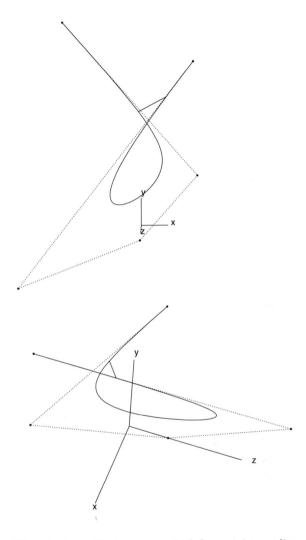

Fig. 11.35. Two different views of spine curve which has minimum distance between two interior points (adapted from [256])

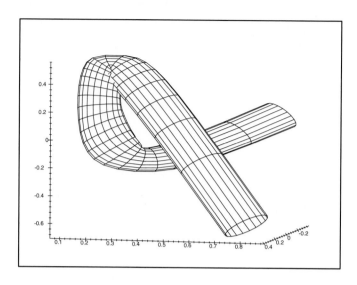

Fig. 11.36. Body to body tangential intersection and local self-intersection ($r = 0.078778$) (adapted from [256])

$$[\mathbf{c}(1) - \mathbf{c}(t)] \cdot \dot{\mathbf{c}}(t) = 0 \ . \tag{11.132}$$

If the spine curve is a rational Bézier curve, (11.132) will become a univariate polynomial equation. This equation contains the trivial solution $t = 1$ and therefore $t - 1$ should be factored out. Similarly, we can substitute $\sigma = 0$ into (11.125) and factor out t. Notice that the line connecting $\mathbf{c}(t)$ and the end point $\mathbf{c}(1)$ is orthogonal to the spine curve at $\mathbf{c}(t)$ but not necessarily orthogonal at $\mathbf{c}(1)$. Therefore, with the radius equal to half the distance between $\mathbf{c}(t)$ and $\mathbf{c}(1)$, the pipe surface may not self-intersect. In the limiting case of tangential self-intersection, at the intersection point, using the parametrization of (11.111), the following equations hold:

$$\mathbf{p}(1, \theta) = \mathbf{p}(t, \phi) \ , \tag{11.133}$$

$$\mathbf{p}_\theta(1, \theta) \cdot [\mathbf{p}_t(t, \phi) \times \mathbf{p}_\phi(t, \phi)] = 0 \ . \tag{11.134}$$

Equation (11.134) comes from the fact that the end circle tangentially self-intersects to the body (see Fig. 11.37). This system consists of four scalar equations with four unknowns, namely r, t, θ and ϕ. We can also form the four scalar equations in terms of polynomials using the rational parametrization of the pipe surface [256]. However we cannot factor out the trivial solution from the system. Maekawa et al. [253] developed a method to handle such a case (see Sect. 11.3.5). But in this specific case we do not need to use this, as we can easily solve the system using Newton's method, since there is only one solution and we can provide a very accurate initial approximation as follows: We consider a circle at $t = t_m$, i.e. $\mathbf{p}(t_m, \theta)$ using the solution of (11.132) as t_m. By considering this circle as one of the end circles, we can use the end circle to end circle global self-intersection technique, that we just introduced, to find the intersection point between the two end circles. From this intersection point we can evaluate the radius r and the two angles θ and ϕ for the initial values, using coordinate transformations. In case when the spine curve is planar, we cannot solve the linear system, since it becomes singular. In such case we will use the solution of (11.132) as t and half the distance between $\mathbf{c}(t)$ and $\mathbf{c}(1)$ (or $\mathbf{c}(0)$) as r, and θ and ϕ as 0 or π as initial approximation. Let us now denote the resulting radius from Newton's method by r_{eb}.

Example 11.6.4. The 3-D quartic spine curve with control points (-0.3, 0.8, 0.1), (0.24, 0.15, -0.45), (0,0,0.2), (-0.24, 0.12, 0.96) and (-2, 0.6, 0) and weights 1, 2, 0.5, 2.5, 1 respectively, has minimum distance 0.0595918 between two points $t = 0.0370295$ and $t = 1$. However with $r = 0.0595918/2 = 0.0297959$, the pipe surface does not self-intersect, since the vector $\mathbf{c}(1) - \mathbf{c}(0.0370295)$ is not orthogonal to the spine curve at $t = 1$. Using Newton's method we obtain the touching radius as $r = 0.041829$. The spine curve also has a global maximum curvature at $t = 0.761006$ with $\kappa{=}31.272916$. Therefore the pipe surface starts to self-intersect locally when $r{=}0.031977$ and globally when $r{=} 0.041829$. Figure 11.37 shows the pipe surface with $r{=} 0.041829$.

A necessary and sufficient condition for nonsingularity. Using the methods of the previous sections we now present a necessary and sufficient condition, in terms of the radius r, for the nonsingularity of a pipe surface. We assume that the spine curve is given by $\mathbf{c}(t) = [x(t), y(t), z(t)]^T$, $0 \le t \le 1$, and that the curve is regular, and $\dot{\mathbf{c}}(t)$ is continuous.

Let κ_a be the maximum curvature of the spine curve, and r_{ee}, r_{bb}, r_{eb} be the maximum possible upper limit radius of the pipe surface such that it does not globally self-intersect between end circle to end circle, body to body and end circle to body of the pipe surface, respectively. Then we have [256]:

Theorem 11.6.1. *Let $\mathbf{p}(r)$ be the pipe surface with spine curve $\mathbf{c}(t)$ and radius r. Then, $\mathbf{p}(r)$ is nonsingular if and only if $r < \delta = \min\{1/\kappa_a, r_{ee}, r_{bb}, r_{eb}\}$.*

Proof (if): It is apparent from the discussion in Sects. 11.6.2 and 11.6.3 that if $r < \delta$ then the pipe surface $\mathbf{p}(r)$ is nonsingular.

(only if): Suppose now that $\mathbf{p}(r)$ is nonsingular. It is enough to show that for all $r \ge \delta$, $\mathbf{p}(r)$ is singular. But this is obvious since if r is as indicated, the pipe surface will either have a singularity due to local self-intersection or one due to global self-intersection, or both. ∎

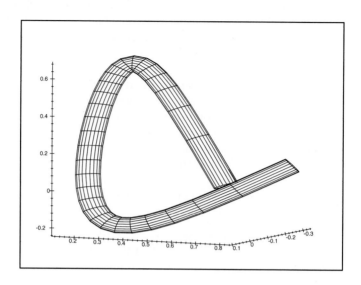

Fig. 11.37. End circle tangentially intersecting the body and the local self-intersection is occurring at $t= 0.761$ ($r = 0.078778$) (adapted from [256])

Remark 11.6.1. When the spine curve is planar, Theorem 11.6.1 can be used to find the maximum offset distance such that the offset of the planar spine curve will not self-intersect.

Example 11.6.5. (2-D spine curve) The quartic spine curve with control points (-0.3, 0.8, 0), (0.6, 0.3, 0), (0,0,0), (-0.3, 0.2, 0) and (-0.15, 0.6, 0) and weights 1, 1, 2, 3, 1 respectively, has minimum distance 0.0777421 between two points $t = 0.0658996$ and $t = 1$. By using Newton's method we obtain the touching radius as $r = 0.055754$. This distance is the maximum offset distance such that the offset of the planar spine curve will not self-interset. Figure 11.38 shows the offset curves when $r = 0.055754$.

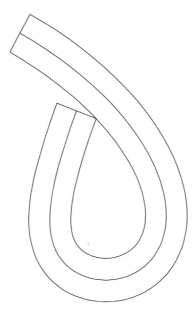

Fig. 11.38. Offset curves of the planar spine curve ($r = 0.055754$) (adapted from [256])

Problems

1. Consider an implicit surface $f(x, y, z) = 0$ where f is a polynomial in x, y, z. Consider the cube $[0, 1]^3$ and the part of the surface inside this cube. The surface can be written in the Bernstein basis as

$$f(x, y, z) = \sum_{i=0}^{n} \sum_{j=0}^{m} \sum_{k=0}^{q} w_{ijk} B_{i,n}(x) B_{j,m}(y) B_{k,q}(z) = 0 .$$

Show the following properties.
   a) The point $(0, 0, 0)$ is on the surface if and only if $w_{000} = 0$. What happens when $w_{00k} = 0$ for all k?
   b) Assuming the condition of question (a) is true, a necessary and sufficient condition for the normal vector of surface f at $(0, 0, 0)$ to be parallel to axis z is that $w_{100} = w_{010} = 0$.
   c) If $w_{ijk} > 0$ or if $w_{ijk} < 0$ for all i, j, k then there is no piece of the surface in the cube under consideration.
   d) Consider a cube $[0, 1]^2 \times [1, 2]$ adjacent to the cube $[0, 1]^3$. Within the new cube define another implicit polynomial surface $g(x, y, z) = 0$ of the same degrees in x, y, z as $f(x, y, z)$. Determine the conditions for the two surfaces to be position continuous at the common face of the two cubes.
   e) Following the condition of question (d), determine the conditions for the two surfaces to be tangent plane continuous at the common face of the two cubes.

2. Show that the derivative of a Bézier curve (also called hodograph) of the form:

$$\mathbf{r}(t) = \sum_{i=0}^{n} \mathbf{b}_i B_{i,n}(t), \quad 0 \le t \le 1 ,$$

is given by:

$$\dot{\mathbf{r}}(t) = \sum_{i=0}^{n-1} n(\mathbf{b}_{i+1} - \mathbf{b}_i) B_{i,n-1}(t), \quad 0 \le t \le 1 .$$

Sketch a cubic Bézier curve and its hodograph and their control polygons.

3. The degree elevation formula for Bézier curves of degree n is given (1.54). Describe a process for approximating a Bézier curve of degree n with a Bézier curve of degree $n-1$ using (1.54) reversely.

4. Show how an explicit polynomial curve $y = y(x)$, where $a \leq x \leq b$ can be converted into a Bézier curve. Provide the control points of the resulting Bézier curve. And show how an explicit polynomial surface $z = f(x,y)$, where $a \leq x \leq b$ and $c \leq y \leq d$ can be converted into a Bézier patch and provide its control points. Extend this to an explicit B-spline patch and provide its control points.

5. Given a planar B-spline curve in the xy plane with a non-uniform knot vector, the control polygon of which is symmetric with respect to the y-axis, find if the curve is also symmetric about the y-axis.

6. What kind of curve is the result of a perspective projection of an integral B-spline curve?

7. Consider the arc of the hyperbola $y = \frac{1}{x}$ for $1 \leq x \leq 2$ and revolve it around the axis $y = 0$ by $\frac{\pi}{2}$, to obtain the quadrant of a surface of revolution, within the first octant of the xyz coordinate system $(x \geq 0, y \geq 0, z \geq 0)$. Express the resulting patch in terms of a rational biquadratic parametric Bézier surface patch.

8. Examine what happens to a cubic B-spline curve in which two, three, or four consecutive vertices, of its control polygon are coincident.

9. Examine what might happen when a rational B-spline curve given by (1.87) or a rational Bézier curve given by (1.88) has weights some of which are positive and some are negative. Examine the validity of the properties of the rational B-spline or rational Bézier curves in such cases.

10. Make plots of the B-spline basis functions of the following order n (degree $= n - 1$) and knot vector $\mathbf{T}$:
   - $n = 4,$ $\mathbf{T} = (0, 0, 0, 0, 1, 1, 1, 1)$
   - $n = 4,$ $\mathbf{T} = (0, 0, 0, 0, 1, 3, 3, 3, 3)$
   - $n = 4,$ $\mathbf{T} = (0, 0, 0, 0, 2, 2, 4, 4, 4, 4)$
   - $n = 4,$ $\mathbf{T} = (0, 0, 0, 0, 1, 1, 1, 3, 3, 3, 3)$
   - $n = 4,$ $\mathbf{T} = (0, 0, 0, 0, 1, 2, 3, 4, 6, 7, 7, 7, 7)$
   - $n = 3,$ $\mathbf{T} = (0, 0, 0, 2, 4, 6, 6, 6)$
   - $n = 2,$ $\mathbf{T} = (0, 0, 1, 2, 2)$

11. Given a list of Cartesian points in 3-D space which represent a non-periodic curve, construct a cubic Bézier curve using least squares approximation of the points. Also, construct a cubic B-spline curve with non-uniform knots using least squares approximation of these points. A user should be able to access your program with an arbitrary number of points and coordinates of points. Include a simple visualization of the results.

12. A cubic planar Bézier curve:

$$\mathbf{r}(t) = \sum_{i=0}^{3} \mathbf{b}_i B_{i,3}(t) \quad 0 \leq t \leq 1,$$

has the following control points

$$\mathbf{r}_0 = (0,0)^T, \mathbf{r}_1 = (1,1)^T, \mathbf{r}_2 = (2,1)^T, \mathbf{r}_3 = (2,0)^T.$$

A designer decides to subdivide (split) the curve at $t_0 = \frac{1}{2}$ and $t_1 = \frac{3}{4}$ in order to be able to modify the curve in the interval $[t_0, t_1]$ and generate a particular shape feature required by his design.

a) Compute the coordinates of the control points of the three curve segments generated by the above subdivision.

b) After the above subdivision, the middle segment of the three curve segments created by subdivision is permitted to be modified by changing the coordinates of its control points. Determine the conditions that the control points must obey so that this curve segment maintains position continuity at its ends with the end and the beginning of the other two curve segments.

c) Assuming position continuity is maintained as in (b), determine the additional necessary conditions to maintain unit tangent vector continuity at the ends of the middle curve segment.

d) Assuming conditions (b) and (c) are satisfied, determine the coordinates of the vertices of the polygonal boundary of the convex hull of the middle curve segment. You need to distinguish several special cases.

13. Derive the monomial or power basis form of curve $\mathbf{r}(t)$ of Problem 12 prior to any subdivision.

14. Derive the (uniform) B-spline form of curve $\mathbf{r}(t)$ of Problem 12 prior to any subdivision. Make a plot of the curve together with its Bézier and B-spline polygon illustrating the principal features of the curve (such as tangencies at the ends etc.). Compare the convex hulls of the curve in the Bézier and the B-spline form in terms of the area they enclose (i.e. determine the ratio of the two areas).

15. We are given a degree (2-1) integral Bézier surface patch

$$\mathbf{r}(u,v) = \sum_{i=0}^{2}\sum_{j=0}^{1} \mathbf{b}_{ij} B_{i,2}(u)B_{j,1}(v), \quad 0 \le u, v \le 1,$$

where the control points $\mathbf{b}_{ij}$ are

$$\mathbf{b}_{00} = (0,0,0)^T, \quad \mathbf{b}_{01} = (-2,0,10)^T,$$
$$\mathbf{b}_{10} = (5,10,5)^T, \quad \mathbf{b}_{11} = (4,12,16)^T,$$
$$\mathbf{b}_{20} = (20,0,0)^T, \quad \mathbf{b}_{21} = (22,0,10)^T.$$

a) Subdivide the surface patch into two patches by the iso-parametric curve $u = 0.5$ and compute the resulting control points of the two patches.

b) Consider the boundary curve of the patch $\mathbf{q}(u) = \mathbf{r}(u,0)$. Provide a tight upper bound for the maximum deviation of the curve $\mathbf{q}(u)$ in the interval $\frac{k}{n} \le u \le \frac{k+1}{n}$ to the straight line passing via $\mathbf{q}(\frac{k}{n})$ and $\mathbf{q}(\frac{k+1}{n})$ for a fixed value of $k(\in 0,1,\cdots,n-1)$, where n is a fixed positive integer.

c) i. Show that the given integral Bézier surface is a developable surface.

   ii. Are there any umbilics on this patch?

16. Consider a curve $u = t$, $v = t^2$ for $0 \le t \le 1$ on a hyperbolic paraboloid $\mathbf{r}(u,v) = (u,v,uv)^T$ where $0 \le u, v \le 1$.

a) Compute the arc length of the curve on the hyperbolic paraboloid for $0 \le t \le 1$.

b) Compute the area of a region of the hyperbolic paraboloid bounded by positive v axis, $v = 1$ and a parabola $v = u^2$.

17. Consider a torus parametrized as follows:

$$\mathbf{r}(u,v) = ((R + a\ cos\ u)cos\ v, (R + a\ cos\ u)sin\ v, a\ sin\ u)^T ,$$

where $0 \le u \le 2\pi, 0 \le v \le 2\pi$ and $R > a$. Derive appropriate formulae for the Gauss, mean and principal curvatures. Sketch the torus and subdivide it into hyperbolic, parabolic and elliptic regions. In a follow-up sketch illustrate the lines of curvature of the torus. Explain the above subdivision and sketches.

18. Show that the curvature of a planar curve is independent of the parametrization. Namely, if

$$\mathbf{r}(t) = (x(t), y(t))^T ,$$

is the curve, then a change of variables

$$t = w(u)\ with\ \dot{w}(u) \ne 0 ,$$

does not affect the curvature.

19. Write a one-dimensional nonlinear polynomial solver based on Projected Polyhedron algorithm. Use the solver to compute the roots of the degree 20 Wilkinson polynomial with different tolerances and discuss robustness issues.

20. Convert an explicit curve $y = x^3$ $(-a \le x \le a)$ into a cubic Bézier curve.

21. Convert the following height function

$$z = h(x,y) = 7.2x^3y^3 - 25.2x^3y^2 + 18x^3y - 10.8x^2y^3 + 37.8x^2y^2 - 27x^2y$$
$$+ 3.6xy^3 - 12.6xy^2 + 9xy, \quad 0 \le x, y \le 1 ,$$

into a bicubic Bézier patch.

22. Compute the characteristic points of the following curve

$$f(u,v) = (x^2 + y^2 - 2x)^2 - (x^2 + y^2) = 0, \quad [x,y] \in [-4,4]^2 ,$$

and trace it.

23. Consider the intersection curve of (11.135) with the plane $15x - 55z + 110 = 0$.

   a) Derive an implicit equation $f(u, v) = 0$ for this intersection curve in the parameter space u, v. Find the characteristic points of this curve, (border, turning, and singular points).

   b) Express this intersection curve as an explicit curve in the u, v parameter space. Indicate the resulting type and degree of this curve. Sketch this curve in the parameter space u, v.

   c) Prove that the above intersection curve is a planar rational Bézier parametric curve of degree 4 in 3D space. Indicate how you would compute its control points (but do not carry out the algebra).

24. Consider the following curves:

$$f(x, y) = - 64y^4 + 128y^3 - 96x^2y^2 + 140xy^2 - 139y^2 + 96x^2y - 140xy + 75y - 96x^4 + 276x^3 - 313x^2 + 165x - 36 = 0 ,$$

and

$$\mathbf{r}(t) = (x(t), y(t))^T = \sum_{i=0}^{3} \mathbf{b}_i B_{i,4}(t) ,$$

where $B_{i,4}(t)$ denotes the ith cubic Bernstein polynomial and $\mathbf{r}_0 = (0.5, 0.5)^T$, $\mathbf{b}_1 = (0.7, 0.6)^T$, $\mathbf{b}_2 = (0.95, 0.1)^T$, $\mathbf{b}_3 = (0.55, 0.25)^T$.

   a) Compute all turning and singular points of $f(x, y)$ to the highest possible accuracy, as well as the tangent lines at all these points.

   b) Using the results of **a** as a guide, sketch $f(x, y)$. Clearly indicate the turning and singular points on your sketch.

   c) Compute the intersections of the two curves given above to the highest possible accuracy. In addition to giving the Cartesian coordinates of the intersection points, also include the parameter values of the points and their multiplicity.

25. Write a program which determines all intersections of two integral planar Bézier curves of arbitrary degrees m and n as accurately as possible, given the control points of the two curves. Your program should report the parametric values of the intersection points as well as the Cartesian coordinates. Give four examples to show how your program works.

26. The following three planar curves are given by:

   1) Implicit curve, $f(x, y) = x^3 + y^3 - 3xy = 0$,
   2) Cubic Bézier curve $\mathbf{r}(t) = (x(t), y(t))^T = \sum_{i=0}^{3} \mathbf{r}_i B_{i,4}(t)$ where $0 \leq t \leq 1$ and with $\mathbf{r}_0 = (0, 0)^T$, $\mathbf{r}_1 = (0, 2)^T$, $\mathbf{r}_2 = (2, 0)^T$, $\mathbf{r}_3 = (0, -2)^T$,
   3) Cubic Bézier curve $\mathbf{q}(u) = (x(t), y(t))^T = \sum_{i=0}^{3} \mathbf{q}_i B_{i,4}(u)$ where $0 \leq t \leq 1$ and with $\mathbf{q}_0 = (-2, -2)^T$, $\mathbf{q}_1 = (-2, 1)^T$, $\mathbf{q}_2 = (4, 1)^T$, $\mathbf{q}_3 = (0, -1)^T$.

a) Compute the characteristic points of the first curve in the rectangle $[-5,5] \times [-5,5]$ and trace it within the same rectangle.

b) Compute the intersections of the first and second curves, and the second the third curves to the highest possible accuracy, and identify their multiplicity.

c) Obtain a parametrization of the first curve in terms of rational polynomials using the transform $y = xt$. Illustrate $x(t)$ and $y(t)$ for all real t. Is this a good parametrization for computer implementation (e.g. tracing of the curve) near $x = y = 0$? Can you suggest better parametrizations for the curve piece in the first quadrant.

27. Compute the intersection curve between the bicubic Bézier patch of Problem 21 and a plane $x - y + z = 0$. Evaluate the curvature of the intersection curve at $u = v = 0.5$.

28. Give the implicit polynomial equation of a torus whose section circle has radius 2, and whose center circle has radius 4 using the implicitization of a surface of revolution. Assume the torus is situated so that it is centered at the origin and the center circle lies entirely in the (x, y)-plane. Using the implicit equation, compute all intersections of the torus with the cubic Bézier curve having control points $\mathbf{r}_0 = (0, 6, 0)^T$, $\mathbf{r}_1 = (-5, 2, -0.5)^T$, $\mathbf{r}_2 = (2, -3, 0.5)^T$, $\mathbf{r}_3 = (6, 0, 0)^T$. Give both the Cartesian coordinates of the intersection points and their associated parameter values on the Bézier curve. Indicate which method you used to solve this problem, and give all answers to at least 5 significant digits.

29. Compute the minimum distance between a point $(0.8, 0.7, 0.2)$ and an iso-parametric line $v = 0.8$ of the bicubic Bézier patch of Problem 21. Also compute the minimum distance between the point and the bicubic patch.

30. Consider two planar Bézier curves which are cubic and quadric with control points: $(0,0)$, $(1, 1)$, $(2,1)$, $(3,0)$ and $(0,1)$, $(\frac{3}{4}, -1)$, $(\frac{6}{4}, 5)$, $(\frac{9}{4}, -1)$, $(3,0)$, respectively. Compute all stationary points of their squared distance function and classify them appropriately into extrema etc. Identify the corresponding Euclidean distances, find the points of intersection of the two curves and the angles between the tangents of the two curves at the intersection points.

31. Consider an ellipsoid of revolution given by (3.81) with $a=b=1$, $c=2$ and a cubic planar Bézier curve with control points $(0,1, -2)$, $(0,0, -1)$, $(0,0,1)$, $(0,1,2)$ on the $x=0$ plane. Compute the stationary points of the squared distance function between the ellipsoid and the curve, classify them into extrema etc. Identify the corresponding Euclidean distances, find the points of intersection and the angles between the surface normals and the Bézier curve tangents at the intersection points.

32. Find the stationary points of the squared distance function between the plane $z=0$ and the wave-like Bézier surface patch of the example in Sect. 8.5.4 and Fig. 8.11. Classify the points into extrema etc., identify the

corresponding distances, and determine the intersections of the two surfaces. Compare the locations of the above extrema with the locations of the various curvature extrema in Sect. 8.5.4.

33. Consider a torus generated by revolving the circle $(x-2)^2 + y^2 - 1 = 0$ around the y axis by a full revolution. Determine the stationary points of the squared distance function between this torus and a) a plane $x = -3$, b) a plane $y = 2$, c) a plane $y = 1$, d) a sphere with center the origin and radius $r = \frac{1}{2}$ and e) a sphere with center the origin and radius $r = 4$.

34. Consider a biquadratic Bezier surface patch whose boundary eight control points are coplanar so that the boundary curves form a square $[0,1]^2$ on the xy plane. The boundary non-corner control points are in the middle of the corresponding boundary edges. The central control point of the patch has coordinates $(\frac{1}{2}, \frac{1}{2}, h)$ where $h = 0$. Determine the surface unit normal vector at the four corners, and at the center, and the extrema of the Gauss, mean and principal curvatures and any umbilics as a function of h and illustrate this for $h = \frac{1}{10}, 1, 10, 100$. Sketch the lines of curvature of the surface patch for these four values of h.

35. Find the range of mean curvature of a hyperbolic paraboloid $\mathbf{r}(u,v) = (u, v, uv)^T$, $(u,v) \in [0,1]^2$ (bilinear surface), and plot four levels of contour lines of mean curvature in the uv-parameter space.

36. Given an implicit surface $f(x,y,z) = 0$, formulate the problem of tracing the lines of curvature and develop an algorithm to do this. Test the resulting implementation for various standard algebraic surfaces (quadrics, torii, cyclides).

37. Given an implicit algebraic surface $f(x,y,z) = 0$, formulate the problem of locating the umbilics of the surface (within a given rectangular box with faces parallel to the coordinate planes).

38. Consider an ellipsoid $\frac{x^2}{a^2} + \frac{y^2}{b^2} + \frac{z^2}{c^2} = 1$ where $a \le b \le c$.

    a) Show that umbilics are located at $\left(\pm a\sqrt{\frac{b^2-a^2}{c^2-a^2}},\ 0,\ \pm c\sqrt{\frac{c^2-b^2}{c^2-a^2}}\right)$.

    b) Show that the patterns of the four umbilics are of the lemon type.

39. Consider a degree (3-1) integral Bézier surface

$$\mathbf{r}(u,v) = \sum_{i=0}^{3}\sum_{j=0}^{1} \mathbf{b}_{ij} B_{i,3}(u) B_{j,1}(v), \qquad 0 \le u,v \le 1,$$

    where

    $\mathbf{b}_{00} = (0,0,0)^T,$ $\mathbf{b}_{01} = (0.5,0,2)^T,$
    $\mathbf{b}_{10} = (1.8,3,0)^T,$ $\mathbf{b}_{11} = (1.895,2.325,2)^T,$
    $\mathbf{b}_{20} = (3.3,-2,1.5)^T,$ $\mathbf{b}_{21} = (3.0575,-1.55,3.1625)^T,$
    $\mathbf{b}_{30} = (4,0,0)^T,$ $\mathbf{b}_{31} = (3.6,0,2)^T.$

    a) Show that the Bézier surface is a developable surface.

    b) Is there an inflection line? If so, find the u parameter which contains the inflection.

40. Derive differential equations for geodesics (10.17) - (10.20) on a parametric surface using Euler's equation (10.24).

41. Write a program which solves differential equations for geodesics (10.17) - (10.20) as a boundary value problem using a shooting method on a parametric surface.

42. For the surface patch of Problem 34 compute the geodesics between two diagonally opposite corners for various values of h. How do these geodesics change as h changes from 0 to large positive values, e.g. in the interval [0,100]. What do you expect in the limit h tends to plus infinity?

43. Let $\mathbf{r}(s)$ be a planar, closed and convex curve (e.g. a circle, an ellipse, etc.) where the arc length s varies in the range $[0, l]$ so that the length of the curve is l. Let

$$\hat{\mathbf{r}}(s) = \mathbf{r}(s) + d\mathbf{n}(s) \, ,$$

be its offset curve, where d is a positive distance and $\mathbf{n}(s)$ is the unit normal vector of the curve $\mathbf{r}$ defined by $\mathbf{t} \times \mathbf{e}_z$ (see convention (**b**) of (2.24) in Table 3.2.)

a) Show that the total length of the curve $\hat{\mathbf{r}}(s)$ exceeds the total length of the curve $\mathbf{r}(s)$ by $2\pi d$.

b) Show that the area enclosed between the two curves is given by $A = d(l + \pi d)$.

c) Show that the curvatures of the two curves are related by

$$\hat{\kappa} = \frac{\kappa}{1 + d\kappa} \, ,$$

where κ is the curvature of $\mathbf{r}(s)$ and $\hat{\kappa}$ is the curvature of the offset curve $\hat{\mathbf{r}}(s)$.

d) Verify your results for questions a to c for a circle of radius R.

44. This problem focuses on the identification of cusps, extraordinary points and self-intersections of offsets of planar curves (use convention (**b**) of (2.24) for the normal vector in Table 3.2.). Consider the ellipse $x^2 + 4y^2 = 1$ or $x = \cos\theta, y = \frac{1}{2}\sin\theta$ and its offset at "distance" d, where d is any real number.

a) Determine all the values of θ for which there can be an extraordinary point on some offset of the ellipse and the values of d at such points. Sketch the offsets at all such values of d.

b) For what range of values of d, are offsets of the ellipse regular curves? Sketch a few such offset curves.

c) Determine a specific offset of the ellipse which includes several cusps and self-intersections but no extraordinary points. Infinite such cases exist. Give the parameter values and coordinates of these cusps and self-intersections. (Hint: Notice that self-intersections are on the axes of symmetry of the ellipse.)

45. Consider the planar cubic integral Bezier curve $\mathbf{r}(t)$ with control points $(0,0), (0,1), (2,1)$, and $(2,-1)$. The offset $\hat{\mathbf{r}}$ of $\mathbf{r}$ at a distance d is given by

$$\hat{\mathbf{r}}(t) = \mathbf{r}(t) + d\mathbf{n}(t) ,$$

where $\mathbf{n}(t)$ denotes the normal to $\mathbf{r}$ at the point $\mathbf{r}(t)$ defined by $\mathbf{t} \times \mathbf{e}_z$ (see convention (**b**) of (2.24) in Table 3.2). Here $\mathbf{r}$ is called the progenitor of $\hat{\mathbf{r}}$.

   a) For values of d between 0 and some critical value d_c, the offset curve resembles its progenitor. At $d = d_c$, however, the offset exhibits a cusp at a parameter value t_c because $\dot{\hat{\mathbf{r}}}(t_c) = \mathbf{0}$. Compute the values of d_c and t_c.

   b) Sketch the progenitor curve, and two offset curves, one at a distance of d_c and one at a distance of around $2d_c$.

46. The evolute of a planar curve is the curve of its center of curvature. Show that cusps and extraordinary points of the offset lie on the evolute. Illustrate the concept by examining the superbola $x = t$, $y = t^4$, $-2 \leq t \leq 2$. Draw the curve, its evolute, and several offsets with offset distance, $d = -0.25$, $d \simeq 0.4648$, $d = -0.8$, $d = -1$, $d = -1.25$ (all on the concave side).

47. Consider a pocket to be machined, bounded by the following four curves:
   1) $x = t$, $y = t^4$, $-2 \leq t \leq 2$,
   2) $y = 16$,
   3) Two circular blends of the first and second curves with radius, $r = 0.25$.

   a) Construct an approximation of the medial axis (skeleton) of the pocket in terms of a set of linear spline curves. The skeleton is the set of points inside the shape with two or more nearest points on the boundary of the shape. The skeleton branches potentially start at the curvature centers corresponding to points of maximum curvature of the boundary. Next, you may specify the skeleton by writing differential equations relating the tangent vector of the skeleton to known functions. For simplicity, write these equations for the specific example. Integrate these differential equations using the Runge-Kutta method.

   b) Assuming you have cylindrical cutters with radii: $0.25, 0.5, 0.75, \cdots$, 2, describe an efficient method to accurately machine the pocket.

48. Write a program which approximates an offset curve of a planar rational curve following the algorithm developed by Tiller and Hanson [421].

49. Give the implicit polynomial equation of a torus with axis in the direction $(0,0,1)$, center circle radius R and section circle radius a where $R > a$. What is the equation of the offset of this torus by $\pm h$, $h > 0$.

50. A pipe surface or canal surface of spheres of constant radius is defined as the envelope of a family of spheres of constant radius r whose centers describe a smooth curve, $\mathbf{c}(t)$ known as the spine. Let $f(\mathbf{p}, t) = |\mathbf{p} - \mathbf{c}(t)|^2 - r^2$, where $\mathbf{p} = (x, y, z)$

a) Show that the canal surface has an implicit equation $g(\mathbf{p}) = 0$ which results from eliminating t from the two equations $f = 0$ and $f_t = 0$.

b) Obtain the implicit equation of a torus by using the approach of part a).

c) Show that canal surfaces can be obtained as generalized cylinders by sweeping a circular cross-section along the spine.

51. In this problem we consider developing a blending surface between the plane $z = 0$ and the right circular cylinder $x^2 + y^2 = 1$, $0 \le z \le 2$. Because the cylinder is a surface of revolution, we will, for simplicity, consider the cross-section of the objects obtained by setting $y = 0$. Our problem is to develop a smooth surface between the cylinder and the plane by creating a cross-section curve starting at height 1 on the cylinder and terminating on the plane 2 units away from the origin.

a) As a first effort, consider using a quadratic Bezier curve (i.e. a parabola) as a blend cross-section. The curve should have the starting and ending points as indicated above, and to ensure a smooth blend, the tangent to the curve at the start (end) point should have the same direction as the tangent to the cylinder (plane). Give the control points of this curve.

b) Using the results of part a), express the blending surface (i.e. the surface of revolution characterized by the cross section obtained in a) as rational B-spline (NURBS) surface.

c) Now suppose we want a cubic Bezier curve as the cross-section of our blending surface. Give the control points of a cubic Bezier curve which generates a "good" blend. To be "good", the curve not only has to satisfy the boundary conditions indicated in part **a**, but also the area under the curve should be between 0.2 and 0.3.

d) Now suppose we want to maintain curvature continuity at the blending surface linkage curves in addition to position and tangent plane continuity. Determine a sufficiently high degree Bezier curve cross section to accomplish this.

A. Color Plates

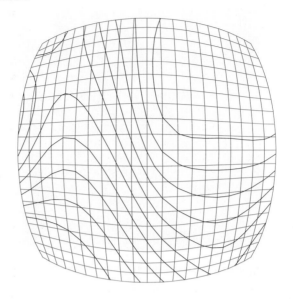

Color Plate A.1. Isophotes for surface of Fig. 8.1

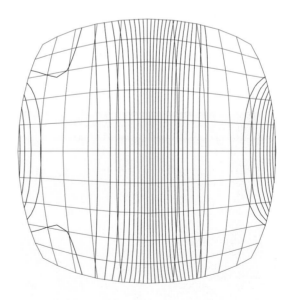

Color Plate A.2. Reflection lines for surface of Fig. 8.1

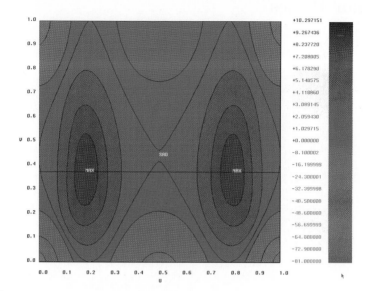

Color Plate A.3. Gaussian curvature color map of surface in Fig. 8.11 (adapted from [255])

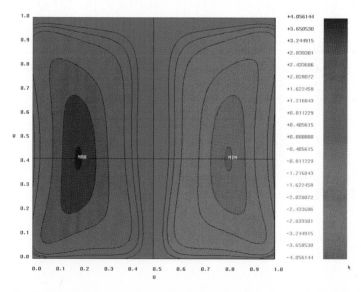

Color Plate A.4. Mean curvature color map of surface in Fig. 8.11 (adapted from [255])

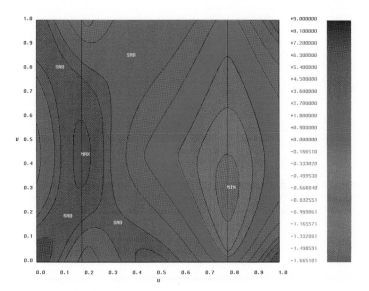

Color Plate A.5. Maximum principal curvature color map of surface in Fig. 8.11 (adapted from [255])

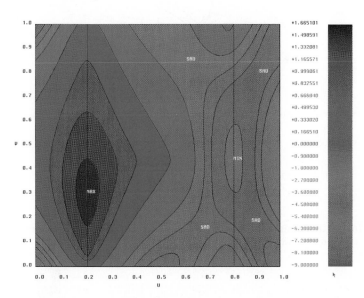

Color Plate A.6. Minimum principal curvature color map of surface in Fig. 8.11 (adapted from [255])

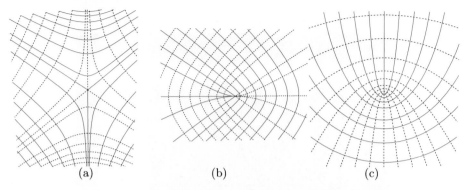

(a)   (b)   (c)

Color Plate A.7. (a) Star pattern (extracted from lower left umbilic of Fig. 9.6), **(b)** monstar pattern (extracted from center umbilic of Fig. 9.6), **(c)** lemon pattern (extracted from lower umbilic of Fig. 9.1) (adapted from [257])

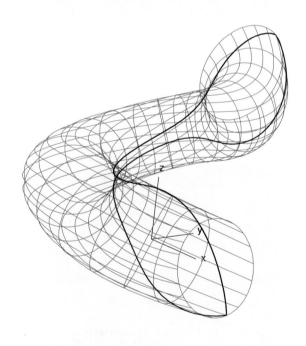

Color Plate A.8. Minimal geodesic paths on the generalized cylinder between two points of two circular edges (adapted from [247])

References

1. K. Abdel-Malek and H.-J. Yeh. On the determination of starting points for parametric surface intersections. *Computer-Aided Design*, 29(1):21–35, January 1997.
2. S. S. Abhyankar. *Algebraic Geometry for Scientists and Engineers*. American Mathematical Society, Providence, RI, 1990.
3. S. L. Abrams, L. Bardis, C. Chryssostomidis, N. M. Patrikalakis, S. T. Tuohy, F.-E. Wolter, and J. Zhou. The geometric modeling and interrogation system Praxiteles. *Journal of Ship Production*, 11(2):117–132, May 1995.
4. S. L. Abrams, W. Cho, C.-Y. Hu, T. Maekawa, N. M. Patrikalakis, E. C. Sherbrooke, and X. Ye. Efficient and reliable methods for rounded-interval arithmetic. *Computer-Aided Design*, 30(8):657–665, July 1998.
5. P. G. Alourdas. *Shape Creation, Interrogation and Fairing Using B-Splines*. Engineer's thesis, Massachusetts Institute of Technology, Department of Ocean Engineering, Cambridge, Massachusetts, 1989.
6. L.-E. Andersson, T. J. Peters, and N. F. Stewart. Selfintersection of composite curves and surfaces. *Computer Aided Geometric Design*, 15(5):507–527, May 1998.
7. R. K. E. Andersson. Surfaces with prescribed curvature I. *Computer Aided Geometric Design*, 10(5):431–452, October 1993.
8. E. V. Anoshkina, A. G. Belyaev, and T. L. Kunii. Detection of ridges and ravines based on caustic singularities. *International Journal of Shape Modeling*, 1(1):13–22, 1994.
9. E. V. Anoshkina, A. G. Belyaev, O. G. Okunev, and T. L. Kunii. Ridges and ravines: A singularity approach. *International Journal of Shape Modeling*, 1(1):1–11, 1994.
10. ANSI/IEEE Std 754–1985. *IEEE Standard for Binary Floating–Point Arithmetic*. IEEE, New York, 1985. Reprinted in *ACM SIGPLAN Notices*, 22(2):9-25, February 1987.
11. S. Aomura and T. Uehara. Self-intersection of an offset surface. *Computer-Aided Design*, 22(7):417–422, September 1990.
12. C. G. Armstrong, T. K. H. Tam, D. J. Robinson, R. M. McKeag, and M. A. Price. Automatic generation of well structured meshes using medial axis and surface subdivision. In G. A. Gabriele, editor, *Proceedings of the 17th ASME Design Automation Conference: Advances in Design Automation, Vol. 2*, pages 139–146, Miami, FL, September 1991. New York: ASME.
13. G. Aumann. Interpolation with developable Bézier patches. *Computer Aided Geometric Design*, 8(5):409–420, November 1991.
14. F. Aurenhammer. Voronoi diagrams — a survey of fundamental geometric data structure. *ACM Computing Surveys*, 23(3):345–405, September 1991.
15. W. Auzinger and H. J. Stetter. An elimination algorithm for the computation of zeros of a system of multivariate polynomial equations. In R. P. Agarwal,

Y. M. Chow, and S. J. Wilson, editors, *Numerical Mathematics, Singapore, 1988, International Series of Numerical Mathematics, Volume 86*, pages 11–30. Birkhäuser Verlag, Boston Basel Berlin, 1988.

16. C. Bajaj, J. Chen, and G. Xu. Modeling with cubic A-patches. *ACM Transactions on Graphics*, 14(2):103–133, April 1995.

17. C. L. Bajaj, C. M. Hoffmann, J. E. Hopcroft, and R. E. Lynch. Tracing surface intersections. *Computer Aided Geometric Design*, 5(4):285–307, November 1988.

18. R. E. Barnhill, G. Farin, L. Fayard, and H. Hagen. Twists, curvatures and surface interrogation. *Computer-Aided Design*, 20(6):341–346, July 1988.

19. R. E. Barnhill, G. Farin, M. Jordan, and B. R. Piper. Surface/surface intersection. *Computer Aided Geometric Design*, 4(1-2):3–16, July 1987.

20. R. E. Barnhill and S. N. Kersey. A marching method for parametric surface / surface intersection. *Computer Aided Geometric Design*, 7(1-4):257–280, June 1990.

21. R. C. Beach. *An Introduction to the Curves and Surfaces of Computer-Aided Design*. Van Nostrand Reinhold, New York, 1991.

22. J. M. Beck, R. T. Farouki, and J. K. Hinds. Surface analysis methods. *IEEE Computer Graphics and Applications*, 6(12):18–36, December 1986.

23. K.-P. Beier and Y. Chen. Highlight–line algorithm for realtime surface-quality assessment. *Computer-Aided Design*, 26(4):268–277, April 1994.

24. M. V. Berry and J. H. Hannay. Umbilic points on Gaussian random surfaces. *Journal of Physics A.*, 10(11):1809–1821, 1977.

25. D. Blackmore, M. C. Leu, and L. P. Wang. Sweep-envelope differential equation algorithm and its application to NC machining verification. *Computer-Aided Design*, 29(9):629–637, September 1997.

26. W. Blaschke. *Kreis und Kugel*. Walter de Gruyter and Co., Berlin, 1956.

27. C. Bliek. *Computer Methods for Design Automation*. PhD thesis, Massachusetts Institute of Technology, Cambridge, MA, July 1992.

28. G. A. Bliss. The geodesic lines on the anchor ring. *Annals of Mathematics*, 4:1–21, October 1902.

29. H. Blum. Biological shape and visual science (part I). *Journal of Theoretical Biology*, 38:205–287, 1973.

30. H. Blum. A transformation for extracting new descriptors of shape. *Models for the Perception of Speech and Visual Form*, pages 362–381, ed: Weinant Wathen-Dunn MIT Press, 1967.

31. H. Blum and R. N. Nagel. Shape description using weighted symmetric axis features. *Pattern Recognition*, 10(3):167–180, 1978.

32. R. M. C. Bodduluri and B. Ravani. Design of developable surfaces using duality between plane and point geometries. *Computer-Aided Design*, 25(10):621–632, October 1993.

33. W. Boehm. Cubic b-spline curves and surfaces in computer aided geometric design. *Computing*, 19:29–34, 1977.

34. W. Boehm. Inserting new knots into B-spline curves. *Computer-Aided Design*, 12(4):199–201, July 1980.

35. W. Boehm. Subdividing multivariate splines. *Computer-Aided Design*, 15(6):345–352, November 1983.

36. F. L. Bookstein. The line skeleton. *Computer Graphics and Image Processing*, 11:123–137, 1979.

37. M. Brady, J. Ponce, A. Yuille, and H. Asada. Describing surfaces. *Computer Vision, Graphics and Image Processing*, 32(1):1–28, October 1985.

38. J. W. Brandt. *Theory and Application of the Skeleton Representation of Continuous Shapes*. PhD thesis, University of California, Davis, CA, December 1991.

39. J. W. Brandt. Describing a solid with the three-dimensional skeleton. In J. D. Warren, editor, *Proceedings of The International Society for Optical Engineering, Volume 1830, Curves and Surfaces in Computer Vision and Graphics III*, pages 258–269. SPIE, Boston, Massachusetts, 1992.

40. J. W. Brandt. Convergence and continuity criteria for discrete approximations of the continuous planar skeleton. *CVGIP: Image Understanding*, 59(1):116–124, January 1994.

41. J. W. Brandt and V. R. Algazi. Continuous skeleton computation by Voronoi diagram. *CVGIP: Image Understanding*, 55(3):329–338, May 1992.

42. J. W. Brandt and V. R. Algazi. Lossy encoding of document images with the continuous skeleton. In P. Maragos, editor, *Visual Communications and Image Processing '92, SPIE 1818*, pages 663–673, 1992.

43. J. W. Brandt, A. K. Jain, and V. R. Algazi. Medial axis representation and encoding of scanned documents. *Journal of Visual Communication and Image Representation*, 2(2):151–165, June 1991.

44. E. L. Brechner. General tool offset curves and surfaces. In R. E. Barnhill, editor, *Geometry Processing for Design and Manufacturing*, pages 101–121. SIAM, 1992.

45. P. Brunet, A. Vinacua, M. Vivo, N. Pla, and A. Rodriguez. Surface fairing for ship hull design application. *Mathematical Engineering in Industry*, 7(2):179–193, 1998.

46. B. Buchberger. *Ein Algorithmus zum Auffinden der Basiselemente des Restklassenringes nach einem nulldimensionalen Polynomideal*. PhD thesis, University of Innsbruck, Innsbruck, Austria, 1965.

47. B. Buchberger. Gröbner bases: An algorithmic method in polynomial ideal theory. In N. K. Bose, editor, *Multidimensional Systems Theory: Progress, Directions and Open Problems in Multidimensional Systems*, pages 184–232. Dordrecht, Holland: D. Reidel Publishing Company, 1985.

48. J. F. Canny. *The Complexity of Robot Motion Planning*. MIT Press, Cambridge, MA, 1988.

49. J. F. Canny and I. Z. Emiris. An efficient algorithm for the sparse mixed resultant. In G. Cohen, T. Mora, and O. Moreno, editors, *Proceedings of 10th International Symposium, Applied Algebra, Algebraic Algorithms and Error-Correcting Codes*, pages 89–104. Springer-Verlag, 1993.

50. J. S. Chalfant. Analysis and Design of Developable Surfaces for Shipbuilding. Master's thesis, Massachusetts Institute of Technology, Department of Ocean Engineering, Cambridge, Massachusetts, 1997.

51. B. W. W. Char, K. O. Geddes, G. H. Gonnet, B. L. Leong, M. B. Monagan, and S. M. Watt. *First Leaves: A Tutorial Introduction to Maple V*. Springer-Verlag, 1992.

52. Y. J. Chen and B. Ravani. Offset surface generation and contouring in computer-aided design. *Journal of Mechanisms, Transmissions, and Automation in Design, Transactions of the ASME*, 109(3):133–142, March 1987.

53. K.-P. Cheng. Using plane vector fields to obtain all the intersection curves of two general surfaces. In W. Strasser and H. Seidel, editors, *Theory and Practice of Geometric Modeling*, pages 187–204. Springer-Verlag, New York, 1989.

54. C.-S. Chiang. *The Euclidean Distance Transform*. PhD thesis, Purdue University, West Lafayette, IN, August 1992.

55. C. S. Chiang, C. M. Hoffmann, and R. E. Lynch. How to compute offsets without self-intersection. In M. J. Silbermann and D. Tagare, editors, *Proceedings*

of The SPIE Conference on Curves and Surfaces in Computer Vision and Graphics II, Volume 1610, pages 76–87, Boston, Massachusetts, 1991. International Society for Optical Engineering.

56. H. Chiyokura. *Solid Modelling with DesignBase*. Addison-Wesley, Reading, MA, 1988.

57. W. Cho, T. Maekawa, and N. M. Patrikalakis. Topologically reliable approximation of composite Bézier curves. *Computer Aided Geometric Design*, 13(6):497–520, August 1996.

58. W. Cho, T. Maekawa, N. M. Patrikalakis, and J. Peraire. Topologically reliable approximation of trimmed polynomial surface patches. *Graphical Models and Image Processing*, 61(2):84–109, March 1999.

59. B. K. Choi and R. Jerard. *Sculptured Surface Machining - Theory and Applications*. Kluwer Academic Publishers, 1998.

60. B. K. Choi, C. S. Lee, and C. S. Jun. Compound surface modelling and machining. *Computer-Aided Design*, 20(3):127–136, April 1988.

61. I. Choi and K. Lee. Efficient generation of reflection lines to evaluate car body surfaces. *Mathematical Engineering in Industry*, 7(2):233–250, 1998.

62. B. Cobb. *Design of Sculptured Surfaces Using the B-spline Representation*. PhD thesis, Computer Science Department, University of Utah, Salt Lake City, Utah, 1984.

63. E. Cohen, T. Lyche, and R. Riesenfeld. Discrete B-splines and subdivision techniques in computer-aided geometric design and computer graphics. *Computer Graphics and Image Processing*, 14(2):87–111, October 1980.

64. G. E. Collins and R. Loos. Real zeros of polynomials. In B. Buchberger, G. E. Collins, and R. Loos, editors, *Computer Algebra: Symbolic and Algebraic Computation*, pages 83–94. Springer-Verlag, Vienna, 1982.

65. S. Coquillart. Computing offsets of B-spline curves. *Computer-Aided Design*, 19(6):305–309, July/August 1987.

66. T. H. Cormen, C. E. Leiserson, and R. L. Rivest. *Introduction to Algorithms*. MIT Press, Cambridge, MA, 1990.

67. M. G. Cox. The numerical evaluation of B-splines. *Journal of the Institute for Mathematics Applications*, 10:134–149, 1972.

68. T. Culver, J. Keyser, and D. Manocha. Accurate computation of the medial axis of a polyhedron. In W. F. Bronsvoort and D. C. Anderson, editors, *In Proceedings of Fifth Symposium on Solid Modeling and Applications, Ann Arbor, Michigan*, pages 179–190. NY: ACM, June 1999.

69. G. Dahlquist and Å. Björck. *Numerical Methods*. Prentice-Hall, Inc., Englewood Cliffs, NJ, 1974.

70. P.-E. Danielsson. Euclidean distance mapping. *Computer Graphics and Image Processing*, 14:227–248, 1980.

71. G. Darboux. *Leçons sur la Théorie Générale des Surfaces, Vol.4*. Gauthier-Villars, Paris, 1896.

72. C. De Boor. On calculating with B-splines. *Journal of Approximation Theory*, 6:50–62, 1972.

73. C. De Boor. *A Practical Guide to Splines*. Springer, New York, 1978.

74. J. C. Dill. An application of color graphics to the display of surface curvature. *ACM Computer Graphics*, 15(3):153–161, August 1981.

75. Q. Ding and B. J. Davies. *Surface Engineering Geometry for Computer-Aided Design and Manufacture*. Ellis Horwood, Chichester, UK, 1987.

76. P. M. do Carmo. *Differential Geometry of Curves and Surfaces*. Prentice-Hall, Inc., Englewood Cliffs, NJ, 1976.

77. T. Dokken. Finding intersections of B-spline represented geometries using recursive subdivision techniques. *Computer Aided Geometric Design*, 2(1-3):189–195, September 1985.

78. D. Dragomatz and S. Mann. A classified bibliography of literature on NC milling path generation. *Computer-Aided Design*, 29(3):239–247, March 1997.

79. A. Dresden. *A Solid Analytical Geometry and Determinants*. Dover, New York, 1964.

80. T. Duff. Interval arithmetic and recursive subdivision for implicit functions and constructive solid geometry. *ACM Computer Graphics*, 26(2):131–138, July 1992.

81. D. Dutta and C. M. Hoffmann. A geometric investigation of the skeleton of CSG objects. In B. Ravani, editor, *Proceedings of the 16th ASME Design Automation Conference: Advances in Design Automation, Computer Aided and Computational Design*, volume I, pages 67–75, Chicago, IL, September 1990. New York: ASME, 1990.

82. D. Dutta and C. M. Hoffmann. On the skeleton of simple CSG objects. *Journal of Mechanical Design, ASME Transactions*, 115(1):87–94, March 1993.

83. D. Dutta, R. R. Martin, and M. J. Pratt. Cyclides in surface and solid modeling. *IEEE Computer Graphics and Applications*, 13(1):53–59, January 1993.

84. H. Edelsbrunner and E. P. Mücke. Three-dimensional alpha shapes. *ACM Transactions on Graphics*, 13(1):43–72, 1994.

85. G. Elber and E. Cohen. Error bounded variable distance offset operator for free form curves and surfaces. *International Journal of Computational Geometry and Applications*, 1(1):67–78, March 1991.

86. G. Elber and E. Cohen. Offset approximation improvement by control points perturbation. In T. Lyche and L. L. Schumaker, editors, *Mathematical Methods in Computer Aided Geometric Design II*, pages 229–237. Academic Press, Boston, 1992.

87. G. Elber and E. Cohen. Second-order surface analysis using hybrid symbolic and numeric operators. *ACM Transactions on Graphics*, 12(2):160–178, April 1993.

88. G. Elber, I.-K. Lee, and M. S. Kim. Comparing offset curve approximation methods. *IEEE Computer Graphics and Applications*, 17(3):62–71, May/June 1997.

89. I. Z. Emiris. *Sparse Elimination and Applications in Kinematics*. PhD thesis, University of California at Berkeley, Berkeley, CA, 1994.

90. W. Enger. Interval Ray Tracing - A divide and conquer strategy for realistic computer graphics. *The Visual Computer*, 9(2):91–104, November 1992.

91. M. Etzion and A. Rappoport. Computing the Voronoi diagram of a 3-d polyhedron by separate computation of its symbolic and geometric parts. In W. F. Bronsvoort and D. C. Anderson, editors, *In Proceedings of Fifth Symposium on Solid Modeling and Applications, Ann Arbor, Michigan*, pages 167–178, NY: ACM, June 1999.

92. G. Farin. *Curves and Surfaces for Computer Aided Geometric Design: A Practical Guide*. Academic Press, Boston, MA, 3rd edition, 1993.

93. R. T. Farouki. Exact offset procedures for simple solids. *Computer Aided Geometric Design*, 2(4):257–279, 1985.

94. R. T. Farouki. The approximation of non-degenerate offset surfaces. *Computer Aided Geometric Design*, 3(1):15–43, May 1986.

95. R. T. Farouki. The characterization of parametric surface sections. *Computer Vision, Graphics and Image Processing*, 33(2):209–236, February 1986.

References

96. R. T. Farouki. Graphical methods for surface differential geometry. In R. R. Martin, editor, *The Mathematics of Surfaces II*, pages 363–385. Clarendon Press, 1987.

97. R. T. Farouki. Hierarchical segmentations of algebraic curves and some applications. In T. Lyche and L. L. Schumaker, editors, *Mathematical Methods in Computer Aided Geometric Design*, pages 239–248. Academic Press, Boston, 1989.

98. R. T. Farouki. On integrating lines of curvature. *Computer Aided Geometric Design*, 15(2):187–192, February 1998.

99. R. T. Farouki and J. K. Johnstone. The bisector of a point and a plane parametric curve. *Computer Aided Geometric Design*, 11(2):117–151, April 1994.

100. R. T. Farouki and J. K. Johnstone. Computing point/curve and curve/curve bisectors. In R. B. Fisher, editor, *The Mathematics of Surfaces V*, pages 327–354. Oxford University, Oxford, 1994.

101. R. T. Farouki and C. A. Neff. Algebraic properties of plane offset curves. *Computer Aided Geometric Design*, 7(1 - 4):101–127, June 1990.

102. R. T. Farouki and C. A. Neff. Analytic properties of plane offset curves. *Computer Aided Geometric Design*, 7(1 - 4):83–99, June 1990.

103. R. T. Farouki and C. A. Neff. Hermite interpolation by Pythagorean hodograph quintics. *Mathematics of Computation*, 64(212):1589–1609, October 1995.

104. R. T. Farouki, C. A. Neff, and M. A. O'Connor. Automatic parsing of degenerate quadric-surface intersections. *ACM Transactions on Graphics*, 8(3):174–203, 1989.

105. R. T. Farouki and V. T. Rajan. On the numerical condition of polynomials in Bernstein form. *Computer Aided Geometric Design*, 4(3):191–216, November 1987.

106. R. T. Farouki and V. T. Rajan. Algorithms for polynomials in Bernstein form. *Computer Aided Geometric Design*, 5(1):1–26, June 1988.

107. R. T. Farouki and V. T. Rajan. On the numerical condition of algebraic curves and surfaces 1. implicit equations. *Computer Aided Geometric Design*, 5(3):215–252, September 1988.

108. R. T. Farouki and T. Sakkalis. Pythagorean hodographs. *IBM Journal of Research and Development*, 34(5):736–752, September 1990.

109. R. T. Farouki and T. Sakkalis. Real rational curves are not 'unit speed'. *Computer Aided Geometric Design*, 8(2):151–157, May 1991.

110. R. T. Farouki and T. Sakkalis. Pythagorean-hodograph space curves. *Advances in Computational Mathematics*, 2:41–46, 1994.

111. R. T. Farouki and T. W. Sederberg. Analysis of the offset to a parabola. *Computer Aided Geometric Design*, 12(6):639–645, September 1995.

112. R. T. Farouki and S. Shah. Real-time CNC interpolators for Pythagorean-hodograph curves. *Computer Aided Geometric Design*, 13(7):583–600, October 1996.

113. R. T. Farouki and R. Sverrisson. Approximation of rolling-ball blends for free-form parametric surfaces. *Computer-Aided Design*, 28(11):871–878, November 1996.

114. R. T. Farouki, K. Tarabanis, J. U. Korein, J. S. Batchelder, and S. R. Abrams. Offset curves in layered manufacturing. *Journal of Manufacturing Science and Engineering, Transactions of the ASME*, 68(2):557–568, 1994.

115. J. C. Faugere, P. Gianni, D. Lazard, and T. Mora. Efficient computation of zero-dimensional Gröbner bases by change of ordering. *Journal of Symbolic Computation*, 16(4):329–344, 1993.

116. I. D. Faux and M. J. Pratt. *Computational Geometry for Design and Manufacture*. Ellis Horwood, Chichester, England, 1981.
117. J. H. Ferziger. *Numerical Methods for Engineering Applications*. Wiley, 1981.
118. J. D. Foley, A. Van Dam, S. K. Feiner, and J. F. Hughes. *Computer Graphics: Principles and Practice*. Addison-Wesley, Reading, MA, 2nd edition, 1996.
119. A. R. Forrest. Computational geometry. *Proceedings of the Royal Society of London A*, 321:187–195, 1971.
120. W. H. Frey and D. Bindschadler. Computer-aided design of a class of developable Bézier surfaces. R&D Publication 8057, General Motors, September 1993.
121. J. Gallier. *Curves and Surfaces in Geometric Modeling: Theory and Algorithms*. Morgan Kaufmann, San Francisco, CA, 1999.
122. J. Gallier. *Geometric Methods and Applications: For Computer Science and Engineering*. Springer-Verlag, New York, 2001.
123. C. B. Garcia and W. I. Zangwill. Global continuation methods for finding all solutions to polynomial systems of equations in n variables. In A. V. Fiacco and K. O. Kortanek, editors, *Extremal Methods and Systems Analysis*, pages 481–497. Springer-Verlag, New York, NY, 1980.
124. A. Geisow. *Surface Interrogations*. PhD thesis, School of Computing Studies and Accountancy, University of East Anglia, Norwich NR47TJ, U. K., July 1983.
125. S. M. Gelston and D. Dutta. Boundary surface recovery from skeleton curves and surfaces. *Computer Aided Geometric Design*, 12(1):27–51, February 1995.
126. C. F. Gerald and P. O. Wheatley. *Applied Numerical Analysis*. Addison-Wesley, Reading, MA, 4th edition, 1990.
127. G. Glaeser, J. Wallner, and H. Pottmann. Collision-free 3-axis milling and selection of cutting tools. *Computer-Aided Design*, 31(3):225–232, March 1999.
128. D. Goldberg. What every computer scientist should know about floating–point arithmetic. *ACM Computing Surveys*, 23(1):5–48, March 1991.
129. M. Golubitsky and V. Guillemin. *Stable Mappings and their Singularities*. Springer-Verlag, New York, 1973.
130. W. J. Gordon and R. F. Riesenfeld. B-spline curves and surfaces. In R. E. Barnhill and R. F. Riesenfeld, editors, *Computer Aided Geometric Design*, pages 95–126. Academic Press, Inc., 1974.
131. T. A. Grandine. Computing zeroes of spline functions. *Computer Aided Geometric Design*, 6(2):129–136, May 1989.
132. T. A. Grandine. Geometry processing and numerical stability. In G. Farin, J. Hoschek, M. S. Kim, and D. Abma, editors, *The Handbook of Computer Aided Design*. Elsevier, 2001.
133. T. A. Grandine and F. W. Klein. A new approach to the surface intersection problem. *Computer Aided Geometric Design*, 14(2):111–134, 1997.
134. J. A. Grant and G. D. Hitchins. An always convergent minimization technique for the solution of polynomial equations. *Journal of Industrial and Mathematical Applications*, 8:122–129, 1971.
135. J. A. Grant and G. D. Hitchins. Two algorithms for the solution of polynomial equations to limiting machine precision. *The Computer Journal*, 18(3), 1973.
136. A. Gray. *Modern Differential Geometry of Curves and Surfaces*. CRC Press, Boca Raton, 1993.
137. L. Guibas and J. Stolfi. Primitives for the manipulation of general subdivisions and the computation of Voronoi diagrams. *ACM Transactions on Graphics*, 4(2):74–123, April 1985.
138. A. Z. Gurbuz and I. Zeid. Offsetting operations via closed ball approximation. *Computer-Aided Design*, 27(11):805–810, November 1995.

139. H. N. Gursoy. *Shape Interrogation by Medial Axis Transform for Automated Analysis*. PhD thesis, Massachusetts Institute of Technology, Cambridge, MA, November 1989.

140. H. N. Gursoy and N. M. Patrikalakis. Automated interrogation and adaptive subdivision of shape using medial axis transform. *Advances in Engineering Software and Workstations*, 13(5/6):287–302, September/November 1991.

141. H. N. Gursoy and N. M. Patrikalakis. An automated coarse and fine surface mesh generation scheme based on medial axis transform, part I: Algorithms. *Engineering with Computers*, 8(3):121–137, 1992.

142. H. N. Gursoy and N. M. Patrikalakis. An automated coarse and fine surface mesh generation scheme based on medial axis transform, part II: Implementation. *Engineering with Computers*, 8(4):179–196, 1992.

143. C. Gutierrez and J. Sotomayor. Lines of curvature, umbilic points and Carathéodory conjecture. *Resenhas IME-UPS*, 3(3):291–322, 1998.

144. J. Hadenfeld. Local energy fairing of B-spline surfaces. In M. Dæhlen, T. Lyche, and L. L. Schumaker, editors, *Mathematical Methods for Curves and Surfaces*, pages 203–212. Vanderbilt University Press, 1995.

145. H. Hagen, S. Hahmann, and T. Schreiber. Visualization and computation of curvature behaviour of freeform curves and surfaces. *Computer-Aided Design*, 27(7):545–552, July 1995.

146. H. Hagen, S. Hahmann, T. Schreiber, Y. Nakajima, B. Wördenweber, and P. Hollemann-Grundstedt. Surface interrogation algorithms. *IEEE Computer Graphics and Applications*, 12(5):53–60, September 1992.

147. G. D. Hager. Constraint solving methods and sensor-based decision making. In *Proceedings of the 1992 IEEE International Conference on Robotics and Automation*, pages 1662–1667. IEEE, 1992.

148. S. Hahmann and S. Konz. Knot-removal surface fairing using search strategies. *Computer-Aided Design*, 30(12):923–930, February 1998.

149. D. G. Hakala, R. C. Hillyard, B. E. Nourse, and P. J. Malraison. Natural quadrics in mechanical design. In *Proceedings of the Autofact West 1, Anaheim, CA in November, 1980*, pages 363–378, 1980.

150. B. Hamann and J. L. Chen. Data point selection for piecewise trilinear approximation. *Computer Aided Geometric Design*, 11(5):477–489, October 1994.

151. R. W. Hamming. *Numerical Methods for Scientists and Engineers*. McGraw-Hill, New York, 1962.

152. H. Hancock. *Theory of Maxima and Minima*. Dover, New York, 1960.

153. A. Hansen and F. Arbab. An algorithm for generating NC tool paths for arbitrarily shaped pockets with islands. *ACM Transactions on Graphics*, 11(2):152–182, 1992.

154. E. Hartmann. G^2 interpolation and blending on surfaces. *The Visual Computer*, 12(4):181–192, 1996.

155. E. Hartmann. Numerical implicitization for intersection and G^n-continuous blending of surfaces. *Computer Aided Geometric Design*, 15(4):377–397, April 1998.

156. R. N. Hawat and L. A. Piegl. Genetic algorithm approach to curve-curve intersection. *Mathematical Engineering in Industry*, 7(2):269–282, 1998.

157. M. Held. *On the Computational Geometry of Pocket Machining*. Springer-Verlag, Berlin, Germany, 1991.

158. P. Van Hentenryck, D. McAllester, and D. Kapur. Solving polynomial systems using a branch and prune approach. *SIAM Journal on Numerical Analysis*, 34(2):797–827, April 1997.

159. P. Van Hentenryck, L. Michel, and Y. Deville. *Numerica: A Modeling Language for Global Optimization*. MIT Press, Cambride, MA, 1997.

160. H.-S. Heo, M.-S. Kim, and G. Elber. The intersection of two ruled surfaces. *Computer-Aided Design*, 31(1):33–50, January 1999.

161. T. Hermann, G. Lukacs, and F. E. Wolter. Geometrical criteria on the higher order smoothness of composite surfaces. *Computer Aided Geometric Design*, 16(9):907–911, October 1999.

162. M. Higashi and K. Kaneko. Generation of high-quality curve and surface with smoothly varying curvature. In D. A. Duce and P. Jancene, editors, *Eurographics '88*, pages 79–92, Nice, France, September 1988. North-Holland.

163. M. Higashi, T. Saitoh, Y. Watanabe, and Y. Watanabe. Analysis of aesthetic free-form surfaces by surface edges. In S. Y. Shin and T. L. Kunii, editors, *Proceedings of the Third Pacific Conference on Computer Graphics and Applications, Pacific Graphics '95*, pages 294–305, Seoul, Korea, August 1995. World Scientific.

164. M. Higashi, H. Tsutamori, and M. Hosaka. Generation of smooth surfaces by controlling curvature variation. *Computer Graphics Forum*, 15(3):187–196, September 1996.

165. D. Hilbert and S. Cohn-Vossen. *Geometry and the Imagination*. Chelsea, New York, 1952.

166. F. B. Hildebrand. *Advanced Calculus for Applications*. Prentice-Hall, Inc., Englewood Cliffs, New Jersey, 1976.

167. C. M. Hoffmann. *Geometric and Solid Modeling: An Introduction*. Morgan Kaufmann Publishers, Inc., San Mateo, California, 1989.

168. C. M. Hoffmann. The problems of accuracy and robustness in geometric computation. *Computer*, 22(3):31–41, March 1989.

169. C. M. Hoffmann. A dimensionality paradigm for surface interrogations. *Computer Aided Geometric Design*, 7(6):517–532, November 1990.

170. C. M. Hoffmann. How to construct the skeleton of CSG objects. In A. Bowyer and J. Davenport, editors, *Proceedings of the Fourth IMA Conference, The Mathematics of Surfaces, University of Bath, UK, September 1990*, pages 421–438, New York, 1994. Oxford University Press.

171. C. M. Hoffmann and G. Vanecek. On alternate solid representations and their uses. Technical Report CSD-TR-91-019, Computer Sciences Department, Purdue University, March 1991.

172. D. H. Hoitsma. Surface curvature analysis. In M. J. Wozny et al., editors, *IFIP TC5/WG5.2 Second Workshop on Geometric Modeling*, pages 21–38, New York, 1988. IFIP, North Holland.

173. M. Hosaka. *Modeling of Curves and Surfaces in CAD/CAM*. Springer-Verlag, New York, 1991.

174. J. Hoschek. Spline approximation of offset curves. *Computer Aided Geometric Design*, 5(1):33–40, June 1988.

175. J. Hoschek and D. Lasser. *Fundamentals of Computer Aided Geometric Design*. A. K. Peters, Wellesley, MA, 1993. Translated by L. L. Schumaker.

176. J. Hoschek and N. Wissel. Optimal approximate conversion of spline curves and spline approximation of offset curves. *Computer-Aided Design*, 20(8):475–483, October 1988.

177. E. G. Houghton, R. F. Emnett, J. D. Factor, and C. L. Sabharwal. Implementation of a divide-and-conquer method for intersection of parametric surfaces. *Computer Aided Geometric Design*, 2(1-3):173–183, September 1985.

178. C. Y. Hu, T. Maekawa, N. M. Patrikalakis, and X. Ye. Robust interval algorithm for surface intersections. *Computer-Aided Design*, 29(9):617–627, September 1997.

179. C. Y. Hu, T. Maekawa, E. C. Sherbrooke, and N. M. Patrikalakis. Robust interval algorithm for curve intersections. *Computer-Aided Design*, 28(6/7):495–506, June/July 1996.

180. C. Y. Hu, N. M. Patrikalakis, and X. Ye. Robust interval solid modeling: Part I, Representations. *Computer-Aided Design*, 28(10):807–817, October 1996.

181. C. Y. Hu, N. M. Patrikalakis, and X. Ye. Robust interval solid modeling: Part II, Boundary evaluation. *Computer-Aided Design*, 28(10):819–830, October 1996.

182. IGES/PDES Organization, U.S. Product Data Association, Fairfax, VA. *Digital Representation for Communication of Product Definition Data, US PRO/IPO-100, Initial Graphics Exchange Specification (IGES) 5.2*, November 1993.

183. C. G. Jensen and D. C. Anderson. A review of numerically controled methods for finish-sculptured-surface machining. *IEE Transactions*, 28:30–39, 1996.

184. R. B. Jerard, R. L. Drysdale, B. Schaudt, K. Hauck, and J. Magewick. Methods for detecting errors in numerically controlled machining of sculptured surfaces. *IEEE Computer Graphics and Applications*, 9(1):26–39, January 1989.

185. H. T. Jessop and F. C. Harris. *Photoelasticity, Principles and Methods*. New York: Dover Publications, 1950.

186. R. A. Jinkerson, S. L. Abrams, L. Bardis, C. Chryssostomidis, A. Clement, N. M. Patrikalakis, and F.-E. Wolter. Inspection and feature extraction of marine propellers. *Journal of Ship Production*, 9(2):88–106, May 1993.

187. J. P. Jouanolou. Le formalisme du resultant. *Advances in Mathematics*, 90(2):117–263, 1991.

188. J. T. Kajiya. Ray tracing parametric patches. *ACM Computer Graphics*, 16(3):245–254, July 1982.

189. K. Kase, A. Makinouchi, T. Nakagawa, H. Suzuki, and F. Kimura. Shape error evaluation method of free-form surfaces. *Computer-Aided Design*, 31(8):495–505, July 1999.

190. E. Kaufmann and R. Klass. Smoothing surfaces using reflection lines for families of splines. *Computer-Aided Design*, 20(6):312–316, July 1988.

191. R. B. Kearfott. Interval Newton/generalized bisection when there are singularities near roots. *Annals of Operations Research*, 25:181–196, 1990.

192. R. B. Kearfott. Decomposition of arithmetic expressions to improve the behavior of interval iteration for nonlinear systems. *Computing*, 47:169–191, 1991.

193. N. Kehtarnavaz and R. J. P. de Figueiredo. A 3-D contour segmentation scheme based on curvature and torsion. *IEEE Transactions on Pattern Analysis and Machine Intelligence*, 10(5):707–713, September 1988.

194. H. B. Keller. *Numerical Methods for Two-Point Boundary Value Problems*. Blaisdell, Waltham, MA, 1968.

195. B. W. Kernighan and D. M. Ritchie. *The C Programming Language*. Prentice-Hall, Englewood Cliffs, NJ, 2nd edition, 1988.

196. J. Keyser, T. Culver, D. Manocha, and S. Krishnan. Efficient and exact manipulation of algebraic points and curves. *Computer-Aided Design*, 32(11):649–662, September 2000.

197. K. I. Kim and K. Kim. A new machine strategy for sculptured surfaces using offset surface. *International Journal of Production Research*, 33(6):1683–1697, 1995.

198. M.-S. Kim, E.-J. Park, and S.-B. Lim. Approximation of variable-radius offset curves and its application to Bézier brush-stroke design. *Computer-Aided Design*, 25(11):684–698, November 1993.

199. T. Kim and S. E. Sarma. Time-optimal paths covering a surface. In R. Cipolla and R. Martin, editors, *The Mathematics of Surfaces IX*, pages 126–143, University of Cambridge, UK., September 2000. London: Springer.

200. R. Kimmel, A. Amir, and A. M. Bruckstein. Finding shortest paths on surfaces using level sets propagation. *IEEE Transactions on Pattern Analysis and Machine Intelligence*, 17(6):635–640, June 1995.

201. R. Kimmel and A. M. Bruckstein. Shape offsets via level sets. *Computer-Aided Design*, 25(3):154–162, March 1993.

202. R. Klass. Correction of local surface irregularities using reflection lines. *Computer-Aided Design*, 12(2):73–76, March 1980.

203. R. Klass. An offset spline approximation for plane cubic splines. *Computer-Aided Design*, 15(4):297–299, September 1983.

204. D. E. Knuth. *The Art of Computer Programming, Vol. 2, Seminumerical Algorithms*. Addison-Wesley, Reading, Massachusetts, 1981. 2nd Edition.

205. J. J. Koenderink. *Solid Shape*. MIT Press, Cambridge, MA, 1990.

206. E. Kreyszig. *Differential Geometry*. University of Toronto Press, Toronto, 1959.

207. E. Kreyszig. *Introduction to Differential Geometry and Riemannian Geometry*. University of Toronto Press, 1968.

208. G. A. Kriezis. *Algorithms for Rational Spline Surface Intersections*. PhD thesis, Massachusetts Institute of Technology, Cambridge, Massachusetts, March 1990.

209. G. A. Kriezis and N. M. Patrikalakis. Rational polynomial surface intersections. In G. A. Gabriele, editor, *Proceedings of the 17th ASME Design Automation Conference, Vol. II*, pages 43–53, Miami, September 1991. ASME, New York, 1991.

210. G. A. Kriezis, N. M. Patrikalakis, and F.-E. Wolter. Topological and differential-equation methods for surface intersections. *Computer-Aided Design*, 24(1):41–55, January 1992.

211. G. A. Kriezis, P. V. Prakash, and N. M. Patrikalakis. Method for intersecting algebraic surfaces with rational polynomial patches. *Computer-Aided Design*, 22(10):645–654, December 1990.

212. S. Krishnan and D. Manocha. Efficient surface intersection algorithm based on lower-dimensional formulation. *ACM Transactions on Graphics*, 16(1):74–106, January 1997.

213. E. Kruppa. *Analytische und Konstruktive Differentialgeometrie*. Springer-Verlag, Wien, 1957.

214. R. Kunze, F.-E. Wolter, and T. Rausch. Geodesic Voronoi diagrams on parametric surfaces. In *Proceedings of Computer Graphics International, CGI '97, June 1997*, pages 230–237. IEEE Computer Society Press, 1997.

215. T. Kuragano. FRESDAM system for design of aesthetically pleasing free-form objects and generation of collision-free tool paths. *Computer-Aided Design*, 24(11):573–581, November 1992.

216. T. Kuragano, N. Sasaki, and A. Kikuchi. The FRESDAM system for designing and manufacturing freeform objects. In R. Martin, editor, *USA-Japan Cross Bridge. Flexible Automation Volume 2*, pages 931–938, 1988.

217. A. Kurosh. *Higher Algebra*. Mir Publishers, Moscow, 1980. Translated by G. Yankovsky.

218. Y. N. Lakshman. *On the complexity of computing Gröbner bases for zero dimensional ideals*. PhD thesis, Rennselaer Polytechnic Institute, Troy, NY, 1992.

219. H. Lamure and D. Michelucci. Solving geometric constraints by homotopy. *IEEE Transactions on Visualization and Computer Graphics*, 2:22–34, 1996.

220. J. M. Lane and R. F. Riesenfeld. A theoretical development for the computer display and generation of piecewise polynomial surfaces. *IEEE Transactions on Pattern Analysis and Machine Intelligence*, 2(1):35–46, January 1980.

221. J. M. Lane and R. F. Riesenfeld. Bounds on a polynomial. *BIT: Nordisk Tidskrift for Informations-Behandling*, 21(1):112–117, 1981.

222. J. Lang and O. Röschel. Developable (1,n) Bézier surfaces. *Computer Aided Geometric Design*, 9(4):291–298, September 1992.

223. C. Lartigue, F. Thiebaut, and T. Maekawa. CNC tool path in terms of B-spline curves. *Computer-Aided Design*, 33(4):307–319, April 2001.

224. D. Lasser. Self-intersections of parametric surfaces. In *Proceedings of Third International Conference on Engineering Graphics and Descriptive Geometry: Volume 1*, pages 322–331, Vienna, 1988.

225. D. Lasser. Calculating the self-intersections of Bézier curves. *Computers in Industry*, 12:259–268, 1989.

226. D. Lavender, A. Bowyer, J. Davenport, A. Wallis, and J. Woodwark. Voronoi diagrams of set-theoretic solid models. *IEEE Computer Graphics and Applications*, 12(5):69–77, 1992.

227. J. D. Lawrence. *A Catalogue of Special Plane Curves*. Dover Publications, Inc., New York, 1972.

228. D. Lazard. Solving zero-dimensional algebraic systems. *Journal of Symbolic Computation*, 13(2):117–131, 1992.

229. D. T. Lee. Medial axis transformation of a planar shape. *IEEE Transactions on Pattern Analysis and Machine Intelligence*, PAMI-4(4):363–369, July 1982.

230. I.-K. Lee, M.-S. Kim, and G. Elber. Planar curve offset based on circle approximation. *Computer-Aided Design*, 28(8):617–630, 1996.

231. K. Lee. *Principles of CAD/CAM/CAE Systems*. Addison-Wesley, 1999.

232. Y. Lee and T. Chang. CASCAM - an automated system for sculptured surface cavity machining. *Computers in Industry*, 16:321–342, 1991.

233. J. Z. Levin. A parametric algorithm for drawing pictures of solid objects composed of quadric surfaces. *Communications of the Association for Computing Machinery*, 19(10):555–563, October 1976.

234. J. Z. Levin. Mathematical models for determining the intersections of quadric surfaces. *Computer Vision, Graphics and Image Processing*, 11:73–87, 1979.

235. M. M. Lipschutz. *Theory and Problems of Differential Geometry*. Schaum's Outline Series: McGraw-Hill, 1969.

236. N. G. Lloyd. *Degree Theory*. Cambridge University Press, Cambridge, 1978.

237. T. Lozano-Perez and M. A. Wesley. An algorithm for planning collision-free paths amongst polyhedral obstacles. *Communications of the ACM*, 25(9):560–570, October 1979.

238. W. Lü. Rational offsets by reparametrization. Technical report, Zhejiang University, December 1992.

239. W. Lü. Offset-rational parametric plane curves. *Computer Aided Geometric Design*, 12(6):601–616, September 1995.

240. W. Lü. Rational parameterization of quadrics and their offsets. *Computing*, 57(2):135–147, 1996.

241. W. Lü and H. Pottmann. Pipe surfaces with rational spine curve are rational. *Computer Aided Geometric Design*, 13(7):621–628, October 1996.

242. R. C. Luo, Y. Ma, and D. F. McAllister. Tracing tangential surface-surface intersections. In C. Hoffmann and J. Rossignac, editors, *Proceedings of the Third ACM Solid Modeling Symposium*, pages 255–262, Salt Lake City, Utah, May 1995. ACM, NY.

243. T. Lyche and K. Mørken. Knot removal for parametric B-spline curves and surfaces. *Computer Aided Geometric Design*, 4(3):217–230, November 1987.

244. Y. Ma and Y.-S. Lee. Detection of loops and singularities of surface intersections. *Computer-Aided Design*, 30(14):1059–1067, December 1998.

245. Y. Ma and R. C. Luo. Topological method for loop detection of surface intersection problems. *Computer-Aided Design*, 27(11):811–820, November 1995.

246. T. Maekawa. *Robust Computational Methods for Shape Interrogation*. PhD thesis, Massachusetts Institute of Technology, Cambridge, MA, June 1993.

247. T. Maekawa. Computation of shortest paths on free-form parametric surfaces. *Journal of Mechanical Design, Transactions of the ASME*, 118(4):499–508, December 1996.

248. T. Maekawa. Self-intersections of offsets of quadratic surfaces: Part I, explicit surfaces. *Engineering with Computers*, 14:1–13, 1998.

249. T. Maekawa. Self-intersections of offsets of quadratic surfaces: Part II, implicit surfaces. *Engineering with Computers*, 14:14–22, 1998.

250. T. Maekawa. An overview of offset curves and surfaces. *Computer-Aided Design*, 31(3):165–173, March 1999.

251. T. Maekawa and J. S. Chalfant. Computation of inflection lines and geodesics on developable surfaces. *Mathematical Engineering in Industry*, 7(2):251–267, 1998.

252. T. Maekawa and J. S. Chalfant. Design and tessellation of B-spline developable surfaces. *Journal of Mechanical Design, Transactions of the ASME*, 120(3):453–461, September 1998.

253. T. Maekawa, W. Cho, and N. M. Patrikalakis. Computation of self-intersections of offsets of Bézier surface patches. *Journal of Mechanical Design, Transactions of the ASME*, 119(2):275–283, June 1997.

254. T. Maekawa and N. M. Patrikalakis. Computation of singularities and intersections of offsets of planar curves. *Computer Aided Geometric Design*, 10(5):407–429, October 1993.

255. T. Maekawa and N. M. Patrikalakis. Interrogation of differential geometry properties for design and manufacture. *The Visual Computer*, 10(4):216–237, March 1994.

256. T. Maekawa, N. M. Patrikalakis, T. Sakkalis, and G. Yu. Analysis and applications of pipe surfaces. *Computer Aided Geometric Design*, 15(5):437–458, May 1998.

257. T. Maekawa, F.-E. Wolter, and N. M. Patrikalakis. Umbilics and lines of curvature for shape interrogation. *Computer Aided Geometric Design*, 13(2):133–161, March 1996.

258. D. Manocha. Solving polynomial systems for curve, surface and solid modeling. In J. Rossignac, J. Turner, and G. Allen, editors, *Proceedings of 2nd ACM/IEEE Symposium on Solid Modeling and Applications*, pages 169–178, Montreal, May 1993. New York: ACM Press, 1993.

259. D. Manocha. Numerical methods for solving polynomial equations. In D. A. Cox and B. Sturmfels, editors, *Proceedings of Symposia in Applied Mathematics Volume 53, Applications of Computational Algebraic Geometry: American Mathematical Society short course, January 6-7, 1997, San Diego, California*, pages 41–66. American Mathematical Society, 1998.

260. D. Manocha and S. Krishnan. Solving algebraic systems using matrix computations. *Sigsam Bulletin: Communications in Computer Algebra*, 30(4):4–21, December 1996.

261. M. Mäntylä. *An Introduction to Solid Modeling*. Computer Science Press, Rockville, Maryland, 1988.

262. K. Marciniak. *Geometric modeling for numerically controlled machining*. Oxford University Press, New York, 1991.

263. R. Markot and R. Magedson. Procedural method for evaluating the intersection curves of two parametric surfaces. *Computer-Aided Design*, 23(6):395–404, July/August 1991.

264. R. P. Markot and R. L. Magedson. Solutions of tangential surface and curve intersections. *Computer-Aided Design*, 21(7):421–429, September 1989.

265. R. R. Martin. Principal patches - a new class of surface patch based on differential geometry. In P. J. W. Ten Hagen, editor, *Eurographics '83, Proceedings of the 4th Annual European Association for Computer Graphics Conference and Exhibition, Zagreb, Yugoslavia*, pages 47–55. Amsterdam: North-Holland, September 1983.

266. J. H. McKay and S. S. Wang. An inversion formula for two polynomials in two variables. *Journal of Pure and Applied Algebra*, 40(3):245–257, May 1986.

267. Z. Michalewicz. *Genetic algorithms + data structures = evolution programs*. Springer-Verlag, Berlin, 1992.

268. J. R. Miller and R. N. Goldman. Geometric algorithms for detecting and calculating all conic sections in the intersection of any two natural quadratic surfaces. *Graphical Models and Image Processing*, 57(1):55–66, January 1995.

269. J. S. B. Mitchell. An algorithmic approach to some problems in terrain navigation. *Artificial Intelligence*, 37:171–201, 1988.

270. K. Mørken. Some identities for products and degree raising of splines. *Constructive Approximation*, 7:195–208, 1991.

271. G. Monge. *Application de l'Analyse à la Géométrie*. Bachelier, Paris, 1850.

272. U. Montanari. Continuous skeletons from digitized images. *Journal of the Association for Computing Machinery*, 16(4):534–549, October 1969.

273. R. E. Moore. *Interval Analysis*. Prentice-Hall, Englewood Cliffs, NJ, 1966.

274. R. E. Moore. *Methods and Applications of Interval Analysis*. SIAM, Philadelphia, 1979.

275. H. P. Moreton. Simplified curve and surface interrogation via mathematical packages and graphics libraries and hardware. *Computer-Aided Design*, 27(7):523–543, July 1995.

276. M. E. Mortenson. *Geometric Modeling*. John Wiley and Sons, New York, 1985.

277. S. P. Mudur and P. A. Koparkar. Interval methods for processing geometric objects. *IEEE Computer Graphics and Applications*, 4(2):7–17, February 1984.

278. G. Müllenheim. On determining start points for a surface/surface intersection algorithm. *Computer Aided Geometric Design*, 8(5):401–408, November 1991.

279. F. C. Munchmeyer. On surface imperfections. In R. Martin, editor, *Mathematics of Surfaces II*, pages 459–474. Oxford University Press, 1987.

280. F. C. Munchmeyer. Shape interrogation: A case study. In G. Farin, editor, *Geometric Modeling*, pages 291–301. SIAM, Philadelphia, PA, 1987.

281. F. C. Munchmeyer and R. Haw. Applications of differential geometry to ship design. In D. F. Rogers, B. C. Nehring, and C. Kuo, editors, *Proceedings of Computer Applications in the Automation of Shipyard Operation and Ship Design IV*, volume 9, pages 183–196, Annapolis, Maryland, USA, June 1982.

282. L. R. Nackman. Curvature relations in three-dimensional symmetric axes. *Computer Graphics and Image Processing*, 20:43–57, 1982.

283. L. R. Nackman and S. M. Pizer. Three-dimensional shape description using the symmetric axis transform I: Theory. *IEEE Transactions on Pattern Analysis and Machine Intelligence*, PAMI-7(2):187–202, March 1985.

284. A. Neumaier. *Interval Methods for Systems of Equations*. Cambridge University Press, Cambridge, 1990.

285. M. Niizeki and F. Yamaguchi. Projectively invariant intersection detections for solid modeling. *ACM Transactions on Graphics*, 13(3):277–299, July 1994.

286. T. Nishita, T. W. Sederberg, and M. Kakimoto. Ray tracing trimmed rational surface patches. *ACM Computer Graphics*, 24(4):337–345, August 1990.

287. M. F. Nittel. Numerically controlled machining of propeller blades. *Marine Technology*, 26(3):202–209, July 1989.

288. M. Noro, T. Takeshima, and K. Yokoyama. Solution of systems of algebraic equations and linear maps on residue class ring. *Journal of Symbolic Computation*, 14:399–417, 1992.

289. H. Nowacki, J. Michalski, B. Oleksiewicz, M. I. G. Bloor, C. W. Dekaski, and M. J. Wilson. In H. Nowacki, M. I. G. Bloor, and B. Oleksiewicz, editors, *Computational Geometry for Ships*. World Scientific, 1995.

290. A. W. Nutbourne and R. R. Martin. *Differenential Geometry Applied to Curve and Surface Design Vol. 1: Foundations*. Ellis Horwood, Chichester, UK, 1988.

291. N. O. Olesten. *Numerical Control*. Wiley-Interscience, 1970.

292. J. O'Rourke. *Computational Geometry in C*. Cambridge University Press, Cambridge, UK, 1994.

293. J. M. Ortega and W. C. Rheinboldt. *Iterative Solution of Nonlinear Equations in Several Variables*. Academic Press, New York, 1970.

294. N. M. Patrikalakis. Shape interrogation. In C. Chryssostomidis, editor, *Proceedings of the 16th Annual MIT Sea Grant College Program Lecture and Seminar, Automation in the Design and Manufacture of Large Marine Systems*, pages 83–104, Cambridge, MA, October 1988. New York: Hemisphere Publishing, 1990.

295. N. M. Patrikalakis. Surface-to-surface intersections. *IEEE Computer Graphics and Applications*, 13(1):89–95, January 1993.

296. N. M. Patrikalakis and L. Bardis. Offsets of curves on rational B-spline surfaces. *Engineering with Computers*, 5:39–46, 1989.

297. N. M. Patrikalakis and L. Bardis. Localization of rational B-spline surfaces. *Engineering with Computers*, 7(4):237–252, 1991.

298. N. M. Patrikalakis and H. N. Gursoy. Shape interrogation by medial axis transform. In B. Ravani, editor, *Proceedings of the 16th ASME Design Automation Conference: Advances in Design Automation, Computer Aided and Computational Design, Vol. I*, pages 77–88, Chicago, IL, September 1990. New York: ASME.

299. N. M. Patrikalakis and G. A. Kriezis. Representation of piecewise continuous algebraic surfaces in terms of B-splines. *The Visual Computer*, 5(6):360–374, 1989.

300. N. M. Patrikalakis and T. Maekawa. Intersection problems. In G. Farin, J. Hoschek, M. S. Kim, and D. Abma, editors, *The Handbook of Computer Aided Design*. Elsevier, 2001.

301. N. M. Patrikalakis and P. V. Prakash. Free-form plate modeling using offset surfaces. *Journal of OMAE, Transactions of the ASME.*, 110(3):287–294, 1988.

302. N. M. Patrikalakis and P. V. Prakash. Surface intersections for geometric modeling. *Journal of Mechanical Design, Transactions of the ASME*, 112(1):100–107, March 1990.

303. N. M. Patrikalakis, T. Sakkalis, and G. Shen. Boundary representation models: Validity and rectification. In R. Cipolla and R. Martin, editors, *The Mathematics of Surfaces IX*, pages 389–409, University of Cambridge, UK., September 2000. London: Springer.

304. J. Pegna and D. J. Wilde. Spherical and circular blending of functional surfaces. *Journal of OMAE, Transactions of the ASME*, 112(2):134–142, May 1990.

305. J. Pegna and F. E. Wolter. Geometrical criteria to guarantee curvature continuity of blend surfaces. *Journal of Mechanical Design, Transactions of the ASME*, 114(1):201–210, March 1992.

306. J. Pegna and F.-E. Wolter. Surface curve design by orthogonal projection of space curves onto free-form surfaces. *Journal of Mechanical Design, ASME Transactions*, 118(1):45–52, March 1996.

307. H. Persson. NC machining of arbitrarily shaped pockets. *Computer-Aided Design*, 10(3):169–174, May 1978.

308. M. Peternell and H. Pottmann. A Laguerre geometric approach to rational offsets. *Computer Aided Geometric Design*, 15(3):223–249, March 1998.

309. T. J. Peters, N. F. Stewart, D. R. Ferguson, and P. S. Fussell. Algorithmic tolerances and semantics in data exchange. In *Computational Geometry '97*, Nice, France, 1997.

310. S. Petitjean. Algebraic geometry and computer vision: Polynomial systems, real and complex roots. *Journal of Mathematical Imaging and Vision*, 10(3):191–220, 1999.

311. F. Pettinati. Private Communication, October 10, 1997.

312. B. Pham. Offset approximation of uniform B-splines. *Computer-Aided Design*, 20(8):471–474, October 1988.

313. B. Pham. Offset curves and surfaces: a brief survey. *Computer-Aided Design*, 24(4):223–229, April 1992.

314. L. A. Piegl and W. Tiller. *The NURBS Book*. Springer, New York, 1995.

315. L. A. Piegl and W. Tiller. Symbolic operators for NURBS. *Computer-Aided Design*, 29(5):361–368, May 1997.

316. L. A. Piegl and W. Tiller. Computing offsets of NURBS curves and surfaces. *Computer-Aided Design*, 31(2):147–156, February 1999.

317. K. G. Pigounakis and P. D. Kaklis. Fairing of 2D B-splines under design constraints. *Mathematical Engineering in Industry*, 7(2):165–178, 1998.

318. K. G. Pigounakis, N. Sapidis, and P. D. Kaklis. Fairing spatial B-spline curves. *Journal of Ship Research*, 40(4):351–367, 1996.

319. T. Poeschl. Detecting surface irregularities using isophotes. *Computer Aided Geometric Design*, 1(2):163–168, November 1984.

320. I. R. Porteous. Ridges and umbilics of surfaces. In R. Martin, editor, *The Mathematics of Surfaces II*, pages 447–458. Oxford University Press, 1987.

321. I. R. Porteous. The circles of a surface. In R. Martin, editor, *The Mathematics of Surfaces III*, pages 135–143. Oxford University Press, 1988.

322. I. R. Porteous. *Geometric Differentiation for the intelligence of curves and surfaces*. Cambridge University Press, Cambridge, 1994.

323. T. Poston and I. Stewart. *Catastrophe Theory and its Applications*. Pitman, San Francisco, CA, 1978.

324. H. Pottmann. Rational curves and surfaces with rational offsets. *Computer Aided Geometric Design*, 12(2):175–192, March 1995.

325. H. Pottmann. General offset surfaces. *Neural, Parallel and Scientific Computations*, 5:55–80, 1997.

326. H. Pottmann and G. Farin. Developable rational Bézier and B-spline surfaces. *Computer Aided Geometric Design*, 12(5):513–531, August 1995.

327. H. Pottmann, W. Lü, and B. Ravani. Rational ruled surfaces and their offsets. *Graphical Models and Image Processing*, 58(6):544–552, November 1996.

328. H. Pottmann and K. Opitz. Curvature analysis and visualization for functions defined on Euclidean spaces or surfaces. *Computer Aided Geometric Design*, 11:655–674, 1994.

329. H. Pottmann and J. Wallner. Approximation algorithms for developable surfaces. *Computer Aided Geometric Design*, 16(6):539–556, June 1999.

330. H. Pottmann and J. Wallner. *Computational Line Geometry*. Springer-Verlag, Berlin, 2001.

331. H. Pottmann, J. Wallner, G. Glaeser, and B. Ravani. Geometric criteria for gouge-free three-axis milling of sculptured surfaces. *Journal of Mechanical Design, Transactions of the ASME.*, 31(1):17–32, 1999.

332. M. J. Pratt. Cyclides in computer aided geometric design. *Computer Aided Geometric Design*, 7(1-4):221–242, June 1990.

333. M. J. Pratt and A. D. Geisow. Surface/surface intersection problems. In J. A. Gregory, editor, *The Mathematics of Surfaces*, pages 117–142. Clarendon Press, 1986.

334. F. P. Preparata. The medial axis of a simple polygon. In G. Goos and J. Hartmanis, editors, *Lecture Notes in Computer Science: Mathematical Foundations of Computer Science*, pages 443–450. Springer-Verlag, 1977.

335. F. P. Preparata and M. I. Shamos. *Computational Geometry: An Introduction*. Springer-Verlag, New York, 1985.

336. W. H. Press, S. A. Teukolsky, W. T. Vetterling, and B. P. Flannery. *Numerical Recipes in C*. Cambridge University Press, 1988.

337. A. Preusser. Computing area filling contours for surface defined by piecewise polynomials. *Computer Aided Geometric Design*, 3:267–279, 1986.

338. M. A. Price, C. G. Armstrong, and M. A. Sabin. Hexahedral mesh generation by medial surface subdivision: I. Solids with convex edges. *International Journal of Numerical Methods in Engineering*, 38(19):3335–3359, 1995.

339. T. Rando and J. A. Roulier. Knot-removal surface fairing using search strategies. *Computer-Aided Design*, 23(7):492–497, September 1991.

340. T. Rausch, F.-E. Wolter, and O. Sniehotta. Computation of medial curves on surfaces. In T. Goodman and R. Martin, editors, *The Mathematics of Surfaces VII*, pages 43–68. Information Geometers, 1997.

341. J. M. Reddy and G. M. Turkiyyah. Computation of 3d skeletons using a generalized Delaunay triangulation technique. *Computer–Aided Design*, 27(9):677–694, September 1995.

342. A. A. G. Requicha. Representations for rigid solids: Theory, methods, and systems. *Computing Surveys*, 12(4), December 1990.

343. A. A. G. Requicha and H. B. Voelcker. Constructive Solid Geometry. Technical Report TM 25, Production Automation Project, University of Rochester, Rochester, NY, November 1977.

344. A. A.G. Requicha and J. R. Rossignac. Solid modeling and beyond. *IEEE Computer Graphics and Applications*, 12(5):31–44, September 1992.

345. R. F. Riesenfeld. *Applications of B-spline Approximation to Geometric Problems of Computer-Aided Design*. PhD thesis, Syracuse University, Syracuse, New York, 1973.

346. J. J. Risler. *Mathematical Methods for CAD*. Cambridge University Press, Cambridge, UK, 1992.

347. D. J. Robinson and C. G. Armstrong. Geodesic paths for general surfaces by solid modellers. In G. Mullineux, editor, *The Mathematics of Surfaces VI, Proceedings of the 6th IMA Conference on Mathematics of Surfaces VI*, pages 103–117, Oxford, UK, 1996. Clarendon Press.

348. D. F. Rogers and J. A. Adams. *Mathematical Elements for Computer Graphics*. McGraw-Hill Inc., 1990. Second Edition.

349. R. F. Rohmfeld. IGB-offset curves - loop removal by scanning of interval sequences. *Computer Aided Geometric Design*, 15(4):339–375, April 1998.

350. A. Rosenfeld. Axial representations of shape. *Computer Vision, Graphics and Image Processing*, 33:156–173, 1986.

351. J. R. Rossignac. *Blending and Offseting Solid Models.* PhD thesis, University of Rochester, July 1985. Production Automation Project Technical Memorandum No. 54.

352. J. R. Rossignac and A. A. G. Requicha. Piecewise-circular curves for geometric modeling. *IBM Journal of Research and Development,* 31(3):296–313, 1987.

353. J. R. Rossignac and A. G. Requicha. Offsetting operations in solid modelling. *Computer Aided Geometric Design,* 3(2):129–148, 1986.

354. M. Sabin. Subdivision surfaces. In G. Farin, J. Hoschek, M. S. Kim, and D. Abma, editors, *The Handbook of Computer Aided Design.* Elsevier, 2001.

355. M. A. Sabin. Recursive division interrogation of offset surfaces. In J. D. Warren, editor, *Curves and Surfaces in Computer Vision and Graphics III, Proceedings of SPIE,* volume 1830, pages 152–161, Boston, MA, November 1992. SPIE.

356. T. Sakkalis. On the zeros of a polynomial vector field. Research Report RC-13303, IBM T. J. Watson Research Center, Yorktown Heights, NY, 1987.

357. T. Sakkalis. The Euclidean algorithm and the degree of the Gauss map. *SIAM Journal on Computing,* 19(3):538–543, June 1990.

358. T. Sakkalis. The topological configuration of a real algebraic curve. *Bulletin of the Australian Mathematical Society,* 43:37–50, 1991.

359. T. Sakkalis and C. Charitos. Approximating curves via alpha shapes. *Graphical Models and Image Processing,* 61(3):165–176, 1999.

360. T. Sakkalis, G. Shen, and N. M. Patrikalakis. Topological and geometric properties of interval solid models. *Graphical Models,* 63, 2001. In press.

361. T. Sakuta, M. Kawai, and Y. Amano. Development of an NC machining system for stamping dies by offset surface method. In *Autofact 87 Conference Proceedings,* pages 2.13–2.27, Dearborn, Michigan, 1987. SME.

362. G. Salmon. *A Treatise on the Analytic Geometry of Three Dimensions, Vol. 1.* Chelsea, New York, seventh edition, 1927.

363. N. M. Samuel, A. A. G. Requicha, and S. A. Elkind. Methodology and results of an industrial part survey. Technical Report Tech. Momo. No. 21, Production Automation Project, University of Rochester, Rochester, NY, 1976.

364. P. T. Sander and S. W. Zucker. Singularities of principal direction fields from 3-D images. In *IEEE Second International Conference on Computer Vision, Tampa Florida,* pages 666–670, 1988.

365. N. Sapidis and G. Farin. An automatic fairing algorithm for B-spline curves. *Computer-Aided Design,* 22(2):121–129, March 1990.

366. R. Sarma and D. Dutta. The geometry and generation of NC tool paths. *Journal of Mechanical Design, Transactions of the ASME,* 119:253–258, June 1997.

367. R. F. Sarraga. Algebraic methods for intersections of quadric surfaces in GMSOLID. *Computer Vision, Graphics and Image Processing,* 22(2):222–238, May 1983.

368. I. Schoenberg. Contributions to the problem of approximation of equidistant data by analytic functions. *Quarterly of Applied Mathematics,* 4:45–99, 1946.

369. L. L. Schumaker. *Spline Functions: Basic Theory.* Pure and Applied Mathematics: a Wiley-Interscience Series of Texts, Monographs, and Tracts. Wiley, New York, 1981.

370. G. L. Scott, S. C. Turner, and A. Zisserman. Using a mixed wave/diffusion process to elicit the symmetry set. *Image and Vision Computing,* 7:63–70, 1989.

371. T. W. Sederberg. *Implicit and Parametric Curves and Surfaces for Computer Aided Geometric Design.* PhD thesis, Purdue University, August 1983.

372. T. W. Sederberg. Planar piecewise algebraic curves. *Computer Aided Geometric Design*, 1(3):241–255, December 1984.

373. T. W. Sederberg. Piecewise algebraic surface patches. *Computer Aided Geometric Design*, 2(1-3):53–59, September 1985.

374. T. W. Sederberg, D. C. Anderson, and R. N. Goldman. Implicit representation of parametric curves and surfaces. *Computer Vision, Graphics and Image Processing*, 28(1):72–84, October 1984.

375. T. W. Sederberg and D. B. Buehler. Offsets of polynomial Bézier curves: Hermite approximation with error bounds. In T. Lyche and L. L. Schumaker, editors, *Mathematical Methods in Computer Aided Geometric Design*, volume II, pages 549–558. Academic Press, 1992.

376. T. W. Sederberg, H. N. Christiansen, and S. Katz. Improved test for closed loops in surface intersections. *Computer-Aided Design*, 21(8):505–508, October 1989.

377. T. W. Sederberg and R. T. Farouki. Approximation by interval Bézier curves. *IEEE Computer Graphics and Applications*, 12(5):87–95, September 1992.

378. T. W. Sederberg and R. N. Goldman. Algebraic geometry for computer-aided geometric design. *IEEE Computer Graphics and Applications*, 6(6):52–59, June 1986.

379. T. W. Sederberg and R. J. Meyers. Loop detection in surface patch intersections. *Computer Aided Geometric Design*, 5(2):161–171, July 1988.

380. T. W. Sederberg and T. Saito. Rational-ruled surfaces: implicitization and section curves. *Graphical Models and Image Processing*, 57(4):334–342, 1995.

381. T. W. Sederberg and J. Zheng. Algebraic methods for CAGD. In G. Farin, J. Hoschek, M. S. Kim, and D. Abma, editors, *The Handbook of Computer Aided Design*. Elsevier, 2001.

382. T. W. Sederberg and A. K. Zundel. Pyramids that bound surface patches. *Graphical Models and Image Processing*, 58(1):75–81, January 1996.

383. U. Shani and D. H. Ballard. Splines as embeddings for generalized cylinders. *Computer Vision, Graphics and Image Processing*, 27:129–156, 1984.

384. D. J. Sheehy, C. G. Armstrong, and D. J. Robinson. Computing the medial surface of a solid from a domain Delaunay triangulation. In C. Hoffmann and J. Rossignac, editors, *Proceedings of the Third Symposium on Solid Modeling and Applications, May 1995, Salt Lake City, Utah*, pages 201–212, New York, 1995. ACM.

385. D. J. Sheehy, C. G. Armstrong, and D. J. Robinson. Numerical computation of medial surface vertices. In G. Mullineux, editor, *The Mathematics of Surfaces VI*, Oxford, UK, 1996. IMA, Oxford University Press.

386. G. Shen. *Analysis of Boundary Representation Model Rectification.* PhD thesis, Massachusetts Institute of Technology, Cambridge, MA, February 2000.

387. G. Shen and N. M. Patrikalakis. Numerical and geometric properties of interval B-splines. *International Journal of Shape Modeling*, 4(1 and 2):35–62, March and June 1998.

388. G. Shen, T. Sakkalis, and N. M. Patrikalakis. Manifold boundary representation model rectification (La rectification des modèles des varietés b-rep). In C. Mascle, C. Fortin, and J. Pegna, editors, *Proceedings of the 3rd International Conference on Integrated Design and Manufacturing in Mechanical Engineering*, page 199 and CDROM, Montreal, Canada, May 2000. Presses internationales Polytechnique.

389. G. Shen, T. Sakkalis, and N. M. Patrikalakis. Boundary representation model rectification. *Graphical Models*, 63, 2001. In press. Also in: *Proceedings of the Sixth ACM Solid Modeling Symposium*. D. Anderson and K. Lee, editors. Ann Arbor, Michigan, June 2001. NY: ACM, 2001.

390. C.-K. Shene and J. K. Johnstone. On the lower degree intersections of two natural quadrics. *ACM Transactions on Graphics*, 13(4):400–424, October 1994.

391. E. C. Sherbrooke. *3-D Shape Interrogation by Medial Axis Transform*. PhD thesis, Massachusetts Institute of Technology, Cambridge, MA, April 1995.

392. E. C. Sherbrooke and N. M. Patrikalakis. Computation of the solutions of nonlinear polynomial systems. *Computer Aided Geometric Design*, 10(5):379–405, October 1993.

393. E. C. Sherbrooke, N. M. Patrikalakis, and E. Brisson. Computation of medial axis transforms of 3-D polyhedra. In C. Hoffmann and J. Rossignac, editors, *Proceedings of the Third Symposium on Solid Modeling and Applications, May 1995, Salt Lake City, Utah*, pages 187–199, New York, 1995. ACM.

394. E. C. Sherbrooke, N. M. Patrikalakis, and E. Brisson. An algorithm for the medial axis transform of 3-D polyhedral solids. *IEEE Transactions on Visualization and Computer Graphics*, 2(1):44–61, March 1996.

395. E. C. Sherbrooke, N. M. Patrikalakis, and F.-E. Wolter. Differential and topological properties of medial axis transforms. *Graphical Models and Image Processing*, 58(6):574–592, November 1996.

396. P. Sinha, E. Klassen, and K. K. Wang. Exploiting topological and geometric properties for selective subdivision. In *Proceedings of the ACM Symposium on Computational Geometry*, pages 39–45. New York: ACM, 1985.

397. S. S. Sinha and P. J. Besl. Principal patches: A viewpoint-invariant surface description. In *IEEE International Robotics and Automation, Cincinnati, Ohio*, pages 226–231, May 1990.

398. J. Sneyd and C. S. Peskin. Computation of geodesic trajectories on tubular surfaces. *SIAM Journal of Scientific Statistical Computing*, 11(2):230–241, March 1990.

399. J. M. Snyder. *Generative Modeling for Computer Graphics and CAD : Symbolic Shape Design Using Interval Analysis*. Academic Press, Boston, MA, 1992.

400. J. M. Snyder. Interval analysis for computer graphics. *ACM Computer Graphics*, 26(2):121–130, July 1992.

401. J. Sone and H. Chiyokura. Surface highlight control using quartic blending NURBS boundary Gregory patch. *Journal of Information Processing Society of Japan*, 37(12):2212–2222, 1996. In Japanese.

402. M. R. Spencer. *Polynomial Real Root Finding in Bernstein Form*. PhD thesis, Department of Civil Engineering, Brigham Young University, August 1994.

403. M. Spivak. *Calculus*. New York: W. A. Benjamin, Inc., 1967.

404. Y. L. Srinivas and D. Dutta. Cyclides in geometric modeling: computational tools for an algorithmic infrastructure. *Journal of Mechanical Design, Transactions of the ASME*, 117(3):363–373, September 1995.

405. V. Srinivasan and L. R. Nackman. Voronoi diagram for multiply connect polygonal domains, I: Algorithm. *IBM Journal of Research and Development*, 31(3):361–372, May 1987.

406. V. Srinivasan, L. R. Nackman, J.-M. Tang, and S. N. Meshkat. Automatic mesh generation using the symmetric axis transformation of polygonal domains. *Proceedings of the IEEE, Special Issue on Computational Geometry*, 80(9):1485–1501, 1992.

407. S. Stifter. *A Medley of Solutions to the Robot Collision Problem in Two and Three Dimensions*. PhD thesis, Johannes Kepler Universität, Linz, Austria, 1989.

408. S. Stifter. An axiomatic approach to Voronoi-diagrams in 3D. *Journal of Computers and System Sciences*, 43(2):361–379, October 1991.

409. P. Stiller. Sparse resultants. Technical Report ISC-96-01-MATH, Texas A & M University, Institute for Scientific Computation, 1996.
410. G. Strang. *Linear Algebra and its Applications*. Harcourt Brace Jovanovich, San Diego, CA, 1988.
411. D. J. Struik. Outline of a history of differential geometry. *Isis*, 19:92–120, 1933.
412. D. J. Struik. *Lectures on Classical Differential Geometry*. Addison-Wesley, Cambridge, MA, 1950.
413. B. Sturmfels. Introduction to resultants. In D. A. Cox and B. Sturmfels, editors, *Proceedings of Symposia in Applied Mathematics Volume 53, Applications of Computational Algebraic Geometry: American Mathematical Society short course, January 6-7, 1997, San Diego, California*, pages 25–39. American Mathematical Society, 1998.
414. B. Sturmfels and A. Zelevinsky. Multigraded resultants of Sylvester type. *Journal of Algebra*, 163(1):115–127, January 1994.
415. A. Sudhalkar, L. Gürsöz, and F. Prinz. Continuous skeletons of discrete objects. In J. Rossignac, J. Turner, and G. Allen, editors, *Proceedings of the Second Symposium on Solid Modeling and Applications, Montreal, Canada*, pages 85–94, New York, 1993. ACM.
416. K. Sugihara. Approximation of generalized Voronoi diagrams by ordinary Voronoi diagrams. *Computer Vision, Graphics and Image Processing: Graphical Models and Image Processing*, 55(6):522–531, November 1993.
417. K. Suresh and D. C. H. Yang. Constant scallop-height machining of free-form surfaces. *Journal of Engineering for Industry, Transactions of the ASME*, 116:253–259, May 1994.
418. T. K. H. Tam and C. G. Armstrong. 2d finite element mesh generation by medial axis subdivision. *Advances in Engineering Software and Workstations*, 13(5/6):313–324, September/November 1991.
419. H. Theisel and G. Farin. The curvature of characteristic surfaces. *IEEE Computer Graphics and Applications*, 17(6):88–96, November/December 1997.
420. W. Tiller. Knot-removal algorithms for NURBS curves and surfaces. *Computer-Aided Design*, 24(8):445–453, August 1992.
421. W. Tiller and E. G. Hanson. Offsets of two-dimensional profiles. *IEEE Computer Graphics and Applications*, 4(9):36–46, September 1984.
422. D. Toth. On ray tracing parametric surfaces. *ACM Computer Graphics*, 19(3):171–179, July 1985.
423. S. T. Tuohy. A visual tool for demonstrating surface curvature. *Computer Applications in Engineering Education*, 5(1):21–27, 1997.
424. S. T. Tuohy, T. Maekawa, and N. M. Patrikalakis. Interrogation of geophysical maps with uncertainty for AUV micro-navigation. In *Engineering in Harmony with the Ocean, Proceedings of Oceans '93, Victoria, Canada*. IEEE Oceanic Engineering Society, October 1993.
425. S. T. Tuohy, T. Maekawa, G. Shen, and N. M. Patrikalakis. Approximation of measured data with interval B-splines. *Computer-Aided Design*, 29(11):791–799, November 1997.
426. S. T. Tuohy and N. M. Patrikalakis. Representation of geophysical maps with uncertainty. In N. M. Thalmann and D. Thalmann, editors, *Communicating with Virtual Worlds, Proceedings of CG International '93, Lausanne, Switzerland*, pages 179–192. Springer, Tokyo, June 1993.
427. S. T. Tuohy, J. W. Yoon, and N. M. Patrikalakis. Reliable interrogation of 3-D non-linear geophysical databases. In J. A. Vince and R. A. Earnshaw,

editors, *Computer Graphics: Developments in Virtual Environments, Procee-dings of CG International '95, Leeds, UK, June 1995*, pages 327–341. London, Academic Press, 1995.

428. G. M. Turkiyyah, D. W. Storti, M. Ganter, H. Chen, and M. Vimawala. An accelerated triangulation method for computing the skeletons of free-form solid models. *Computer-Aided Design*, 29(1):5–19, January 1997.

429. U. S. Product Data Association. *ANS US PRO/IPO-200-042-1994: Part 42 – Integrated Geometric Resources: Geometric and Topological Representation*, 1994.

430. M. E. Vafiadou and N. M. Patrikalakis. Interrogation of offsets of polyno-mial surface patches. In F. H. Post and W. Barth, editors, *Eurographics '91, Proceedings of the 12th Annual European Association for Computer Graphics Conference and Exhibition*, pages 247–259 and 538, Vienna, Austria, Septem-ber 1991. Amsterdam: North-Holland.

431. P. J. Vermeer. *Medial Axis Transform to Boundary Representation Conver-sion*. PhD thesis, Purdue University, May 1994.

432. A. Verroust and F. Lazarus. Extracting skeletal curves from 3D scattered data. *The Visual Computer*, 16(1):15–25, 2000.

433. K. J. Versprille. *Computer Aided Design Applications of the Rational B-Spline Approximation Form*. PhD thesis, Syracuse University, Syracuse, New York, February 1975.

434. H. B. Voelcker et al. An introduction to PADL: Characteristics, status, and rationale. Technical Report Tech. Momo. No. 22, Production Automation Project, University of Rochester, Rochester, NY, December 1974.

435. M. N. Vrahatis. CHABIS: A mathematical software package for locating and evaluating roots of systems of nonlinear equations. *ACM Transactions on Mathematical Software*, 14(4):330–336, December 1988.

436. M. N. Vrahatis. Solving systems of nonlinear equations using the nonzero value of the topological degree. *ACM Transactions on Mathematical Software*, 14(4):312–329, December 1988.

437. R. J. Walker. *Algebraic Curves*. Princeton University Press, Princeton, New Jersey, 1950.

438. L. Wang, M. C. Leu, and D. Blackmore. Generating sweep solids for NC verification using the SEDE method. In *Proceedings of the Fourth Symposium on Solid Modeling and Applications*, pages 364–375, Atlanta, Georgia, May 14-16 1997.

439. W. P. Wang. Integration of solid geometric modeling for computerized process planning. In C. R. Liu, T. C. Chang, and R. Komanduri, editors, *Computer-Aided/Intelligent Process Planning, ASME, Winter Annual Meeting*, pages 177–187, 1985.

440. Y. Wang. Intersection of offsets of parametric surfaces. *Computer Aided Geometric Design*, 13(5):453–465, 1996.

441. C. E. Weatherburn. *Differential Geometry of Three Dimensions, Vol. 1*. The University Press, Cambridge, 1939.

442. H. S. Wilf. A global bisection algorithm for computing the zeros of polynomials in the complex plane. *Journal of the Association for Computing Machinery*, 25(3):415–420, July 1978.

443. I. Wilf and Y. Manor. Quadric-surface intersection curves: shape and struc-ture. *Computer-Aided Design*, 25(10):633–643, October 1993.

444. T. J. Willmore. *An Introduction to Differential Geometry*. Clarendon Press, Oxford, 1959.

445. F. Winkler. *Polynomial Algorithms in Computer Algebra*. Springer-Verlag, New York, 1996.

446. S. Wolfram. *The Mathematica Book*. Wolfram Media, Champaign, IL, 3rd edition, 1996.

447. F.-E. Wolter. Distance function and cut loci on a complete Riemannian manifold. *Archiv der Mathematik*, 32:92–96, 1979.

448. F.-E. Wolter. Interior metric, shortest paths and loops in riemannian manifolds with not necessarily smooth boundary. Master's thesis, Free University of Berlin, Berlin, Germany, 1979.

449. F.-E. Wolter. *Cut Loci in Bordered and Unbordered Riemannian Manifolds*. PhD thesis, Technical University of Berlin, Department of Mathematics, December 1985.

450. F.-E. Wolter. Cut locus and medial axis in global shape interrogation and representation. Memorandum 92-2, Cambridge MA: MIT Ocean Engineering Design Laboratory, January 1992.

451. F.-E. Wolter and K.-I. Friese. Local and global geometric methods for analysis interrogation, reconstruction, modification and design of shape. In *Computer Graphics International, GCI 2000. (Invited paper)*, pages 137–151, Geneva, Switzerland, June 2000. IEEE Computer Society Press. Los Alamitos, CA: IEEE, 2000.

452. F. E. Wolter and S. T. Tuohy. Approximation of high degree and procedural curves. *Engineering with Computers*, 8(2):61–80, 1992.

453. F.-E. Wolter and S. T. Tuohy. Curvature computations for degenerate surface patches. *Computer Aided Geometric Design*, 9(4):241–270, September 1992.

454. S.-T. Wu and L. N. Andrade. Marching along a regular surface/surface intersection with circular steps. *Computer Aided Geometric Design*, 16(4):249–268, May 1999.

455. F. Yamaguchi. *Curves and Surfaces in Computer Aided Geometric Design*. Springer-Verlag, NY, 1988.

456. F. Yamaguchi. A shift of playground for geometric processing from Euclidean to homogeneous. *The Visual Computer*, 14(7):315–327, 1998.

457. Y. Yamaguchi. Differential properties at singular points of parametric surfaces. In P. Brunet, C. M. Hoffmann, and D. Roller, editors, *CAD-Tools and Algorithms for Product Design*, pages 211–221. Springer, 2000.

458. X. Ye and T. Maekawa. Differential geometry of intersection curves of two surfaces. *Computer Aided Geometric Design*, 16(8):767–788, September 1999.

459. W. I. Zangwill and C. B. Garcia. *Pathways to solutions, fixed points, and equilibria*. Prentice-Hall, Englewood Cliffs, NJ, 1981.

460. C. Zhang and F. Cheng. Removing local irregularities of NURBS surfaces by modifying highlight lines. *Computer-Aided Design*, 30(12):923–930, October 1998.

461. J. Zhou, E. C. Sherbrooke, and N. M. Patrikalakis. Computation of stationary points of distance functions. *Engineering with Computers*, 9(4):231–246, Winter 1993.

Index

absolute curvature, 201
affine parameter transformation, 79
algebraic curve, 115
algebraic distance, 116
algebraic numbers, 114
arbitrary speed, 40
arc length, 35
arc length parametrization, 37
artificial singularity, 51
auxiliary variable method, 89

B-spline basis function, 20
– derivative, 21
B-spline curve
– algorithms, 24
– derivative, 22
– properties, 21
B-spline surface, 29
Bézier curve
– algorithms, 13
– derivative, 12
– properties, 12
Bézier point, 12
Bézier surface, 18
basis conversion
– B-spline to Bézier, 24
– monomial to Bernstein
– – one variable, 79
– – two variables, 81
Bernstein polynomial
– derivative, 7
– arithmetic operation, 7
– properties, 7
binormal vector, 43
Boehm's algorithm, 26
border point, 142
Boundary Representation, 109
boundary value problem
– finite difference method, 274
– relaxation method, 274
– shooting method, 273
bounding box, 118

bounding wedge, 131

calculus of variations, 269
Christoffel symbols, 268
clamped curves, 22
clamped knots, 22
class of function, 1, 4
collinear normal point, 149, 184
condition number of root, 10
conic sections, 1
conjugate point, 266
Constructive Solid Geometry, 109
continuity
– geometric, 15
– parametric, 17
contouring, 197, 225
control net, 18
control polygon, 12
convex hull, 13
convex hull property, 12, 23
curvature, 40
– tangential intersection point, 173
– transversal intersection curve, 167
curvature map, 202
curvature plots, 200
curvature vector, 40
cusp, 215
cyclide, 232, 293

Darboux vector, 209
de Boor algorithm, 24
de Boor points, 21
de Casteljau algorithm, 13, 87
degenerate patch, 19
denormalized number, 96
developable surface, 195, 257, 351
development, 289
directrix, 256
distance function, 181
– stationary points, 185
Dupin's cyclide patch, 232
Dupin's indicatrix, 70, 172

ellipsoid, 4
elliptic cone, 4, 51
elliptic cylinder, 4
elliptic paraboloid, 4
– umbilic, 219
elliptic point, 59
end point geometric property, 12, 21
essential singularity, 51
Euler's equation, 270
Euler's theorem, 69, 259
evolute, 203
explicit curve, 2
explicit quadratic surface, 217
exponent, 90

first fundamental form, 53
– coefficients, 53
flat point, 59, 61
floating point arithmetic, 90, 95
focal curve, 203
focal surface, 203
Folium of Descartes, 2
Frenet-Serret formulae, 47, 162, 307

Gaussian curvature, 61, 64
– ellipsoid, 68
– elliptic cone, 68
– explicit surface, 64
– hyperbolic cylinder, 68
– implicit surface, 66
generator, 256
geodesic, 205, 265, 266
– curvature, 205, 266, 267
– curvature vector, 56, 267
– equation of, 268, 271
– offset, 284
– on developable surface, 287
geometry invariance property, 12, 21
Gröbner bases, 77
Greville abscissa, 21

helix, 46
highlight line, 199
hodograph, 12, 130
hyperbolic cylinder, 4
hyperbolic paraboloid, 4, 55, 65, 270
hyperbolic point, 59
hyperboloid of one sheet, 4
hyperboloid of revolution, 4
hyperboloid of two sheets, 4

IGES, 30
implicit curve, 1
implicit surface, 4

implicitization, 118, 119
incidence intransitivity, 114
inflection line, 259
inflection point, 41, 200, 203, 204
initial value problem, 272
interrogation, 195
– first-order, 197
– fourth-order, 208
– second-order, 200
– third-order, 205
– zeroth-order, 196
intersection
– curve to curve, 126
– curve to surface, 134
– lattice method, 148
– marching method, 148
– point to curve, 114
– point to point, 114
– point to surface, 121
– subdivision method, 148
– surface to surface, 137
interval arithmetic, 91, 92
– algebraic properties, 94
– rounded interval arithmetic, 95
interval Newton's method, 92
Interval Projected Polyhedron (IPP)
    algorithm, 105
intrinsic equation, 47
inverse function theorem, 240
inversion, 120
isophotes, 198

knot insertion, 26
knot removal, 27
knot vector, 21
– clamped, 22

Lagrange multiplier, 221
line of curvature, 62, 205, 231
linear precision, 7, 79
linkage curve theorem, 203
local support property, 24

mantissa, 90, 95
marching method, 190
maximum principal curvature, 61
– ellipsoid, 68
– elliptic cone, 68
– hyperbolic cylinder, 68
mean curvature, 61, 64
– ellipsoid, 68
– elliptic cone, 68
– explicit surface, 64

– hyperbolic cylinder, 68
– implicit surface, 66
medial axis, 299
meridians, 63
Meusnier's theorem, 57
minimum principal curvature, 61
– ellipsoid, 68
– elliptic cone, 68
– hyperbolic cylinder, 68
Monge form, 64, 233

NC machining, 293
Newton's method
– modified, 74
– n variables, 275
– one variable, 74
node, 21
non-algebraic distance, 116
non-arc-length parametrization, 40
Non-Uniform Rational B-Spline
    (NURBS), 30
– curve, 30
– surface, 31
nonparametric form, 1
normal curvature vector, 56
normal plane, 169
normal pyramid, 343
normal vector
– curve, 40
– surface, 50
normalized number, 96

offset curve, 307
– approximation, 312
– cusp, 311
– extraordinary point, 308
– irregular point, 308
– isolated point, 311
– ordinary cusp, 308
– self-intersection, 309, 311
– singularity, 308
offset surface, 316
– approximation, 345
– Gaussian curvature, 317
– implicit surface, 320
– irregular point, 318
– mean curvature, 317
– principal curvature, 317
– self-intersection, 318
– – explicit quadratic surface, 328
– – implicit quadratic surface, 319
– – parametric surface, 335
– – tracing, 343

– singularity, 318
ordinary point, 37, 50
orthotomics, 204
osculating plane, 40
Oslo algorithm, 26
overlapping, 155

parabolic cylinder, 4
parabolic point, 59
paraboloid of revolution, 4
parallels, 63
parametric curve, 1
parametric speed, 37
parametric surface, 4
pick feed, 299
pipe surface, 353
planar point, 59, 61
polynomial solver
– algebraic technique, 76
– balanced system, 83
– homotopy method, 78
– hybrid technique, 77
– overconstrained system, 84
– subdivision method, 78
– underconstrained system, 84
principal direction, 61
principal normal vector, 40
principal patch, 231
procedural curve, 110
procedural surface, 110
Projected Polyhedron (PP) algorithm,
    78
Pythagorean hodograph, 349

quadric, 4, 66

radial curve, 201
radius of curvature, 40
rational arithmetic, 90
rational number, 114
ray tracing, 198
rectifying plane, 43
reflection line, 198
regular point
– on curve, 37
– on surface, 50
resultant, 76, 120
root mean square curvature, 202
rotation matrix, 238
ruled surface, 256
– conical, 124
– cylindrical, 123
ruling, 256

secant method, 280
second fundamental form, 57
– coefficients, 57
self-intersection, 157, 158
semi-cubical parabola, 39
shaded image, 197
shooting method, 273
significant digits, 90
singly curved surface, 257
singular point, 37, 50, 143
span, 21
STEP, 30
surface inflection, 258
surface of revolution, 63, 122

tangent plane, 49
tensor product surface, 18, 29
tolerance region, 306
tool driving plane, 299
torsion, 44, 163, 205
– transversal intersection curve, 168
torus, 32, 123
total curvature, 209

trip algorithm, 226
triple scalar product, 44
turning point, 142

umbilic, 61, 215, 231
– criterion, 249
– generic, 233
– index, 236
– lemon pattern, 233
– lemon type, 219
– monstar pattern, 233
– non-generic, 233
– star pattern, 233
unit in the last place (ulp), 95
unit speed, 37
unit tangent vector, 36
– tangential intersection points, 170
– transversal intersection curve, 164

variation diminishing property, 13, 24

wireframe, 196

Printing: Mercedes-Druck, Berlin
Binding: Stürtz AG, Würzburg